Naturally Selective

T0143328

Researchers of human behavior have identified a so-called "orgasm gap": Men usually orgasm during intercourse, whereas women often do not. This book addresses this mystery. The two leading explanations are either that women are "psychologically broken"—Freud's theory—or badly designed—the "by-product theory." However, there is a much more compelling third explanation. Evolutionary biology, anatomy, physiology, and direct sex research suggest women have evolved under their own selection pressures, and orgasm is a fitness-increasing consequence of such selective factors. This is revealed in their patterns of orgasmic response, which are neither random nor inexplicable.

Key Features

- Synthesizes decades of peer-reviewed sex research in anatomy, biology, physiology, and behavior.
- Engagingly written based on feedback from students, peers, and interested lay persons.
- Makes sense of the "orgasm gap" between men and women.
- Provides a wider context of human sexual dimorphism and mutual sexual selection.
- Balances sex research and real-world research and practical applications.

Naturally Selective
Evolution, Orgasm, and Female Choice

Robert King

CRC Press
Taylor & Francis Group
Boca Raton London New York

CRC Press is an imprint of the
Taylor & Francis Group, an **informa** business

Designed cover image: Shutterstock

First edition published 2024
by CRC Press
2385 NW Executive Center Drive, Suite 320, Boca Raton FL 33431

and by CRC Press
4 Park Square, Milton Park, Abingdon, Oxon, OX14 4RN

CRC Press is an imprint of Taylor & Francis Group, LLC

© 2024 Taylor & Francis Group, LLC

Reasonable efforts have been made to publish reliable data and information, but the author and publisher cannot assume responsibility for the validity of all materials or the consequences of their use. The authors and publishers have attempted to trace the copyright holders of all material reproduced in this publication and apologize to copyright holders if permission to publish in this form has not been obtained. If any copyright material has not been acknowledged please write and let us know so we may rectify in any future reprint.

Except as permitted under U.S. Copyright Law, no part of this book may be reprinted, reproduced, transmitted, or utilized in any form by any electronic, mechanical, or other means, now known or hereafter invented, including photocopying, microfilming, and recording, or in any information storage or retrieval system, without written permission from the publishers.

For permission to photocopy or use material electronically from this work, access www.copyright.com or contact the Copyright Clearance Center, Inc. (CCC), 222 Rosewood Drive, Danvers, MA 01923, 978-750-8400. For works that are not available on CCC please contact mpkbookspermissions@tandf.co.uk

Trademark notice: Product or corporate names may be trademarks or registered trademarks and are used only for identification and explanation without intent to infringe.

Library of Congress Cataloging-in-Publication Data
Names: King, Robert (Psychologist), author.
Title: Naturally selective : evolution, orgasm, and female choice / Robert King.
Description: First edition. | Boca Raton, FL : CRC Press, 2024. | Includes bibliographical
    references and index.
Identifiers: LCCN 2023040587 (print) | LCCN 2023040588 (ebook) | ISBN 9781032444765
    (hardback) | ISBN 9781032444758 (paperback) | ISBN 9781003372356 (ebook)
Subjects: LCSH: Female orgasm. | Generative organs, Female—Evolution. | Sex (Biology) |
    Evolution (Biology) | Sexual selection.
Classification: LCC QH481 .K56 2024 (print) | LCC QH481 (ebook) | DDC 574.5/248—dc23/
    eng/20240111
LC record available at https://lccn.loc.gov/2023040587
LC ebook record available at https://lccn.loc.gov/2023040588

ISBN: 9781032444765 (hbk)
ISBN: 9781032444758 (pbk)
ISBN: 9781003372356 (ebk)

DOI: 10.1201/9781003372356

Typeset in Times
by Apex CoVantage, LLC

To my parents, Don and Hilary, without whose mutual sexual selection, I would not have existed.

Contents

Preface

What is this book about, why now, and why am I writing it?

LIONS, TIGERS, AND BEARS: OH MY!

All the animals have started having sex again.

I mean, of course, that all the animals in zoos and wildlife parks have been having sex. Over the last few years, all over the world, there has been a veritable orgy of menagerie-based shenanigans, from species previously thought coy. Once-shy cheetahs have reproduced in droves. Hippos have frolicked. Even those notoriously undersexed pandas have joined in the debauch.

What changed? Did the heating up of the climate similarly heat up their blood? Was this sexing-up the result of those alleged hormones in the water supply that so excite the conspiracy theorists? Not a bit of it. There is one single factor that explains all the sudden passion in those parts of the animal kingdom that humans do not reach: Humans had stopped watching them.

The Covid-19 pandemic denuded zoos of human visitors for months, thus giving the animals what they needed—privacy—to do their, well, what can one say, other than, their "thing"?

We humans often pride ourselves as being very different from other animals, but we are not so different in that respect, at least. We also, unless we have unusual tastes (about which I will have more to say later), normally prefer privacy for our sexual behavior. Not completely, of course, else we would not count voyeurs and exhibitionists among our ranks, but enough so that we might not expect the full range of human sexual behaviors to occur while we are being watched. And that is an issue for scientists, like me, who study sex.

This is a book about sex research—about applying scientific methods to human sexual behavior—and what I have just said to you raises an immediate problem. Science is about observing things. Some (idealized) versions of science imply that *only* observing events under the most controlled conditions—experiments—counts as science at all (these purists appear to have forgotten all about mathematics and astronomy, which have few, if any, controlled experiments). But, as I have just pointed out, we sex researchers have a problem because even other animals hardly have sex when being observed.

Although, of necessity, I will be discussing sex research as it pertains to both men and women, this book focusses on the mechanisms of female sexual selection. This is because the bulk of what I study is the nature and function of female orgasm, but it turns out that you cannot study that without studying how sexual selection—especially female sexual selection—works in both the general cases of the animal kingdom and the specific ones of human beings.

The raw mechanics of sex change when complex social mammals watch one other—and humans are the most complex social mammals of all. It isn't just what an external observer can record that matters to humans, it is what the events mean to the participants. Anyone who has ever asked if it is okay to fantasize about others during sex or, more darkly, thought about the implications of issues such as consent already knows this, of course, but apart from some sex therapists, there have been few attempts to integrate this insight into sex research per se.

WHY NOW?

Over fifteen years ago, a best-selling book about the nature and function of female orgasm ended with the provocative words "The time is ripe to do some good science."[1] Even more provocatively, the rest of the book argued strongly that all of our most reliable sex research implied the conclusion that female orgasm had no discernible function whatsoever, that it only existed at all as a by-product of selection on male anatomy.

To say I was intrigued by this claim was an understatement. Echoing the great evolutionary biologist Robert Trivers, I found myself thinking, "One has to wonder how often [the scientist in question] has been witness to such a blessed event to regard it as a by-product!" When I started looking at the research in question, it turned out that a surprisingly large amount of it involved nothing like sexual behavior as practiced by actual human beings across time and place. It was, in other words, as if we had put humans in zoos and gone, "They don't seem to do much, do they?" This so astonished me that it formed the basis of my PhD research and resulted in a career change. In order to try to answer the questions about the nature and function of female orgasm, I pretty soon realized that I had to understand the fundamentals of how sexual selection operates and because, in common with all other primates, females are the key drivers of sexual selection, this meant studying the mechanisms of that.

I soon discovered that, although a lot of sex research took place in very limited observed settings, not all of it did. Despite this, ignorance about basic female anatomy and physiology bedeviled the field. For example, I found that fully half the textbooks in my local specialist bookshop, which were used to train the next generation of medical students, mislabeled the clitoris as a tiny, functionless, entirely external organ. No wonder so many people had been led into thinking it plausible that female sexuality was likewise tiny, functionless, and external. On the contrary, female sexuality— and the orgasms that lie at the heart of it—is rich, complex, but often somewhat cryptic in nature. Men and women are not the same, and neither are their sexualities. Any discomfort with what this implies about our being "just another critter," and subject to the same rules, needs to be faced in an adult fashion.

Not only have we done a large amount of research into the nature and function of female orgasm over the last fifteen years, but there is also a rich (but neglected) body of research into this phenomenon going back nearly a century that deserves to be better known. I hope to put the record straight.

THE MOST IMPORTANT SEX ORGAN

Gynecologists have told me that they get bored of being asked if the activities of their day jobs put them off sex. One answer I have heard such a colleague give is, "Does being a Ferrari mechanic put you off driving Ferraris?" And, when it is put that way, it is obvious that the answer is "no" because the state of mind (problem solving in the gynecological/mechanics cases) is totally different from the other activity. Not only in terms of raw mechanics, but also in terms of what particular actions mean to the participants.

This implies that attention to the mechanics of acts, while important, is not the whole story when it comes to describing human behavior. Humans ascribe meanings to actions and, when it comes to sex, this utterly crucial element is, too often, absent from behavioral descriptions. As one therapist put it, "A woman's biggest erogenous zone is between her ears." Humans care a lot about meanings, and this is true of other supposedly basic functions besides sex. For example, no other animal cares about how food is presented, but we do—unless we are almost dead of starvation. Ethology is the study of behavior in natural settings, and the human setting includes social worlds that are rich in meaning. Meaning-making might even be said to be our species-typical behavior.

WHO DO I THINK I AM?

Some colleagues told me that it would be a more than usually hard sell for me to write this book. One reason was that I was a man writing about female sexual selection and this had become "problematic."[2] The second was that, beyond a very tight circle of specialists, potential readers would be put off by some of the necessary technical details.

My response to the first point is that if we have really reached a point where people can only write from within their own personal experiential bubble, then behavioral scientists may as well *all* pack up and go home. Science is, minimally, an attempt to correct for biases in a collaborative group exercise. But even if that were not the case, why wouldn't a heterosexual man, who is, after all, the recipient (or at least the witness) of female sexual selection, be capable of studying it? A female doctor would not be prevented from writing about prostate cancer. That aside, a number of eminent people are on record as actually calling for such analysis as I am offering here.

For instance, the (then) head of the American Psychological Association, Lisa Feldman Barrett, recently said (with some asperity) that she was tired of reading about the effects of sexual selection on the female of the species and that evolutionary theorists (leaving aside that these are not all men) should turn that perception on themselves: "I've always felt there should be a special place in hell, filled with mirrors, reserved for people who suggest that waist or hip size predicts anything important about a woman.[3]

This is an explicit demand that male (and, one hopes, female) scientists of sexual behavior look in that mirror that all of our evolved bodies, and minds, hold up to us. And I take her point. Female sexual selection is that mirror, and this book is about that. Whether men will like what they see is a different matter.

As to the second point: I want this book to be of use to professional scientists, but I always have in mind also Steven Pinker's writing ideal of "My smart college room-mate who just did not happen to study my subject" while writing. Some colleagues have warned me that the gory details of scientific method and philosophy—especially the vexed but central question of what actually constitutes causation in behavioral sciences—would be over the heads of a general readership.

I can only say that I hope the pessimists are wrong. One reason for my thinking that they *are* wrong is the proliferation of podcasts, often offering long-form discussions of some pretty technical science and philosophy issues, that boast audiences in the millions. My own experience in public lectures, podcasts, occasional writing for non-technical outlets such as *Quillette*, and my *Psychology Today* column, which has been running for over a decade now, matches this thought. I think there is a public appetite for technicalities, and a concentration span to match. I also think that the general public is being underestimated by much mainstream media, including some mainstream publishing media.

Maybe I am wrong. Let us see.

NOTES

1 Lloyd, E. A. (2005). *The case of the female orgasm: Bias in the science of evolution*. Harvard University Press.
2 "Problematic" is a magic word invoked by moral relativists, who want to say that something is morally wrong but cannot bring themselves to admit that their relativism has painted them into an intellectual corner.
3 Barrett, L. F. (2019). Blog post *Zombie ideas*. www.psychologicalscience.org/observer/zombie-ideas last accessed 22/08/2023.

Acknowledgments

Maria Dempsey, without whose support, feedback, and inspiration this book would never have gotten finished.

People who read, commented on, reviewed, and disputed various sections. All errors are my own.

Olivia Judson
Geoffrey Miller
Jay Belsky
Alice Dreger
Raegan Murphy
John Dempsey for specialist consultation

The Darwin at the LSE group, especially Tom Dickins, Richard Webb, Fatima Felisberti, Diana Fleischman, Max Steur, and Andrew Wells.

Helena Cronin
Dan Nettle
Hanny Lightfoot-Klein
Maggie McNeil

All my co-authors, but especially Kat Valentine, Ken Mah, and Yitzchak ("Irv") Binik (whose kind collaboration started me off).

Mike Bailey and the other sexnetters, but especially David Puts and David Schmitt, who were always quick to recall specific papers from my feeble recollections and whose generosity with their time saved me so much of my own.

Roy Levin for taking the time at IASR in 2010 to try and patiently explain to me why I was wrong about everything.

About the Editor

Robert King is a psychology lecturer at University College Cork, in the school of Applied Psychology. He lectures on social and biological psychology—especially in the field of human sexual behavior—as well as on the history and philosophy of science, research methods, statistics, and behavioral genetics. He is the co-director of the Masters in Mental Health, which has close ties to local health care providers, and he participates in the clinical courses. King writes an (almost) monthly column for *Psychology Today*, which typically gets over 30,000 reads when he blogs about sexual behavior (www.psychologytoday.com/blog/hive-mind). He also writes for *Quillette* online journal about the same topics. *Quillette* articles typically get more than 2 million hits and has a subscriber base of 70,000 (https://quillette.com/author/robert-king/). King has published in the peer-reviewed literature on human sexual behavior, especially female orgasm, in several journals, including the highest impact factor journal in the field of human sexual behavior—namely *Archives of Sexual Behavior*.

1 Introduction

1.1 WHAT IS THIS BOOK ALL ABOUT?

What did Jennifer Aniston mean when she said that the best smell in the world "Is that man that you love"?[1] Why did Germaine Greer tell a luckless paramour, "I'm not frigid, you're boring"?[2] What can we learn from the Mangaians of Polynesia, where the teenage boys are forbidden to have sex with girls of their own age, until they have been properly trained in appropriate techniques by the older women of the tribe?[3]

The answers are to be found by studying female orgasms.

You have probably heard of the so-called orgasm gap: That men usually orgasm easily during sex, whereas women often have a more difficult time of it. Therapeutic careers have been built on trying to reduce the orgasm gap, while comedic careers have been built on making jokes about it. Would it surprise you to learn that many leading scientific figures argue that the orgasm gap exists because women are, intrinsically, either poorly designed by nature, or are too psychologically damaged to enjoy sex?

This is all wrong. Integrating new insights from evolutionary biology, human sex research has advanced to a point where we can see clearly that women are neither psychologically dysfunctional nor anatomically poorly designed. On the contrary, women are miracles of their own unique history of evolutionary design. Female sexuality—of which their orgasms deserve to form a central part—makes sense only in the light of these understandings.

To start answering the sorts of questions I opened with (and I promise you that each of those questions does have a scientifically respectable answer), I will have to start with an even more basic question: What is the most valuable substance on the planet? Not platinum. Not diamonds even. Certainly not gold. In terms of cost to weight ratio, the most valuable substance—currently estimated at one quadrillion dollars per pound—is the human egg.[4]

On the other hand, sperm are cheap. Every human ejaculation contains a quarter of a billion of them. Every second of every day there are 18,000 ejaculations and 4.4 births. That is quite a ratio, and it underscores an important fact about human sexuality: Women are built to be choosy. They have to be, because what they carry is the most valuable thing in the world.

This choosiness manifests itself in a myriad number of interesting ways. For example, it deserves to be a better-known fact that, when women are in egalitarian cultures—and thus freer to choose their paths in life—on average they end up being *more* different—behaviorally and psychologically—from the men around them, than do their sisters in more restricted cultures.[5] The less the environment explains differences, the more that genetics does. The evidence is that these differences include a complex psycho-physical sexual system, engineered to be a rigorous gatekeeper for that valuable egg. Women are thus choosy, rather than passive, and this will be the topic of chapter two.

Sex was never meant to be a merely mechanical doddle for humans; it was always meant to be a challenge. Furthermore, that challenge has been largely set by women, for men. Putting this point less provocatively, the challenging nature of sex arises directly from women's choosiness, or to give it its more technical terminology: sexual selection.[6]

The fundamental thesis of this book is that, when it comes to explaining female orgasm, many people—of both sexes—have mischaracterized this female choosiness. They have mistaken it for women's being either psychologically broken (Freud's theory, in a nutshell)[7] or being badly designed in the first place (the theory that female sexuality is mostly a by-product of male sexuality).[8] Both of these mistakes are born of the same conceptual error: androcentrism.

DOI: 10.1201/9781003372356-1

Androcentrism is the error of seeing the male of a species as the norm, with females as the deviations from this norm. But, with current research, another—and much richer—perspective is possible: Women are neither damaged nor are they pale reflections of men. Their evolutionary path is unique to them. Furthermore, while this is not a sexual self-help book, knowledge is power. Knowledge of human sexual nature can increase our relationship satisfaction and mutual pleasure.

1.1.1 "Just Another Critter"?[9]

Why are all these things I have mentioned not already widely known? Mark Twain memorably said that humans are "The only animal that blushes—or needs to."[10] We are capable of uneasy responses to learning the truth about ourselves. Sex research renders humans—the only animal to wear clothes—both literally, and metaphorically, naked. It strips away layers of comforting illusions. Chief among these illusions is the notion that humans are uniquely free of the laws governing the rest of the animal kingdom. Humans are *special*, we want to say. Humans are self-creating. Surely, we cannot be, in biologist Martin Daly's memorable phrase, "just another critter"?

Humans *are* special, of course. All species are special—that is why they are species. Humans are rarely content to be merely one species among many, however. We have a rich inner life, we insist. We transmit culture. We have ideas.

Those thoughts are not controversial, but they have extreme versions that pander to human vanity. Chief among these is the step that insists that our thoughts—our ideas—*alone* make the world. The ancient Greek philosopher Protagoras gave an early formulation of this *idealism* when he declared, "Man is the measure of all things, of the things that are—that they are, of the things that are not, that they are not."[11] Protagoras' argument was roundly defeated at the time by Socrates, but his core conceit—that thoughts themselves alone are real—is a very persistent and ancient one. Idealism may even be a cross-culturally universal conceit. For instance, at roughly the same time Protagoras was expressing that western version of idealism, Buddha opened the *Dhammapada* with the words: "With our thoughts, we make the world."[12]

Does anyone think such an extreme form of idealism today? Thirty million people bought *The Secret*, a book that peddles the idea that the universe will align itself to your desires, if only you wish hard enough.[13] Less extreme, though no less delusional, is a modern version of Protagoras' idealism that says, "All sex differences are social constructions." There are various ways in which this could be meant. There are desirable versions, which point out that humans can outgrow socially restrictive gender roles. However, at least one version is to insist that sex differences exist only in our heads, and that, therefore—the thought continues—we could wish them all away if we wanted to.[14]

As sci-fi author Philip K. Dick puts it, "Reality is that which, when you stop believing in it, doesn't go away."[15] On one level sex differences do, of course, exist "only" in our heads. For one thing, brains exist in our heads, and male and female brains—whatever you may have heard to the contrary—can be readily and reliably distinguished.[16] This topic—of what sense biologists make of what is technically called *sexual dimorphism*—is such a foundational one that it is the subject of chapter three. However, somebody viewing from the sidelines might be forgiven for being puzzled as to why this is even a contentious issue.[17]

Quite often, so the thought goes, the moment we admit to innate human differences, we mandate treating some people—typically women—worse. I need to say something about this form of social constructionist reflex to sex differences right away, to persuade some readers that admitting sexual dimorphism may not be quite the political bogey-person that they fear.

The moral and political case for sexual egalitarianism I take to have been already, and decisively, won in our culture, but I do not believe it can, or should, be maintained through lies. To echo the historian and activist Alice Dreger, truth and justice go hand in hand, and it is therefore impossible to build a lastingly just society on the shifting sands of untruths.[18] To say even that, true though it is, under-describes the problem, and ignores one of the principal reasons why applying scientific reasoning to humans has been so fraught up until now.

If generals are notorious for tending to fight the last war, then philosophers are even worse, and philosophers of science perhaps the worst of all. They are prone to re-fight battles that have already been decisively won by other people, and then turn up in sharply creased military uniform for the victory parade, claiming to have been in the vanguard all along. The philosophers posing as victors in this case are the existentialists, and their descendants are the post-modernists and social constructionists. At the core of their philosophy is the famous Sartrean claim that "existence precedes essence."[19] Sartre thereby thought to free human souls from what he saw as the twin tyrannies of science and religion—both of which he felt had nefarious plans for humans, based on their falsely claimed essential natures.

Liberated from these tyrannies by existentialism, humans would then be free to construct themselves as they wished.[20] There are some things that we can grant post-modernism, and its offshoot—social constructionism. One concession is that it is most certainly true that modern science has discarded the idea of essences. We did not need the post-modernists to point this out. Rejecting essences is the mark of the Enlightenment, characterized by the change from magical thinking to modern science. For example, alchemy became chemistry through the abandonment of magical essentialist thinking, such as the doctrine of signatures.[21] Chemistry was not alone. Biology also has no use for ideas of an essence of male, or essence of female, any more than it has need of the concept of a magical essence—an élan vital—to explain life. But the alternative to belief in essences is not the unrestricted flight of the human imagination, or that more extreme idea that human thought creates all reality.

Scientific advance has been characterized by replacing *essences* with *processes*. No one now believes that fire burns because flammable substances contain an essence—once called "phlogiston." Instead, we now understand the processes of combustion. Functionalism—the idea that what something *is*, is what it *does*—is the guiding principle of physics, chemistry, and biology.

Functionalism is the guiding principle of modern psychology too, although this is far too seldom made explicit.[22] These psychological and behavioral processes that behavioral scientists study are not randomly scattered throughout the population. Many of them are patterned according to someone's biological sex. Furthermore, the fascinating exceptions that test the boundaries of sexual dimorphism—the people with intersex conditions, transsexual folk, the sex-role reversed, gender-atypical people, and all the fascinating kaleidoscope of human variation—turn out to provide profound and illuminating insights into the underlying processes that unite us all.

1.1.2   BEHAVIORAL SCIENCE AND PHYSICAL SCIENCE: DOES EVERYBODY LIE?

How do we explore these processes in humans? In chapter four, I outline some of the main principles of the discipline of ethology: Specifically, of how we observe and describe behaviors—in this case, sexual behaviors—in natural settings, inflected by the knowledge that evolutionary theory brings to this pursuit. Achieving this goal when it comes to human sexual behavior is not easy. Some of this is for practical reasons; humans (like a lot of animals) tend to mate in private. However, there are also conceptual reasons why sex research poses unique challenges.

Daniel Dennett, in *Darwin's Dangerous Idea*, reminds us that "There is no such thing as philosophy-free science; there is only science whose philosophical baggage is taken on board without examination."[23] The issue of human interests affecting objectivity in science becomes especially acute when it comes to sex research. This is because sex research is carried out by actual men and actual women, and there exists an actual battle of the sexes which, as renowned biologist Olivia Judson puts it, is "Eternal, insoluble, and inevitable."[24] In other words, men and women have reproductive interests, and these interests are always in danger of biasing their perception of the data, because they do not always align perfectly. Put more simply—we tend to lie to both ourselves, and to others, on the topic of sex.

Is there a way to solve this epistemological problem; that we are trying to investigate sexual behavior while remaining sexually reproducing beings, at the mercy of our own sexually reproductive urges? Can we ever be objective about these things, or are we as doomed as someone trying to open a box with a key locked inside? I think part of the solution is the least bad one that humans have come up with to solve large-scale coordination problems: namely, democracy. In this case, democracy does not mean that majority rules, because the majority cannot decide on matters of scientific fact. However, it can mean enshrining certain democratic principles in one's work. A key democratic principle is freedom of speech. Freedom of speech makes a society a better place in which to live, of course, but it is even more fundamental than that.

At the core of the liberal totem of free speech is the belief that a multiplicity of views—a heterodoxy—will create a marketplace of ideas in which the best ideas can win out through mutual correction. This means that we need the correcting mechanisms of other people, perhaps with very conflicting viewpoints, to scrutinize data. This approach is also part of the answer to the postmodernist challenge to science—namely, that (allegedly) no one can be objective, because ultimately we all see the world through a subjective lens. On the contrary. According to the view I am sketching here, behavioral scientists do not see the world through *a* lens at all. They see it through a multiplicity of lenses including, but not limited to, independently converging lines of inquiry using vastly different methodologies from disparate fields. These techniques are used by a wide variety of individual scientific practitioners who are on the lookout for one another's errors.

This means that, in a book about female orgasm, I am going to have to address some issues of how we go about studying these things systematically, in other words, to reflect on issues of scientific methodology. Science is somewhat akin to a detective inquiry, and there is rarely a smoking gun solution to any problem. This is certainly true in the field of female orgasm research. What there is, is a compelling picture built up from interlocking scientific approaches and the rejection of some promising leads that turned out to be dead ends. If people want to skip these issues, and get right to the sex research, they could skip to chapter five and refer back if any methodological details seem unclear.

Heterodox values lie at the core of the scientific method. One example of heterodoxy in action is the necessity of passing peer review to be able to publish. Achieving this benchmark in the collaborative effort that is science does not mean that your scientific colleagues all agree with you. Best of luck with your dreams of that. It means only that your work has passed the minimal requirements of demonstrating awareness of relevant other work in the field; that your sums add up; that alternative hypotheses have been duly considered, and so on. That is quite a steep "only," but it is not enough for some critics. There have been some recent and noisy calls to the effect that peer review is "broken." However, proposals for replacing peer review are invariably a lot sketchier than the cries of pain at its inevitable imperfections.[25] What do people want to replace peer review with? Asking god politely whether she will allow us a peek into the teachers' edition of the Textbook of the Universe?

As a reviewer of peers, I can assure everyone that none of us has the answers in the back of the book when we review our colleagues' work. This is not school anymore. It is science, and "peer reviewed" just means your piece of work is good enough to get out in the open scientific playground, to be kicked around by your colleagues. And kick it they surely will. Repeated resilience in the face of that kicking is one of the best guides we have in science. The philosopher of science, Karl Popper, called this kickable quality *falsifiability*.[26] Falsifiability is the requirement that your hypotheses be subjectable to relevant tests, with the condition that if they fail to predict results, they are abandoned. Sometimes reluctantly—but abandoned they are.

What all this means is that in this book, I am going to need to examine a number of good ideas about female orgasm that are, to the best of our knowledge, wrong. To put this in more scientific terms, we have performed as many tests as our imaginations have allowed us (to date), and these theories have come up short. Testability, which is really what falsifiability is all about, is another cornerstone of science. An absence of testability is what makes such fringe activities as pseudo-science, alternative facts, and conspiracy thinking, not so much false as meaningless. Those approaches are simply not playing the same game of knowledge building through the competitive sieving of ideas, refining them, without ever perfecting them. On the other hand, some key lines of inquiry into sex have followed the scientific rules, so that we can build on their work—even if it is flawed.

Some TV shows have made a wider public familiar with Masters and Johnson, those giants in the field of sex research.[27] They tested some hypotheses about female orgasm under laboratory conditions and concluded that it serves no adaptive function. I detail their studies in chapter five, but here I want to point out that their lab-based studies are not the final word on sex because sex, unless one has some very unusual predilections, does not reach its fullest expression in laboratories. As I said at the outset: Mating in captivity is different for humans, just as it is for other species. Drawing on the ethological insights outlined in chapter four, the fifth chapter will also detail a more naturalistic line of sex research, that of the Fox team in the UK. Their research was conducted at the same time as that of Masters and Johnson, but they lacked the media savvy of the American team. The Fox team found a number of interesting effects of female orgasm and gave birth to a rich line of inquiry, which a large number of scientists, including our own team, have inherited and developed.

1.1.3   An Outlaw Discipline?

Ken Zucker, the editor of the flagship scientific journal in sex research, once described the field, with a touch of gloomy relish, as an "outlaw discipline."[28] He meant, at least, this much: That the field of sex research draws together what might appear to be the most motley collection of scientists of any field. This ragtag posse of disciplines needs a leader to effectively hunt down the truth about female orgasm. I contend that this leader is ethology, and I will try to show the value of measuring behavior in natural settings, because I am convinced that it is a missing piece of the puzzle in explaining female orgasm.

Ethology is not just a methodology that focuses on behavior in natural settings, however. It also offers a conceptual framework that unifies different types of questions and answers in the behavioral sciences. The fourth chapter addresses this conceptual framework in some detail, but I will not limit myself to reporting on the results of ethology alone. The answer to the riddle of female orgasm requires a multiplicity of evidence-gathering methods, including (but not limited to) anthropology, psychology, endocrinology, anatomy, physiology, and sex therapy. Chapter six focuses on some crucial elements of clitoral anatomy, physiology, and the attendant neuronal system, which have gained surprisingly scant attention in the relevant literature—popular and scientific—up until now. This integrated female reproductive system has a set of functions, whose nature spans several scientific disciplines.

The sheer complexity of this highly specific integrated system will, I hope, help to convince you that the female sexual response system is not some pale reflection of the male one. On the contrary, the female system is just as complex as the male one, if not more so, and female sexual physiology has an integrated neural system to support it, along with specific brain structures to regulate it. Androcentrism is, therefore, demonstrably scientifically wrong, as well as morally suspect. Is it not nice that truth and justice align?

With a deeper knowledge of the anatomical, physiological, and neurological details at our disposal, we will be able to reconsider those two traditions of sex research I detailed in chapter five. The research that both includes and extends that of the husband-and-wife team of the Foxes makes human sexual physiology and behavior congruent with that of other mammals. In addition, this line of research has firmly resisted multiple and spirited attempts at refutation, justifying a strengthening of our conviction that we are on the right track, just as Popper said science progresses. Both the developments, the attempted refutations of them, and what can be learned by responses to these are the focus of chapter seven.

1.1.4   She Sells Sea Shells on the Sea Shore

Why does sex research routinely find that women get more orgasmic with age? In the preface I mentioned the Mangaians, who have a custom where the older women train the younger men in techniques to generate orgasms in their partners. There are a number of factors at play here. For one thing, technical skills take time to master, but it also seems likely that, over time, women become more in touch with what they want from sexual partners and become more direct about communicating this desire. It is thus doubly important that scientists pay close attention to women's voices—both those in sexual science and those who can report on their experiences. Women's voices have often been absent from science, and this is a sorry state of affairs.

For instance, the great paleontologist Mary Anning could not make a living directly in science, so she had to sell her finds ("sea shells," or sometime ichthyosaur fossils) to make ends meet. The early part of the scientific revolution was mainly for gentlemen of leisure, not for the poor, and certainly not for women.

It is a well-known fact that the word "scientist" was first coined by a man: William Whewell. What is less well known was that he coined it of a woman: Mary Sommerville, whose magnificent *On the Connexion of the Physical Sciences* (1834)[29] he was then reviewing. Up until that point, people working in this field had called themselves "Men of Science," and Whewell wanted to mark the fact that this term was now outmoded. Up until then, women and their work had largely been excluded from recognition.[30]

Of course, it matters in terms of social, and individual, justice that women are not excluded from the sciences. There are purely (and distinctly) factual reasons for valuing more inclusivity as well. Diversity of representation is not just a value to aspire to, it's a procedural necessity for getting to the truth in behavioral science.

This is because behavioral science is not precisely like the other physical sciences. It did not make a scientific difference that Jocelyn Bell-Burnell was a woman, in order for her to discover pulsars. However, it may well have mattered as being a factor in her not gaining the Nobel prize that she deserved. But who now remembers the de facto winner (and co-author on the key paper) of the Nobel for pulsar discovery as anyone other than "that guy who denied the Nobel prize to Jocelyn Bell-Burnell"? Pulsars do not much care about the sex of the person who detected them. Radium could have been discovered by a man, but as it happens, it was not.

However, it probably *did* matter that a generation of female biologists were needed to upend the biological orthodoxy on the importance of female choice mechanisms in evolution. Conservative Victorian scientists (and some later ones) simply missed the fact that the females of any species were active sexual strategists, rather than being the coy recipients of male ardor. This is partly because one of the best strategies can sometimes be to appear to have no strategy. The male scientists had, somewhat embarrassingly, fallen for this ploy. Many of them still do. Victorian notions of the female of any species—let alone our own—as being sexually coy are still alive and well and exert subtle (and not so subtle) influences. Furthermore, it has taken some time to integrate these insights into the behavioral sciences. This integration will be a major theme throughout this book.

Part of the problem in getting traction on the nature and function of female orgasm is that the experience of it is not univocal. That said, remarkable similarities in experience exist cross-culturally. What sense can be made of this apparent paradox? Part of the solution is to appreciate that not all female orgasms are the same. Converging evidence from a multiplicity of cultures—including ones where the mere admittance of female sexual pleasure is taboo—supports the findings of sex researchers and sex therapists in the western tradition. Many of these sources have been insisting for years that there is not *one* female orgasm at all—but several. The insights from this realization will be integrated into the book in chapter eight.

1.1.5 Orgasms as Signals and Signalers

Jennifer Aniston averred that the best smell in the world was that man that you love. It turns out that smell is a critical signal of genetic compatibility—of having babies with strong immune systems—and our team were the first to find that it was the best predictor of female orgasmic response.[31] The theme of chapter nine is that there are patterns of these responses, and those patterns are not random.

On the contrary, they track certain salient partner characteristics and behaviors. Attractive smell, advertising a compatible immune system, is one such characteristic, but there are many others.

One important side effect of the post-modernist (that is, idealist) line of thought has been to overplay human cultural differences and to downplay human biological similarities. Anthropology involves the comparative study of human culture, and I will make a lot of references to it in this book. However, while studying female orgasm, this emphasis on comparison led to some anthropologists being a tad hasty in their accepting the words of (for instance) male elders of the studied groups that, naturally, their wives and daughters were sexually inert, in-orgasmic beings.

For instance, some anthropologists reported—with straight faces—that even the concept (let alone the experience) of female orgasm did not exist in Irish folk communities as late as the 1970s.[32] We would caution against mistaking the unwillingness to report private matters to strangers, with a decisive proof that a phenomenon has been shown to not exist. We are also happy to report that female orgasm has made a comeback in this century, in Ireland at least.[33] Closer examination of other cultures reveals that the evidence for female orgasm being a social construction is less than paper thin, and that, even in cultures where the discussing of sexual response is taboo, sensitive questioning reveals a remarkable similarity in women's reported experiences.[34] This topic will require a chapter of its own, the tenth.

Some prominent scholars and sex therapists have argued that female orgasm occurs inefficiently. My response to that is that efficiency is not the hallmark of good sex. At least, not of good *human* sex. Some of these are issues of technical competence and partner attentiveness, but human sex is rarely a function of mere mechanistic rehearsal. If it were, we would never bother going to the trouble of finding partners and learning their preferences and desires. Indeed, even given the increase in technological solutions to unrequited sexual desire—such as sex dolls and immersive pornography—humans still desire to pair up with other real humans.[35] This getting *in sync* is more than mere metaphor; it seems to have measurable neural correlates. A growing body of research suggests that we learn to be better at sex with particular people, and this manifests in a variety of interesting ways. Female orgasm can be seen as part of the general sexual choosiness of women, and we have a growing body of evidence for the sort of things that they are choosing—not necessarily consciously—on the basis of. This will involve addressing that part of human sexuality that reveals itself in fantasy and desire.

There is at least this much truth in the idealist impulse: Our imaginations *do* have power. But what sort of power? Some things exist only because enough of us believe in them. Money is a good example. But, on an individual level, human sensuality, fantasy, and eroticism are not optional add-ons to an understanding of our sexual natures.

Germaine Greer famously retorted to a critical lover, "I'm not frigid, *you're* boring." Multiply converging lines of evidence suggest that sexual encounters act as (literal) test beds for partner qualities. This is true of both sexes, but it is especially true that men are on the spot here. The logic of evolutionary signaling prompts consideration of the potential recipients of said signals. The pattern of female sexual response outlined in the book up until this point has prompted a range of male reactions to try to either align with, or subvert, female choice mechanisms. For example, there is now evidence that male interest in female orgasm is no more accidental than male interest in lots of other things that have fitness-related consequences, such as status. However, one set of male strategic responses is to attempt to subvert those female choice mechanisms. Females are also often in competition with one another, and patterns of subversion of choice mechanisms exist within each sex, as well as between the sexes. These strategic responses—often manifest in cultural practices, cognitions, and social norms—are also part of the complete picture of a mutually sexually selecting species, and I will address these issues in chapter eleven.

I bring together all these themes in that final chapter. Humans attempting to explore nature, while it is acting through them, is a dominant pattern, when I examine the various ways in which men and women have maneuvered around their key rivals (members of their own sex) and potential collaborators (members of the opposite sex) to maximize their fitness goals. These have manifested themselves in those large, but finite, proximate strategic responses of physiology and culture.

1.1.6 HUMANS AND NATURE

Once the implications of the fact that humans are a mutually sexually selected (and selecting) species are appreciated, then many hitherto unfathomable mysteries become puzzles to be solved. One puzzle in particular—as to the nature and function of the female orgasm—is the reason for this book. But why has the application of science to human behavior taken so long in any case? Our species has been studying physics for half a millennium and biology for three centuries, but psychology is barely a hundred years old. Some even contend that psychology is not, cannot be, or possibly even, *should* not be a science.

"We live on a placid island of ignorance in the midst of black seas of infinity, and it was not meant that we should voyage far."[36] H. P. Lovecraft feared that, faced with the universe and its utter indifference to our values and desires, humans would go mad. With all due respect to the dark prince of horror, reality is not quite as bad as he feared. But, for interesting reasons, we humans have found the turning of the lenses of science back on ourselves to be an especially difficult exercise.

The objective methods that we have used to view the worlds of physics, chemistry, and biology can—when used on ourselves—show us images of humanity with which we are not totally comfortable. Caliban, the monster of the Tempest, raged at seeing his own face in the glass.[37] This is part of the reason why the human sciences are still in their infancy. It has not been mere ignorance that has held us back, but fear of what we might find, fear that Caliban had a point. Fears exist that a science of human nature offers a stark brutal picture, without the consolations of art, or the redemptions of religion.[38] I hope to show that many of these fears have been overblown. However, these fears did not come from nowhere. They arise because humans have moral and image-rich cognitions, and they rebel at certain pictures of themselves that science has appeared to paint. A richer conception of behavioral science as enhancing and deepening our other understandings of ourselves, rather than replacing them, is now available.

As a final thought for this introductory chapter, Stephen J. Gould reminds us that: "Scientific images are not frills or summaries; they are foci for modes of thought."[39] How right he was. Here he was referring to the iconic "March of Progress" illustration of evolution (Figure 1.1). We are all familiar with it. It has something a bit like a gibbon at one end, loping

FIGURE 1.1   Zallinger, *The March of Progress*, from Wikipedia Commons.[40]

FIGURE 1.2 *Evolution of Man and Woman* by Tom Rhodes.

Source: (With permission of Tom Rhodes, the artist)

up from the phylogenetic mists, and then a progression of more (than less) simian ancestors, leading up to the (secretly, of course, we all believe this) "Real Reason for the Whole Affair": namely, "Us"! (Fanfare optional).

Zallinger's illustration is justly celebrated, but it tends to carry with it a few somewhat unfortunate implications, along with its elegance. Two are of particular importance for what I need to say.

1.1.7 LOVERS *AND* FIGHTERS

For the first erroneous implication—there are no women in that early evolutionary picture—it is always depicted with men. Men striding confidently. Men carrying weapons or (sometimes) fire. But always, men without genitals, as if the inclusion of these would somehow lower the tone. "Man (and, here, I do mean *Man*) The Hunter." "Man, the Warrior." Never, "Humans, the Lovers." This emphasis matters greatly for the picture it builds of ourselves.

The second issue is that the linear progression can be taken to imply that we humans are the main event, as if our hairier ancestors were proto-humans, mere supporting cast. But those other apes are not also-rans. Our ancestors (and sometimes contemporaries) were highly successful hominids in their own right, very well adapted to their particular ecological niches. We could profit by their example.

The artist Tom Rhodes offers an alternative to this parochial picture, which captures much of the spirit of this book (Figure 1.2). Namely, that proper attention to our highly successful evolutionary lineage, with men and women both represented as equal partners (and even enjoying one another's company) is a crucial part of understanding humanity. We are a mutually sexually selecting species, and this matters a lot because, to a large extent, we have created one another. Sexual selection has made us what we are.

NOTES

1 In an interview to *People Magazine*, August 1998.
2 Said in interview in *The Rebels of Oz: Germaine, Clive, Barry and Bob.* BBC Four Programme, aired 1 July 2014, Serendipity Productions.
3 Reported on by Davenport, W. H. (1977). Sex in cross-cultural perspective. *Human Sexuality in Four Perspectives*, 115–163.
4 Miller, G. (2009). *Spent.* Viking Penguin.

5 Schmitt, D. P., Realo, A., Voracek, M., & Allik, J. (2008). Why can't a man be more like a woman? Sex differences in Big Five personality traits across 55 cultures. *Journal of Personality and Social Psychology*, *94*(1), 168. See also *Science Daily*. (2018). Blog post *Countries with greater gender equality have a lower percentage of female STEM graduates*. www.sciencedaily.com/releases/2018/02/180214150132.htm last accessed 22/08/2023.
   for a good discussion of a replication of the gender equality paradox.

6 A recent work of popular science, Prum, R. O. (2017). *The evolution of beauty: How Darwin's forgotten theory of mate choice shapes the animal world-and us*. Anchor argued that female sexual selection was simultaneously important, because it explains the baroque, but that the baroque was itself unimportant. This is a big topic area, and I will address it in chapter nine.

7 Freud, S. (2017). *Three essays on the theory of sexuality: The 1905 edition*. Verso Books.

8 The by-product theory of female orgasm is first expressed in Symons, D. (1979). *The evolution of human sexuality*; and then staunchly defended in (e.g.) Gould, S. J. (1987). Freudian slip. *Natural History*, *96*(2), 14; and Lloyd, E. A. (2009). *The case of the female orgasm: Bias in the science of evolution*. Harvard University Press.

9 The great biologist Martin Daly used to use this phrase in lectures. Thanks to Catherine Salmon for bringing it to my attention.

10 Twain, M. (1897). *More tramps abroad* (p. 238). Chatto & Windus. Other animals show inadvertent shame responses as well, of course. However, the particular pattern that we display to other humans appears unique. Shearn, D., Bergman, E., Hill, K., Abel, A., & Hinds, L. (1990). Facial coloration and temperature responses in blushing. *Psychophysiology*, *27*(6), 687–693.

11 *The Theaetetus of Plato*. Jackson, Wylie & Company, 1928. (Trans. M. J. Levett). Section 152a.

12 Babbitt, D. D. (Ed.). (1936). *The Dhammapada*. Oxford University Press.

13 Byrne, R. (2006). *The secret*. Simon and Shuster.

14 This is one meaning of "social construction." It is not the only one. For one thing, the term could also usefully be deployed to mean "path dependent." For example, being a man or a woman includes a certain historicity—perhaps of hopes, privileges, and expectations that the other sex knows little or nothing of. There is nothing objectionable and, indeed, much to be learned from this usage.

15 Dick, P. K. (1985). *I hope I shall arrive soon*. Doubleday Books.

16 This is a very big topic and a large part of chapters 3 and 4. The short answer to the brain dimorphism challenge is that male and female brains are somewhat like male and female faces. There may not be any such thing as a male nose or a male chin per se, but people can distinguish the patterns of difference most of the time and recognize a male or female face. In fact, when it comes to brains, the differences are stronger than that because the male brain contains a map of the male body in it, and (more crucially for the purposes of this book) the female brain contains a map of the female body—especially the sexual organs—in it. This brain map is called the somatosensory cortex and is discussed in Di Noto, P. M., Newman, L., Wall, S., & Einstein, G. (2013). The hermunculus: What is known about the representation of the female body in the brain? *Cerebral Cortex*, *23*(5), 1005–1013.

17 And it is contentious. For example, Rippon, G. (2019). *Gender and our brains: How new neuroscience explodes the myths of the male and female minds*. Pantheon; received a favorable review in Nature, of all places (Eliot, L. (2019). The gendered brain: The new neuroscience that shatters the myth of the female brain. *Nature*, *566*(7745), 453–454. There were furious rebuttals to this, which I discuss in more detail in chapter 3.

18 Dreger says this in various places, but chiefly in Dreger, A. (2016). *Galileo's middle finger: Heretics, activists, and one scholar's search for justice*. Penguin Books.

19 Sartre, J. P. (1948). Existentialism and humanism (1947). *Philosophy: Key texts, 115*.

20 The essence (if such a pun be permitted) of the idea that "existence precedes essence" is to be found in Heidegger (whom Sartre credits with it) and also Kierkegaard. Heidegger, M. (1996). *Being and time: A translation of Sein und Zeit*. SUNY Press.
   Kierkegaard, S. (2013). *Kierkegaard's writings, VII, volume 7: Philosophical fragments, or a fragment of philosophy/Johannes Climacus, or De omnibus dubitandum est* (Two books in one volume) (Vol. 22). Princeton University Press.
   More pessimistic thinkers like Foucault thought we were doomed to be at the mercy of others' constructions. See, e.g., Foucault, M. (1980). *Power/knowledge: Selected interviews and other writings, 1972–1977*. Pantheon.

21 Sagan, C. (2011). *The demon-haunted world: Science as a candle in the dark*. Ballantine Books.

22 See Dennett, D. C. (2001). Are we explaining consciousness yet? *Cognition*, *79*(1–2), 221–237, for a more extended discussion of this topic.

23 Dennett, D. C. (1995). *Darwin's dangerous idea* (p. 2). Penguin.

24 Judson, O. (2002). *Dr. Tatiana's sex advice to all creation: The definitive guide to the evolutionary biology of sex*. Macmillan.

25 McCook, A. (2006). Is peer review broken? Submissions are up, reviewers are overtaxed, and authors are lodging complaint after complaint about the process at top-tier journals. What's wrong with peer review? *The Scientist, 20*(2), 26–35.

26 Popper, K. (1953). *Science: Conjectures and refutations*.

27 For example, the period drama, *Masters of Sex* (2013). TV Show, distributed by CBS Television, produced by Round Two and Timberman/Beverly productions.

28 Zucker, K. (2002). From the editor's desk: Receiving the torch in the era of sexological renaissance. *Archives of Sexual Behavior, 31*, 1–6. Quote is from page 1.

29 Somerville, M. (1858). *On the connexion of the physical sciences*. J. Murray.

30 Whewell, W. (1834). Mary Somerville: On the connexion of the physical sciences. *Quarterly Review, 51*, 54–68.

31 King, R., & Belsky, J. (2012). A typological approach to testing the evolutionary functions of human female orgasm. *Archives of Sexual Behavior, 41*(5), 1145–1160.

32 Messenger, J. C. (1971). Sex and repression in an Irish folk community. In *Human sexual behavior: Variations in the ethnographic spectrum* (pp. 3–37).

33 We surveyed attitudes toward and opinions of a range of sexual behaviors to help understand knowledge of and attitudes toward assistive reproductive technology. Modern Irish populations, religious or not, appear to have rediscovered female orgasms, if they had lost them previously. Dempsey, M., King, R., & Nagy, A. (2018). A pot of gold at the end of the rainbow? A spectrum of attitudes to assisted reproductive technologies in Ireland. *Journal of Reproductive and Infant Psychology, 36*(1), 59–66.

34 Obermeyer, C. M. (1999). Female genital surgeries: The known, the unknown, and the unknowable. *Medical Anthropology Quarterly, 13*, 79–106. Arguably, Mead, M. (1963). *Sex and temperament in three primitive societies*. Morrow, makes a similar claim. Messenger, J. C. (1971). Sex and repression in an Irish folk community. *Human Sexual Behavior: Variations in the Ethnographic Spectrum*, 3–37.

35 Although it is certainly possible that this is in something of a decline.

36 Lovecraft, H. P. (1999). *The call of Cthulhu and other weird stories*. Penguin.

37 "The nineteenth century dislike of Realism is the rage of Caliban seeing his own face in a glass." So says Oscar Wilde in the preface to (1891) *The picture of Dorian Gray*. Wilde goes on to say that "The nineteenth century dislike of Romanticism is the rage of Caliban not seeing his own face in a glass." The tension between realism and romanticism is not quite the tension between science and art—because great art also aspires to conditions of truthfulness—but there are important parallels.

38 This squeamishness can sometimes manifest in calls for absurd levels of confirmation in human behavioral sciences, but only these, making one suspect that the scientific probity being alluded to is less than authentic. For example, such levels of required evidence should also apply to human biomedical science, and they rarely do.

39 Gould, S. J. (1991). *Bully for Brontosaurus. Further Reflections in Natural History*. This quote comes from (p. 171).

40 Zallinger, R. (1970). *Illustration of the march of progress*. Time Life.

2 Coy Females?

"I feel sorry for straight men. The only reason women will have sex with them is that sex is the price they are willing to pay for a relationship with a man, which is what they want," he said. "Of course, a lot of women will deny this and say, 'Oh no, but I love sex, I love it!' But do they go around having it the way that gay men do?"

(Stephen Fry, *Attitude* Magazine, November 2010)[1]

Alice: Millions of years of evolution, right? Right!? Men have to stick it in every place they can, but for women it's just about security, and commitment, and—and whatever the fuck else!
Bill: A little oversimplified, Alice, but yes, something like that.
Alice: If you men only knew.

Nicole Kidman (as Alice Harford) and Tom Cruise (as Bill Harford) in *Eyes Wide Shut*[2]

The heart of this book is the scientific implications of female-driven sexual selection on human nature. In this chapter, I explore some of the practical and visible ways that you can see this happening around you every day. The fact that some of the things I am going to say will surprise at least some of you (most, but not all, of the surprises will come to the men reading this) is testament to the oft-hidden nature of these female choice mechanisms. Once you develop an eye for it, it becomes like one of those ambiguous optical illusions where you suddenly see a hidden figure. You will wonder how you ever missed it. There is a reason for this: You were meant to miss it.

The quotes that start this chapter capture two popular, yet polar opposite, conceptions of female sexual desire. The first view is that heterosexual sex is rarely (if ever) pleasurable to women; it is something that they endure (lying back and thinking of England, perhaps) in order to get what they really want, namely children (unless they really enjoy those patriotic musings, of course). The second view captures the insight that the reason men have been inclined to think that women are sexually coy is that they (the men) have been subjected to one of the most successful strategic hoodwinkings since (literally) the dawn of recorded time. And, like many cons, it is more effective if the con artist sometimes believes it herself.

When such differing interpretations of the same data set, in this case human sexual behavior, are held so strongly, this is an opportunity to investigate the separate slices of perceived reality that humans occupy. To give away the ending, I think the evidence is that the answer to this question is overwhelmingly that women are neither sexually passive nor are they coy. What they are is comparatively *choosy*, which is a somewhat different thing.

Relative female choosiness is a function of differential minimum parental investment, rather than deriving from some medieval notion of essences. But rather than trot out the phrase "that sex which, in our species, is the one that has the highest minimum potential investment in child-producing" every time, I am just going to say "women," and trust that this does not cause undue confusion.

A large part of the reason that men have been willing to be fooled is that admittance of active female sexual agency taps into one of the most basic fears of a species like ours with internal gestation—paternity certainty.[3] Women can never be in any doubt as to the mother of their child, and thus (otherwise costly) genes that alerted them to being fooled about this would not be selected for. On the other hand, fathers can be wrong about paternity, and we have developed a raft of cultural and psychological adaptations (such as sexual jealousy) that speak to this genetic threat. This topic is so important that it will be dealt with in detail in a later chapter, but for the moment I have a different question. How would an internal willingness to deny female sexual agency fit this need for

DOI: 10.1201/9781003372356-2

genetic certainty? Would it not be better for men to be obsessively jealous of their partners at each and every moment?

Sometimes, and in some places, men do develop, and even enshrine in culture, obsessive sexual jealousy. An obvious example would be the *crime passionel*, where, in France, even as late as the 1970s, it was considered a borderline legal defense for a man to murder a lover found with another man.[4] But, as with many things that are the result of evolutionary pressures that pull in slightly different directions, a balance is likely to generate the highest average fitness benefit. The most plausible reason for the common (individual) blindness about female sexual agency—when it does occur—seems to be because once a man has decided—or is in the process of deciding—to invest in a woman long term, then he needs to silence some of these fears and take the commitment plunge. How is this trade-off manifest at the proximate level—the level of daily experience? Women have long (and not unreasonably) complained of the *Madonna/Whore* stereotypes that get foisted upon them by social convention, but the ethologist asks, "Where do such stereotypes originate, and why do they persist often in the teeth of socio-political push-back?"

Madonnas—those women who could be reliably invested in long term, because their lack of sexual agency guarantees paternity certainty—would benefit from being seen as sexually naïve and thereby unlikely to stray. It is also strategically advantageous for some women to project this image under certain circumstances. Such female judiciousness is easy to mistake (by some men) for a general lack of interest in sex. As renowned biologist Robert Trivers (and Sigmund Freud before him) has pointed out, the easiest way to *appear* sincere is to *be* sincere.[5] Like spies in the field operating on a "need to know" basis, humans have a raft of mechanisms to prevent themselves from having transparent access to their own underlying motives. Nothing I say about sexual strategizing implies that it must be conscious strategizing. It might be—and interestingly, often is—but it need not be.

This hoodwinking insight—that females may appear sexually coy but are, in fact, sexually choosy—is true, not just of humans, but of many species. Across taxa, females thought (by human observers—and very often male observers—to begin with at least) to have been coy turned out to be active sexual strategists.[6] A range of complex and sophisticated behaviors and preferences have been observed to have evolved to enhance their (really their *genes*) fitness.[7] When it comes to humans, the level of delusion is even higher. Humans are quite possibly the most mutually sexually selecting species in existence. So, how did the story that males across nature are naturally promiscuous, while females are naturally monogamous, get traction in the scientific community? This part of our tale goes back to 1948.

2.1 HOGAMOUS, HIGAMOUS

The story goes that one night, after experimenting with nitrous oxide (laughing gas), someone believed they had written down a deep insight into the universe. Upon awakening, they read what they had written:

> *Hogamous, Higamous,*
> *Man is polygamous,*
> *Higamous, Hogamous,*
> *Woman is monogamous.*

This piece of doggerel has been variously attributed to (among others) William James, Ogden Nash, and Dorothy Parker, but in biology, the idea that this captures a deep insight about nature goes back to the geneticist Angus Bateman.

In 1948, Bateman did an important series of experiments with fruit flies, in which he appeared to establish the fact that multiple matings improved male fitness, but had no effect on female fitness.[8] Since then, people have used "Bateman's principle" as an explanation for the (apparent) prevalence of male sexual promiscuity, and female sexual coyness, across nature. However, biologists have done a lot of work since 1948, and no mainstream scientist thinks that Bateman's principle is true in anything like its original formulation. A little logic reveals that every promiscuous act by any male of any species must be matched by one by a female.[9] However, it does not follow that the desires and goals underlying each act are the same.

Indeed, the truth is much more interesting. Multiple matings do (perhaps unexpectedly) benefit females in all sorts of ways, across all sorts of species. Surprisingly, they even helped the fruit flies in Bateman's original experiments (Figure 2.1). Why then did he not say this at the time? It turns out that he did. It is simply that Bateman could not explain his own findings at the time, but—proper scientist that he was—he faithfully reported them all the same.

At the time Bateman was experimenting, no one could think of ways that multiple matings could help females of any species. It seemed (and still seems to some) obvious that humans (or lady fruit flies) cannot be a "little bit pregnant." However, since Bateman's time, we have discovered a whole host of ways in which multiple matings can increase female fitness. How do males show their love? Let me count the ways. Nuptial gifts, sperm selection mechanisms, pair-bonding, and a host of other cryptic (and not so cryptic) female choice mechanisms.

Some evolutionary theorists have been slow to catch up to this, insisting that women are a limiting resource, thereby forcing men to do all the competing for them. However, more recently there has been some sign of a welcome correction to this simplification, by including some facts that ethologists—and common wisdom—have known for some time. Women feel raw sexual desire, compete over access to men, test levels of partner commitment, and denigrate rivals. I will have more to say about the latter two points in subsequent chapters. My goal in this chapter is to kill, or at least knock down for the count, the myth of the sexually coy female human.[10]

Let me start this by thinking of some obvious ways to test the claim of female sexual coyness in humans. If females were really sexually coy, then we might predict them to be uninterested in physical arousal in the absence of partners, to be incapable of generating their own sexually oriented fantasy scenarios, and to passively wait for singular male sexual attention, which they could then

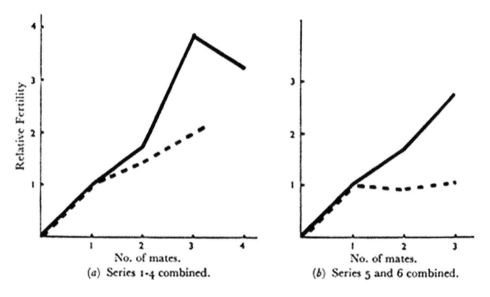

(a) Series 1-4 combined.   (b) Series 5 and 6 combined.

FIGURE 2.1   Redrawing of Bateman's original graphs of six series of experiments, in two groups, plotting relative fertility against number of mates for male and female fruit flies. The dotted lines show the rise of female relative fertility, the solid lines male. As can be clearly seen in the first series of four experiments (a), female relative fertility also increased with the number of mates, although not quite as sharply as male (b).

Source: Courtesy of the *Heredity Journal* (redrawn)

finesse into commitment to offspring and an exclusively monogamous relationship. To put it briefly, the data are in, and none of those things is true. Let me detail each in turn.

2.2 FEMALE-DESIGNED SEX TOYS

"Desire is the essence of man," wrote Spinoza,[11] and I think I am safe in saying that he would have been happy to add "and of woman." So, let us start with the most basic question: Do women enjoy raw sexual stimulation? The most direct and obvious proof that they certainly do and that they enjoy such stimulation in the absence of men (or other women) is that they pay for it: The sex toy industry, which *Forbes* estimated as being worth $15 billion in the United States in 2016, is predicted to rise to $50 billion by 2020, and a huge element of this is vibrators for women.[12]

Human desire—when markets are given free rein—becomes more and more visible in the products we choose over others.[13] Advertisers may be able to get us to associate their products with other things we desire—like sex, status, and security—but they have no mechanisms for creating those basic desires from nothing. Increasingly, sex toys are being designed by women and for women, with profit-driven feedback mechanisms. This allows us an interesting window into our evolved psychologies and physiologies.

One of Darwin's inspirations was to notice that evolution—descent with modification from common ancestry—can work even faster by *artificial* selection than by the natural kind.[14] We see such changes in consumer-led alterations to products such as teddy bears[15] and Mickey Mouse™.[16] Both became noticeably more altricial—the technical word for cute—over a century of production, resulting from audience/customer feedback. Another self-consciously evolved popular example would be the Teletubbies™. These brightly colored, repetitively chattering, baby-friendly gonks were the result of a deliberate, iterative, evolutionary process that favored and replicated those physical elements that a team of specially recruited babies paid the most attention to. At each stage, the designer consciously discarded the duller elements (judged as so by measuring levels of baby attention) and then re-presented newly evolved Proto-tubbies to the waiting baby audience. Their creator, Anne Wood, is a child psychologist who knew exactly what she was doing.[17] Natural selection is slower than the artificial kind, but no less unforgiving of failures.

In each of these cases, small changes—in some cases unconsciously driven—led to increased sales, more positive audience response, happy baby stares, or some other form of positive feedback. As rapper Baba Brinkman puts it, "Performance, Feedback, Revision" is the essence of both natural and sexual selection.[18] But this is also true of artificial selection, and here it can happen with much greater speed than someone might have expected. For instance, the great Russian biologist Belyaev et al. (1985) managed to selectively breed aggression out of Arctic foxes in only eight generations. Not only did the foxes become domesticated, but a process technically called *linkage disequilibrium* (genes that really like to hang out with one another) meant that the foxes also became cuter with each generation—developing adorable floppy ears and eye-patches.[19]

In artificially evolving human cultural artifacts, we have one big advantage over blind evolutionary processes. We can tear up existing designs, and go back and produce the thing we *really* wanted all along. And then, we can improve on that—like the Teletubbies™. The resultant products, which includes both artifacts and ideas, are then windows into human desires, relatively free from constraints. When we produce physical artifacts, these are windows into human needs. Recently, a leading sex toy manufacturer allowed a team of women to directly engineer in this way with sex toys. None of these women was a professional engineer, and none had had any other relevant design experience. What they came up with was very interesting to a sex researcher.

Prior to this current female-driven engineering renaissance, for a considerable time in the twentieth century, sex toys for women orbited male expectations about female desires. In consequence, dildos tended to be large, surgical pink, and somewhat rigid.[20] This has changed, with many stores and manufacturers catering solely to women—indeed, some stores do not even let unaccompanied men into them. The same is true of female sex toy parties. Studying the dynamics of them, one set of researchers wrote: "Rather than stroking the male ego, women can focus on their own pleasure. This exclusionary narrative conscientiously constructs the party as a safe place."[21]

So, what about these modern sex toys? The recent ones created by women, for women. What lessons in female desire can be drawn from them?[22]

1) *Penetration is sometimes, but not always, desired.* Five out of six of these winning toys were penetrative ones. This is tough to explain if we believe the notion that women are not sensitive inside their vaginas, as some scientists have claimed and as we will explore more thoroughly in chapter six. Indeed, the women in question typically reported multiple areas of internal sensitivity.

2) *Size. Does it matter?* The answer appears to be that size matters somewhat. None of these toys were huge—unlike (for example) gay male penetrative toys whose size can tend towards the alarming for those unprepared. However, all the female penetrative toys had significant girth, albeit that this was well within the range of most human males.

3) *Polymorphous perversity.* Some toys were also designed to stimulate other areas, such as the external parts of the clitoris and the nipples. Women clearly do not have one single erogenous zone—they have many. The new toys often cater to these.

4) *Flexibility.* Modern silicone has allowed the creation of sex toys with significant ability to bend. In nature, the human penis bends to the shape of the vagina during intercourse, with the possibility of interacting with the full extent of the internal parts of the clitoris, as well as other sensitive areas detailed in chapter six. Fascinating fMRI studies, done of coupled sex in the late 1990s, revealed this surprising fact. If there is a lesson there, it is that female choice is driving male anatomy. Many of the preferred sex toys bend in exactly the same way as the human penis, using modern silicone to facilitate this design.[23]

This is all interesting information and tends to undermine notions both of a generalized female sexual coyness and also the curious notion that has entered some academic discourse, that even aroused women have little or no internal sexual sensitivity. But that is not the only reason to reject claims that sexual penetration is something that is only undertaken to please men.

Hard to explain on the "man-pleasing" view is the existence of toys aimed specifically at lesbian clients. *Strap-ons* is the usual term for these items, but a brief internet search (possibly done while not at work, unless you also happen to be a sex researcher) will reveal a bewildering variety of such devices, all designed to produce vaginal penetration as part of an all-female sexual encounter. In our own research in 2012 we found, without explicitly looking for this, that vaginal penetration was reported as leading up to or occurring during orgasm, in most of the lesbian encounters reported to us.[24] We are not alone in finding this sort of response. In addition, plenty of sex toys aimed specifically at the lesbian market specifically involve vaginal penetration.[25] Furthermore, this is neither a new nor a specifically western phenomenon. For example, a strikingly large collection of Japanese erotic art from the Edo period depicts lesbian sexual encounters, some of which use an impressive and imaginative range of insertable sex toys.[26] As I explore in more detail in chapter ten, the walls of many ancient Hindu temples have similarly instructive themes.[27]

These observations need to be set against an oft-repeated insistence that the fact that only 20% of women use penetration during digital masturbation is somehow a hugely important fact about human sexuality.[28] There is nothing wrong with solo unaided masturbation, to be sure. On the contrary, it is natural and healthy, and humans who care about their sexual partners would benefit greatly by studying one another's preferred methods for achieving orgasm through it. However, its goals and opportunities are different from those of partnered sex. People do not typically kiss objects during masturbation either, and men rarely use toys that they penetrate, to achieve orgasm during their masturbation. What of it? I submit that precisely nothing of importance follows from these utterly commonplace and unsurprising facts. For one thing, not all orgasms feel the same. I will have much more to report on this in chapter eight, but it is worth noting, even here, that women routinely and readily distinguish masturbatory and coital orgasms.[29] Second, rapidity and efficiency at achieving orgasm, using minimal stimulation, are routinely achieved by *both* sexes during masturbation; there is nothing special or unique about women in this respect.

Sex toys have also undergone revisions in the last few thousand years.[30] That is not a typo. We have good evidence of dildo-like objects (which are often rather coyly termed "ritual artifacts" in museum cabinets, such as the British Museum's "Cabinet 55"[31] and Naples' "Forbidden Rooms")[32] dating back to the ice age. Some of these allow double penetration—e.g., of two women facing one another. Some authors even make claims of ancient vibrating toys—that insects trapped in tubes were used in ancient Rome or by such notables as Cleopatra. There do not seem to be any references to these before 1992, however, making it inherently unlikely. And, speaking as a sometime bee-keeper, I cannot think of any material thin enough to feel a bee's buzz through that would not also make you distractingly vulnerable to stings.[33] All that aside, there is some evidence that the use of insertable sex toys has a venerable history. For instance, Aristophanes was making ribald references to *olizboi* to his all-male (yet surprisingly au fait with female sex toys) audiences, 2,500 years ago.[34]

Rachel Maines advanced the intriguing hypothesis that the use of vibrators to bring women to orgasm was a central part of neurology practice in the nineteenth century, with specialist tables used for the purpose.[35] Such a hypothesis would have fitted well with the Freudian claim that sexual neurosis and dissatisfaction lay at the heart of many mental health issues of the time.[36] However, subsequent research by Kate Lister and Fern Riddell[37] cast serious doubt on this tempting interpretation. Lister notes that while many sex toys are detailed in Victorian pornography, a diligent search turned up no mention of vibrators used in this context. This would have been an odd omission, given what was *not* omitted. For example, a famous work of Victorian erotica was not remotely coy about dildos, as a representative quote will illustrate: "As we are five to two you will find I have a stock of fine, soft, firmly made dildoes to make up the deficiency in males, which alternated with the real article will enable us to thoroughly enjoy ourselves."[38]

Tellingly, it transpires that the apparent reticence of Victorians in describing sex was rather like our own modern reticence: In other words, context was all. Kate Lister, a scholar of Victorian erotica, puts it thusly, "Victorian doctors knew exactly what the female orgasm was; in fact, it's one of the reasons they thought masturbation was a bad idea."[39]

In summary, far from being uninterested in raw sexual pleasure, women have always been, and remain today, demanding and multiply sensitive beings, who use complex (and sometimes penetrative) toys in sex play, whether men are present or not.

Are these women just "lying back and thinking of England" while all this occurs? Not hardly. Fortunately, many of them have told us exactly what is in their heads, and patriotic thoughts are not prominent. What lessons can be learned from female-driven sexual fantasy?

2.3 PORNOGRAPHY WRITTEN AND FILMED BY WOMEN FOR WOMEN: WHAT LESSONS CAN WE LEARN?

There is something psychologically and sociologically fascinating about the fact that each generation finds new ways to deny female sexual agency. There have been, to take an absurd example, eras and cultures where women had to eat bananas in private because of the supposed shocking effect this might have, if they did so publicly.[40] However, when we turn to the erotica written by and for women, it becomes increasingly hard to maintain the notion that women are sexually coy.

Homo Sapiens' active imaginations must exist, at least in part, to reflect on past and plan for future experiences. The products of those imaginations—especially when they are freely generated and shared—therefore offer interesting insights into our natures. Imagination can even be a form of planning, unconstrained by messy detail and compromise, especially compromise with the desired sex.

It is commonly reported that men's sexual imaginations are visual, whereas women's are conceptual. The truth is somewhat more nuanced than that, but there certainly do seem to be trends in that direction.[41] That said, there is a thriving industry in visual—i.e., film—pornography written by, directed by, and targeted at women. Some examples would include *Our Porn Ourselves*, or the work of self-consciously feminist pornographers such as Petra Joy or Anna Arrowsmith. Such pornography does tend to have more of a storyline than a lot of the pornography directed at men, and the market for it is comparatively small. Still, it suggests that the old industry slogan that "men crave simulation while women crave stimulation" requires some nuanced unpacking. And, since any mention of pornography tends to raise defensive hackles in some, I will have to say a few things about that as well, even though I am only going to be discussing female-written and targeted pornography, not male.[42]

The Internet is a window into human nature.[43] We are being given the chance to get much closer than we would have previously thought possible to reading one another's minds. Is what we see in there always nice? Not so much. It is a fact that needs to be faced that many humans desire things in the bedroom at night that they would march in protest against during the day.[44] It is also a fact that there is no pressing contradiction in acknowledging this point. Fantasy—as long as it is reliably distinguished from reality—is healthy. In addition, there is decent evidence that those with even somewhat kinky fantasies are psychologically healthier than those without, as long as consenting adults are their objects of desire.[45] Despite these findings, some would like to repress all such fantasies. Some even make a connection linking fantasy to reality. One of the most famous of these would be Susan Brownmiller, who boldly stated, "*Pornography, like rape*, is a male invention, designed to dehumanize women, to reduce the female to an object of sexual access, not to free sensuality from moralistic or parental inhibition."[46]

While it may be true that some people cannot distinguish fantasy from reality, that in itself is the cause for worry, independent of the content of the fantasies.[47] For the rest of the species, it had better not be true that fantasy and reality are so closely linked as Brownmiller implies. Think, for example, about how many times you have fantasized about killing your boss. Did you do it? Did you appear to know the difference between fantasy and planning?[48] And, if female pornographic fantasy cannot be distinguished from reality, then our species is going to be in just as much trouble as if this was the case with men. In any case, part of the "pornography is patriarchal oppression" model can be tested directly. If male pornography is to be understood as nothing more than part of a pattern of oppression of women (rather than the expression of specifically male evolved desires), then we would expect gay male pornography to be noticeably different in themes and styles from its heterosexual equivalent. However, this prediction is conclusively refuted. Gay porn has just the same proportion of roughness and dominance, with just the same absence of relationship focus, that heterosexual male porn typically has.[49]

In any case, it can hardly be said that themes of dominance, hierarchy, and objectification are absent from female-generated pornography, now that women have had a chance to express these desires more freely, and we have all had a chance to access them. Fantasy might well be the only reliably true thing on the internet. Grass-roots support generates much of the massive amount of female-oriented pornographic content that we find there. A notable recent example is the incredibly successful book and movie franchise, *Fifty Shades of Grey*, which originated as fan fiction about the *Twilight* characters, but that is just the famous tip of a very deep iceberg.[50]

A prominent example of similar work, which is usually below the icy surface of polite society, would be slash fiction. This is sexual fantasy starring famous folk (usually two men) in highly sexualized encounters. Thus, one might read about Holmes/Watson or Kirk/Spock. This fiction is written almost exclusively by women and for women. Catherine Salmon and Donald Symons, in the evocatively titled *Warrior Lovers*, exhaustively documented the details of how such a fictionalized world details the desires of many women for complex, highly sexual, but committed partnerships.[51]

Typically, the more masculine protagonist in these stories is someone who could choose from a raft of desirable females, and the end goal of the slash story is frequently their discovery that they will forgo these options for the commitment to the female (or more feminine) hero. For example, in the *Man From UNCLE* slash novel, *City of Byzantium*, Ilya says to Napoleon, "I don't mind if you look, I don't mind if you flirt. Both are as natural to you as breathing. But, you unzip and it's over."[52]

It should be added that the desirable males in these encounters have properties such as strength, courage, and physical prowess, and the psychological characteristics to attain and maintain status and power.

Helen Hazen has drawn attention to the fact that romance fiction, almost exclusively consumed by women, all has these same basic themes: Namely, the tensions and pitfalls that are attendant on searching out a suitable partner, attracting their attention, acquiring them, and retaining them, possibly in the teeth of rivals, or the partner's own inclinations and temptations.[53] These are not the actions of coy passivity; they are exceptionally difficult games of social and sexual strategizing. The fact that women are intensely interested in fictional accounts of them implies either that this is an enjoyable by-product of their highly evolved abilities in these areas, or that the stories are a way to dry-run these skills. The jury is still out as to which of these accounts has the most empirical support, but the only thing we can be sure of is that humans in general, and women in particular, are not mere passively brainwashed dupes. They actively seek out such stories, and generate them independently, via things like fan fiction.

A key difference between the pornography aimed at men (both heterosexual and homosexual) and the romance, even the highly sexualized romance, aimed at women is that in the latter, sex is never pursued as an end in itself. If anything counts as a robust difference between the sexes in terms of sexual fantasy, it is this finding. Men will pursue sex, even bad sex, for its own sake (until they learn better, perhaps). On the other hand, women are intensely interested in sex, but it serves as a test-bed (pun intended) for other important qualities of the partner. Some romance novels do not even mention sex at all, but many do, and it is done through the eyes of the heroine, who often has the job of taking a cad (who could have had his pick of a number of women) and turning him into a dad, with all the qualities of loyalty and commitment this implies.

Such heroes also need to be tall, dark, and handsome (of course) and have other indicators of good genes (such as strength and stamina) and good character (generous, brave, and so on). On this last point, it is worth noting that anti-heroes, or even downright villains, can also be highly desirable, just so long as they are strong and (apparently) tameable by the heroine (some wag described *Fifty Shades of Grey* as an "extended contractual negotiation," which is not far from the truth). The sort of cruelty that can only come with power can also be sexy and, in our own research, two of the best predictors of female orgasmic response were sexual dominance and penetrative vigor.[54] That research reflects real sexual experiences, but it has deep roots that appear in fantasy forms. Noted feminist scholar Germaine Greer recently got herself into trouble by pointing out that the bulk of the market for sexually violent TV fiction was women, but she was just reporting a commonplace fact. This fact does not imply that women are actively seeking violent deaths in real life, just that the male qualities portrayed—power, dominance, and sexual aggression—are sexy qualities.[55] And female desire in these areas can take darker paths. Recently we documented the different types of female fetishization of actual killers, something I will discuss in more detail later.[56]

2.3.1 DON'T DO IT IN PUBLIC AND FRIGHTEN THE HORSES

Why did Greer get in trouble for saying these things? The Japanese have a pair of incredibly useful words: *Tatemae*, meaning the things we will admit in public, and *Honne*, to refer to things we know to be true but only say in private. The moment I mentioned the meaning of those words, you could probably think of dozens of relevant examples of *Honne*, from urinating in public swimming pools to not always observing the speed limit. Here is another piece of nearly universal *Honne*: female desire for and response to sexually dominant men. I mentioned earlier that one of the most successful pieces of fan fiction erotica is the famous *Fifty Shades* series, which has now sold something like 150 million copies.[57] It is one of many such series.

Every generation seems to produce its own exemplar version of this tale of female submission to a dominant, high-status man. Furthermore, each generation's version is often published anonymously (and in the case of *Fifty Shades*, needing huge underground support, before going public).

These publications are often accompanied by howls of protest from concerned people, assuming that such strong erotica could only have been penned by a man, and by a thoroughly misogynistic man at that. It does not seem to matter how often this phenomenon repeats itself, each generation credulously swallows the idea (at least publicly) that women are passively coy recipients of male sexual aggressiveness and that, therefore, erotic representations of female sexual responsiveness to these patterns must be written by men, for men.

These critics are wrong. Some of the most iconic erotic fiction in the BDSM[58] genre was written by women. Examples abound, but most readers, even if such works are not to their personal taste, are likely to have heard of *Delta of Venus* (Anais Nin),[59] *The Story of O* (Pauline Reage),[60] *Emmanuelle* (Emmanuelle Arsan),[61] *9 ½ Weeks* (Elizabeth McNeil),[62] and *Secretary* (Mary Gaitskill).[63] The reason most readers will have heard of these works is (of course) that they were all made into (more or less) mainstream movies—movies, incidentally, to which women took their men in droves. These stories all feature sexually submissive (but by no means socially, intellectually, or emotionally submissive) women in strongly sexually charged relationships.

One response to these stories is to call these representations "objectifying," but a bit of reflection reveals the story to be not so simple. If the morally valenced charge of "objectification" is to be consistent, then it must apply equally to the men presented in these fantasies. Such men are universally tall, broad-chested, sexually dominant, and independently wealthy. Interestingly, there seems to be a non-random scatter of male names in these stories. A "Mr. Gray" or "Grey" is the male lead in *9 ½ Weeks*, *Secretary*, and the eponymous *Fifty Shades*. In other notable works, the male leads often do not have full names at all but are identified solely through aristocratic or militaristic titles, such as "The Baron" in *Delta* or "The Count," "The Commander," and "Sir Steven" in *Story of O*. "Rene" does get a name in "O," but not a full one. It seems that objectification in sexual matters might be an equal opportunities affair.

Such fantasies may seem very different from mainstream romances, but the themes of being sexually irresistible, and compelling investment from a specifically chosen male, is at the core of at least half of female sexual fantasies according to more than one extensive review.[64] It might seem that such fetishistic elements are incompatible with what the rest of us might deem as true love. To the contrary. High-cost, hard-to-fake indicators lie at the heart of sexual selection for those species where partnered investment is a typical part of the reproductive pattern. And, for humans, this often means the hard-to-fake signs of love. Even, perhaps especially, of obsessive love.

One important thing seems worth considering seriously as arising from the data. It is the possibility that adult fantasy is a controlled way to express desires that we might otherwise be in denial about or would be horrified by in reality. However, it *is* play. At least, for psychologically healthy people, it is play. As one professional dominatrix pointed out to me, "spanking benches are padded." People may, for example, be able to eroticize certain forms of pain, in controlled circumstances, but that does not mean that they are pathologically addicted to pain in some general sense. Masochists do not get aroused by stubbing their toes. There will be more to say about the role of fantasy and the role of human imagination in sex in later chapters.

The great biologist Amotz Zahavi drew attention to the fact that the concept of *self-handicapping* made sense of many otherwise baffling behaviors and morphologies across nature.[65] Lots of advertising may lie, but some behaviors are hard to fake—producing a truth in advertising. The peacock's famous train is a hard-to-fake sign of genetic quality. Weaker peacocks do not have the resources to spare to grow one to impress the females.

Viewed this way, symbols of submission (in *Story of O* they include branding, chains, ritual clothing, etc.) can be framed as incontrovertible proof of devotion and, it could be easily argued that such extreme submissiveness creates an appropriate countervailing response in the male recipient. Framed thus, it becomes obvious why humans are caught between *tatemae* and *honne* over this. For women to publicly state that they desire sexual dominance from men in general would undercut the very choice mechanisms that lie at the heart of female sexual selection. In short, women are exceptionally picky about who—if anyone—gets to sexually dominate them.

To see how all this might work ethologically, consider the actions of the hornbill.[66] In this African bird species, the female makes herself dependent on the male by walling herself up in a tree before laying their eggs. The male hornbill—whose extravagantly colored beak and loud calls have made him irresistible—now has no choice but to provision both her and *their* eggs. If he were to split his resources with another female, then the first one would die, taking his genetic inheritance with her. This extreme dependence on the part of the female thus compels investment from the male.

Incidentally, it is worth noting that this superficially submissive female hornbill is no pushover in other ways. For example, female hornbills regularly take on snakes and monitor lizards, that the males of the species back away from, and they actively connive at their own walling up and dependence. Not all female (human)-created erotica has such themes of dependence, dominance, and devotion, of course, and I am far from suggesting that this pattern is the only one for human mating. However, these themes are perfectly easy to understand and accommodate within the mainstream of biology. Not only that, but it should be clear, even from the prior sketch, that simplistic, unidimensional analyses of power relations will struggle to make sense of them.[67] Finally, there is nothing inherently pathological about any of them.[68]

2.3.2 WOMAN IS MONOGAMOUS?

Finally, in this section, we should consider the phenomena of swinging, polyamory, and similar explicitly openly multi-male pairings, driven by female desire.[69] These are by no means universal characteristics of our species, but they are not that rare either.[70] We are only just beginning to study them, partly for reasons that I will discuss in later chapters. But, for the moment, it is worth noting that an increasing body of evidence suggests that a number of heterosexual, homosexual, and bisexual relationships in mainstream society have consensual, female-driven, extra-pair sexual behaviors.

When questioned, the women (as well as the men) involved in these relationships frequently feel that these activities enhance their sexual experience and can benefit the primary relationship.[71] That admitted, it is also possible that polyamory and swinging can be preparatory to mate-switching: These are potentially risky activities in that regard, but it is also possible that openness and honesty about the fact that humans are, in Dan Savage's memorable coinage, "monogamish" could even strengthen the primary mateship.[72]

For humans to engage in such activities, they have to deal with the fact that we have an evolved mate-guarding mechanism of jealousy.[73] However, there is some evidence that some people do overcome these barriers, or even consciously invoke them, to add spice and excitement in some cases, and such actions can be highly stimulating to female orgasm. For example, one study noted that nearly three-quarters of women "always" or "often" reached orgasm during swinging exchanges.[74] As noted sex advice columnist Dan Savage puts it, this is "varsity level" sexual activity and not for everyone.[75] However, it is another demonstration of the thinness of the traditional conception of coy female sexuality.

In summary, women enjoy physical sexual stimulation, and they enjoy fantasizing about males—with a pattern of qualities indicating good genetic features, commitment, and a willingness to invest in offspring. Frequently, though not universally, they actively seek out multiple sexual pairings, apparently for comparison purposes. What have been all too often condemned as pathological forms of erotic desire can be easily re-framed as fitness-maximizing behaviors, when we apply mainstream ethological understanding to human activity and the selection of suitable males. Thus equipped, do women just passively wait for such Prince Charmings to arrive and sweep them off their feet? In a word, *no*. Once again, here, as elsewhere, women are active sexual strategists.

2.4 WALLFLOWERS?

Súil mhealltach a chaitheamh le duine'.
(When an Irish woman throws someone a tantalizing eye)[76]

The final question to be addressed in this chapter is—given that women do enjoy sexual pleasure, enough to simulate it with toys and fantasy—do they also actively solicit sexual encounters from potential partners? Once again, the evidence is unequivocal. When it comes to dating, men may think that they are making all the moves, but if any of those moves are to count in the sexual game, then it turns out that women typically made the first one.

One of the heroes of human ethology was Eibl-Eibesfeldt.[77] In a career spanning seven decades, he documented, often using ingenious hidden cameras, the myriad ways in which women signaled potential receptivity to specific men's sexual advances. While he would tend to frame these behaviors as fixed action patterns, or in terms of drives, a more modern ethological formulation would be in terms of functional cognitive/emotional modules, designed by natural selection to increase fitness.[78]

However, the key elements of behavior he observed remain unchanged. These "courtship signals," as they are technically known, form a cross-culturally universal catalogue of human—especially *female* human—mating behavior. One of these, which he found across the continent of Africa all the way to Polynesia, was the "coy glance." "Turning toward a person and then turning away," he wrote, "are typical elements of human flirting behavior." There was also the eyebrow flash, opening the eyes wide, lowering the eyelids, tilting the head down and to the side, and looking away. In summary, make eye contact and then remove it, a ritualized avoidance that is not followed up, and see if the target gets in sync with you. It is worth noting that humans, unique among primates, have white sclera—the areas surrounding the pupil and iris.[79] Lots of research has stressed the importance of this adaptation to gaze synchronicity in social interactions such as mother/baby attention.[80] But this (proximate) mechanism can also be recruited to serve other purposes, such as generating field-based hypotheses about who is sexually interested in you.

As an aside, in no instance of his field studies did Eibl-Eibesfeldt, or any of his researchers, ever witness anything that they could describe as a successful move on the part of a man that was not preceded by some sort of signal, on the part of the woman. Not that such signals from women guaranteed success for the man—far from it. However, absence of such interest from the women pretty much guaranteed failure. This choosiness is also underscored by the selective and easily withdrawn nature of such signals. They have a built-in level of plausible deniability. The man is being given an opportunity to flirt, nothing more. Then, the real decisions will be made, but mostly not by him, whatever he may feel about his moves, chat-up lines, casually name-dropped alma mater, or subtly exposed Rolex.

Monica Moore followed up these early ethological observations with supporting ones, identifying many subsets of courtship behaviors, which were interestingly found to be more important than physical attractiveness in generating male attention.[81] Using naïve observers to confirm interpretations, she found a set of behaviors that was highly predictive of prompting an approach from the men (although the behaviors originated in the women) at dances. First, the woman would glance around the room, not directing the gaze to any particular recipient. This specific glance type was not usually repeated throughout the evening. The second type of glance was identical to Eibl-Eibesfeldt's coy glance detailed previously, the gaze was briefly held, and then the woman looked away. This "darting glance" could be repeated to see if the recipient had caught on. Finally, there could be the direct gaze.

Once eye contact had been made, there were a raft of possible attention-gathering strategies on the part of the women. These included the eyebrow flash, head toss, hair flip, face to face, lipstick application, lip lick, lip pout, smiling, laughing, giggling, kissing, whispering, arm flexion, tapping, palming, gesticulation, hand hold, primp, skirt hike, object caress, caress (hair, leg, buttock, arm, torso, back), lean, brush, breast touch, thigh touch, foot to foot, placement, lateral body contact, parade, approach, promenade, pinching, and finally, tickling.

The highly ritualized nature of many of these courtship behaviors led at least one researcher to make an explicit comparison between humans and wild turkeys, and to call these human behaviors a "courtship dance."[82] There is also a huge degree of cross-cultural stability about these behaviors.

However, as many evolutionarily trained scientists are at pains to point out, humans manage to adapt their courtships to new technologies (such as mobile phones and dating apps), so a simplistic "fixed action pattern model" of instinct is not to be preferred.[83] More fruitful is a computational model of instincts, where patterns of behavior are linked with some degree of local flexibility to outputs, which are assessed in terms of salient goals.[84]

Male sexual displays seem to be more ad hoc than female ones and less ritualized. These can be displays of dominance (such as entering physical space and making noise), resource-based (e.g., ostentatiously paying for things), or even submissive displays (like shrugging).[85] No ethologist has yet found a consistent ritual set that works for men to attract women's attention, although now people can earn good livings trying to tell men how to achieve these goals in online dating forums, with indifferent success. The comedian Jerry Seinfeld put the situation less kindly, "Men want to attract women's attention but are really bad at it. Honking car horns, shouting from building sites—these are actually the best ideas we have had so far."[86] This is an interesting observation, and it probably speaks to the need for truth in advertising that the Zahavian handicap principle draws attention to. There is likely more to it than just this, however. It may not just be that men's signaling is overt, but that women's signaling needs to be covert, in that it has to achieve the twin goals of attracting one specific individual while screening out less desirable others, and not appearing to be generally sexually available to all. There is some support for this interpretation. In a review of the sorts of ethological observations I have been discussing, Andrew Gersick and Robert Kurzban found that potential costs to reputation constrained female sexual signaling.[87] It seems to be that individuals are adept at matching risks to particular social situations, and thereby minimizing the possibility of reputational damage.

Many commentators complain of a social double standard when it comes to male and female sexual behaviors. They are right, of course. However, as is often the case, without some background about *why* such double standards arise and are maintained, criticisms of them can appear to have an almost magical quality—as if we could simply wish such things away. Mere wishing has an uphill task against the forces of sexual selection. That said, attention to why such proximate differences between attitudes toward the sexes keep arising, at least contains the possibility of prompting objectively fairer, kinder, and more decent human social behavior.

2.5   SUMMARY

The model of coy females and indiscriminate males was always an over-simplification. Women's sexual desire is discriminating, rather than passive, in relation to that of men. Bringing together diverse strands of evidence from other species, and our own behaviors around sexuality, makes it increasingly clear that the coy female model needs serious revision. If the Standard Social Science Model—sometimes called the Blank Slate[88] model of human nature—cannot explain the differences that do exist between the sexes, then what can? It is time to ask what scientists mean when they talk of sexual dimorphism.

NOTES

1   Fry, S. (2010). *Interview for attitude* [Magazine]. Stream Publishing Ltd.
      Stephen Fry does raise an interesting point. When we consider homosexual behaviors of both sexes, we do get a unique window into our natures, with behaviors that do not have to be constrained by compromises and complementarity with the opposite sex. And, many studies have shown that the average number of lovers, both real and desired, for gay men is radically different from that of lesbian women. Both of these sets of numbers are different again from those of both sets of heterosexuals. These differences are not best explained by a difference in interest in sex per se, but rather by a difference in interest in consequence-free sex. A classic study is presented in Bell, A. P., & Weinberg, M. S. (1978). *Homosexualities: A study of diversity among men and women.* Among those sampled here, nearly half of the men had had more than 500 sexual partners in a lifetime, and over 90% had had at least 25. By

contrast, most of the lesbians had had ten same-sex partners or fewer and rarely sought casual sex. This implies that the putative consequences are not just pregnancy (which cannot be the result of lesbian sex) but also emotional commitment.

2 Kubrick, S. (1999). *Eyes wide shut* [Film]. Warner Bros.

3 For more on paternity uncertainty and adaptations to it, see especially Buss, D. M., & Schmitt, D. P. (1993). Sexual strategies theory: An evolutionary perspective on human mating. *Psychological Review*, *100*(2), 204.

4 Ferguson, E. E. (2006). Judicial authority and popular justice: Crimes of passion in fin-de-siècle Paris. *Journal of Social History*, 293–315.

5 On self-deception see Trivers, R. (2011). *The folly of fools: The logic of deceit and self-deception in human life*. Basic Books (AZ). Anna Freud (Sigmund's daughter) is the one who deserves credit for the most robustly testable formulations of the concept of ego defenses; see especially Freud, A. (1936). *Das ich und die Abwehrmechanismen*. A good example of the ways that humans lie to themselves can be demonstrated by the work of Leon Festinger via the concept of cognitive dissonance. Festinger, L. (1962). *A theory of cognitive dissonance* (Vol. 2). Stanford University Press.

6 The classic case of our realizing (once we could do genetic tests) that apparently monogamous behavior was nothing of the sort was the case of those unassuming birds (or "little brown jobs") the Dunnocks. See Davies, N. B. (1985). Cooperation and conflict among dunnocks, *Prunella modularis*, in a variable mating system. *Animal Behaviour*, *33*, 628–648. Since then, many other socially monogamous species have been discovered to be not genetically monogamous.

7 In other species, the range of post-copulatory mechanisms that the females of the species have adopted to increase the chances that a particular mate will be the father of the offspring include, but are not limited to, controlling the depth of penetration to optimize either storage or fertilization; maintaining longer copulation; directly transporting sperm to fertilization sites; modifying internal conditions to make them less hostile to sperm; directly nourishing sperm; discarding sperm from sub-optimal males; increasing egg supply; ovulating; allowing male manipulations that result in spermatophore being discharged; increasing chance of implantation; keeping copulatory plug (a sort of chastity belt preventing further implantation from other males) in place when it could be removed; assisting male in production of copulatory plug; aborting prior zygotes (fertilized eggs); refraining from aborting selected zygotes when this is possible; avoiding other males; and investing more in offspring. If anyone detects coyness or lack of agency in this list, then they are engaging in willful delusion. See Eberhard, W. G. (2009). Postcopulatory sexual selection: Darwin's omission and its consequences. *Proceedings of the National Academy of Sciences*, *106*(supplement_1), 10025–10032, for a fuller discussion and the source of this list (plus other examples).

8 Bateman, A. J. (1948). Intra-sexual selection in Drosophila. *Heredity*, *2*(Pt. 3), 349–368.

9 This is a direct consequence of the equal ratios of the sexes. Fisher, R. A. (1930). *The genetical theory of natural selection* (Mimicry). Dover.

10 A very welcome correction to extreme forms of the "Males compete, females choose" model for humans was the focus of a target article in *Psychological Enquiry* in 2013. See Stewart-Williams, S., & Thomas, A. G. (2013). The ape that thought it was a peacock: Does evolutionary psychology exaggerate human sex differences? *Psychological Inquiry*, *24*(3), 137–168, for the target article, and the rest of the issue for responses to it. Most of the responses to this conceptual re-adjustment were positive. This re-adjustment brings contemporary evolutionary scholarship more into line with classical ethological observations of humans, as depicted in the rest of this chapter.

11 Spinoza, B. (1992). *Ethics: With the treatise on the emendation of the intellect and selected letters*. Hackett Publishing. (Original work published 1677). Ethics book 3, definition of emotion 1.

12 Burns, J. (2016). *Forbes report on the sex toy industry*. www.forbes.com/sites/janetwburns/2016/07/15/adult-expo-founders-talk-15b-sex-toy-industry-after-20-years-in-the-fray/#2329ff675bb9 last accessed 9/9/2018.

13 For more on how free markets are windows into human desire, see Saad, G. (2007). *The evolutionary bases of consumption*. Psychology Press; and Miller, G. (2009). *Spent: Sex, evolution, and consumer behavior*. Penguin. Both these books review the extensive literature.

14 Darwin, C. (1859). *The origin of species*. Murray. Darwin joined pigeon clubs to explore, at first hand, the myriad ways in which artificial selection could modify a species.

15 Hinde, R. A., & Barden, L. A. (1985). The evolution of the teddy bear. *Animal Behaviour*, *33*(4), 1371–1373.

16 Gould, S. J. (2008). A biological homage to Mickey Mouse. *Ecotone*, *4*(1), 333–340.

17 Carter, M. (2013). *Interview: Anne Wood, the co-creator of the Teletubbies | Society.* Guardian.co.uk. Retrieved 2015–04–06. If you want to explore the process in detail, her website can be found at www. ragdoll.co.uk/ (link is external) (last accessed 13/2/2013).

18 Baba Brinkman. His song "Performance, Feedback, Revision" is available at www.youtube.com/ watch?v=hod20AzYB4o (link is external) last accessed 14/11/2018.

19 For more on turning wild foxes into tame dogs, see Belyaev, D. K., Plyusnina, I. Z., & Trut, L. N. (1985). Domestication in the silver fox (*Vulpes fulvus*): Changes in physiological boundaries of the sensitive period of primary socialization. *Applied Animal Behaviour Science, 13*(4), 359–370; Belyaev, D. K., & Trut, L. N. (1964). Behaviour and reproductive function of animals. II. Correlated changes under breeding for tameness. *Bulletin of the Moscow Society of Naturalists Biological Series (in Russian), 69*(5), 5–14; Belyaev, D. K., & Trut, L. N. (1982). Accelerating evolution. *Science in the USSR*, 5, 24–29, 60–64.

    For a great introduction to Belyaev's team, including up-to-date interviews with surviving members, all written in a lively yet authoritative style, see Dugatkin, L. A., & Trut, L. (2017). *How to tame a fox (and build a dog): Visionary scientists and a Siberian tale of jump-started evolution.* University of Chicago Press.

20 Lieberman, H. (2017). *Buzz: The stimulating history of the sex toy.* Pegasus Books.

21 McCaughey, M., & French, C. (2001). Women's sex-toy parties: Technology, orgasm, and commodification. *Sexuality & Culture, 5*, 77–96. The quote is from p. 85.

22 Joshi, P. (2014). The best vibrators and sex toys to buy now (Magazine article, updated August 2023). *Good Housekeeping.* www.goodhousekeeping.co.uk/health/health-advice/the-great-good-housekeeping-vibrator-test%20 accessed 18/4/2014.

    For *Anne Summer*'s "O-Team," a group of citizen designers given free rein to have their creations made real, see www.annsummersproducts.co.uk/Sex-Toys/O-Team last accessed 13/2/2013.

23 Schultz, W. W., van Andel, P., Sabelis, I., & Mooyaart, E. (1999). Magnetic resonance imaging of male and female genitals during coitus and female sexual arousal. *British Medical Journal, 319*, 1596–1600. If you access the article, pay particular attention to the way that the penis bends to interact with all the sensitive internal areas talked about in chapter six.

24 King, R., & Belsky, J. (2012). A typological approach to testing the evolutionary functions of human female orgasm. *Archives of Sexual Behavior, 41*(5), 1145–1160.

25 *Sh* is a good example of a sex shop that caters only to women. They were kind enough to stretch their rules enough to allow the author in on several occasions to sit in on seminars and talk to people. www. sh-womenstore.com/ One of those seminars was given by the highly informative Deborah Sundahl, whose book [Sundahl, D. (2003). *Female ejaculation and the G-spot.* Hunter House] is well worth a read for details of practical consequences of internal sensitivity.

26 For more on *shunga* art, see Screech, T. (2009). *Sex and the floating world: Erotic imagery in Japan, 1720–1810.* Reaktion Books. As well as rather imaginatively repurposing a *Noh* theater mask of a Tengu spirit for such encounters, there were also purpose-made dildos from this period, called *harigata*.

27 For entire friezes dedicated to the production of sexual pleasure, including (but not limited to) stimulation of g-spots, collection of female ejaculate, and insertable sex toys, the Karnatak temple carvings are an education. They are discussed in Anand, M. R. (1958). *Kama Kala: Some notes on the philosophical basis of Hindu erotic sculpture.* London: Skilton.

28 The most enthusiastic proponent in the modern era of the "sexual penetration pleases only men" approach was Anne Koedt, as expressed inRef_385_FILE150644625BM1Koedt, A. (1968). *The myth of the vaginal orgasm* (First Published in *Notes from the First Year*) New York Radical Women. Available at www.atu.edu/lfa/brucker/amst2003/Texts/Orgasm.pdf last accessed 14/2/2010.

    However, this approach was taken up with almost equal enthusiasm by Elizabeth Lloyd in Lloyd, E. (2005a). *The case of the female orgasm: Bias in the science of evolution.* Harvard University Press, and especially on her website supporting the book: Lloyd, E. (2005b). http://mypage.iu.edu/%7Eealloyd/ Reviews.html#FAQs (Website with questions and answers and clarifications of her (2005a) work. Last accessed March 2008, although the specific sections detailing support for Koedt's model appear to have been moved now.) The aforementioned enthusiasm matches the idea that enjoyment of heterosexual intercourse is one-sided, and that this appears to buttress claims that female orgasm only occurs as a by-product of male orgasm. Elizabeth Lloyd also usefully summarizes sex research that suggests a somewhat low figure for penetration during female masturbation. For the reasons stated, I find this claim to be with little merit in understanding interactions between real humans. Unless, that is, we are to take male masturbatory habits as somehow settling the selection conditions for the male anatomy and physiology. These things are *clues* to sexual behavior, but if they were the be-all and end-all of sexual behavior, then none of us would ever go to the trouble of finding partners. It might get that far some day with artificial sex partners, but we are not there yet. In any case, it is not true. As discussed in chapter six, there are lots of internal sites for sexual sensitivity in aroused women, and this is reflected in the sex toys they buy and use.

29 For work on different types of orgasm identified by women, see Bentler, P. M., & Peeler, W. H. (1979). Models of female orgasm. *Archives of Sexual Behavior, 8*, 405–423; Fisher, S. (1973). *The female orgasm.* Basic Books; King, R. J., Belsky, J., Mah, K., & Binik, Y. (2011). Are there different types of female orgasm? *Archives of Sexual Behavior*; King, R., & Belsky, J. (2012). A typological approach to testing the evolutionary functions of human female orgasm. *Archives of Sexual Behavior, 41*(5), 1145–1160; Levin, R. (1981). The female orgasm—a current appraisal. *Journal of Psychosomatic Research, 25*, 119–133; Levin, R. (1998). Sex and the human female reproductive tract-what really happens during and after coitus. *International Journal of Impotence Research, 10*(Suppl. 1), S14–S21; Levin, R. (2002). The physiology of sexual arousal in the human female: A recreational and procreational synthesis. *Archives of Sexual Behavior, 31*, 405–411; Levin, R. (2004). An orgasm is . . . who defines what an orgasm is? *Sexual and Relationship Therapy, 19*, 101–107; Levin, R. (2008). Critically revisiting aspects of the human sexual response cycle of Masters and Johnson: Correcting errors and suggesting modifications. *Sexual and Relationship Therapy, 23*, 393–399; Singer, J., & Singer, I. (1972). Types of female orgasm. *Journal of Sex Research, 8*, 255–267; Sundahl, D. (2003). *Female ejaculation and the G-spot.* Hunter House. Obviously, I will have a lot more to say about this in subsequent chapters, particularly chapter nine.

30 For more on ice age dildoes, see http://carnalnation.com/content/55700/4/worlds-oldest-sex-toy-still-rock-hard-after-28000-years last accessed 15/2/2018.

31 For more on Cabinet 55, see Bailey, M. (1991, July 7). Cupboard love: Sex secrets of the British Museum. *The Observer*, p. 7.

32 For more on the Forbidden Rooms (*Gabinetto degli oggetti osceni*, the last word changing euphemistically to *riservati* in 1823) of the Naples museum—that used to require Papal permission to view—see Grant, M., & Mulas, A. (1982). *Eros in Pompeii. The secret rooms of the national museum of naples.* Bonanza Books.

33 Brenda Love makes the first mention that I can find of the bee vibrator in Love, B. (1992). *The encyclopedia of unusual sex practices.* Barricade Books. However, for the reasons stated, I do not find it likely.

34 For a typical ancient Greek reference to women's use of leather dildoes as a sex aid, see Aristophanes (431 BC/trans. 1973). *Lysistrata.* Penguin.

35 Maines, R. (1999). *The technology of orgasm. Hysteria", the vibrator, and women's sexual satisfaction.* Johns Hopkins Press.

36 Freud, S. (1953). Three essays on the theory of sexuality (1905). In *The standard edition of the complete psychological works of Sigmund Freud, volume VII (1901–1905): A case of hysteria, three essays on sexuality and other works* (pp. 123–246).

37 Riddell, F. (2014). *No, no, no! Victorians didn't invent the vibrator.* www.theguardian.com/commentis-free/2014/nov/10/victorians-invent-vibrator-orgasms-women-doctors-fantasy last accessed 16/02/2018.

38 Anon. (1968). Sub-umbra, or sport among the she-noodles. In *The Pearl: A journal of voluptuous reading, the underground magazine of Victorian England.* Grove. (Original work published 1879)

39 More from Kate Lister can be found at www.thewhoresofyore.com/kates-journal/buzzkill-vibrators-and-the-victorians last accessed 07/10/2018.

40 For China's ban on camgirls eating fruit in an erotic manner, see http://shanghaiist.com/2016/05/06/dont_eat_that_banana.php last accessed 16/3/2018.

41 In particular, Magnanti, B. (2012). *The sex myth: Why everything we're told is wrong.* Hachette UK, takes this idea to task.

42 I am discussing female-oriented porn rather than male, but the evidence is that liberal access to pornography (as long as it is ethically produced) is good for society in general, but particularly so for women. There are many studies about this issue, but I am just going to concentrate on one because it gives us the closest we get in social science to a controlled experiment. In the Czech Republic, after the fall of communism, the iron grip of authority was relaxed. This also meant more liberal access to porn, as well as laxer crime suppression generally. Crime rates in every single area *except* sex crime—all types of them—increased. See Diamond, M., Jozifkova, E., & Weiss, P. (2011). Pornography and sex crimes in the Czech Republic. *Archives of Sexual Behavior, 40*(5), 1037–1043. For more details. This natural experiment in the Czech Republic has been replicated in Japan, Croatia, and the USA, where the exact same effects are detectable at a state-by-state comparison level. Landripet, I., Štulhofer, A., & Diamond, M. (2006). Pornography, sexual violence and crime statistics: A cross-cultural perspective. In *IASR thirty-second annual meeting.*

43 Gottschall, J. (2012). *The storytelling animal: How stories make us human.* Houghton Mifflin Harcourt.

44 I think the first person to make the fantasy at night versus politics during the day remark was Ester Perel in her TED talk: www.ted.com/talks/esther_perel_the_secret_to_desire_in_a_long_te... last accessed 12/04/2018 but if there is an earlier reference, then please let me know.

45 For the psychological health of those with somewhat kinky fantasies, see, for example, Pawlowski, W. (2009). BDSM: The ultimate expression of healthy sexuality. In W. J. Taverner & R. W. McKee (Eds.), *Taking sides: Clashing views in human sexuality* (11th ed., pp. 70–75). McGraw-Hill, and Richters, J., De Visser, R. O., Rissel, C. E., Grulich, A. E., & Smith, A. M. (2008). Demographic and psychosocial features of participants in bondage and discipline, "sadomasochism" or dominance and submission (BDSM): Data from a national survey. *The Journal of Sexual Medicine, 5*(7), 1660–1668.

46 Brownmiller, S. (2013). *Against our will: Men, women and rape.* Open Road Media (Original work published 1975). The quote is from pp. 394–395. I am not going to have much to say about male pornography, only that directed at a female audience. However, if her point is to be valid, then it would apply to female-created pornography as well. If all and only males that are tall, dark, handsome, status-bearers are attractive, then how is this not objectification, if voluptuous, willing, orgasmic females (the staple of male porn) is to count as such? Perhaps what healthy humans need is the ability to distinguish externalization of desire—which both sexes do—from treating people purely as a means to an end. Both sexes are apt to do this as well, of course (though in slightly different ways), as pointed out by Immanuel Kant in Kant, I. (1987). *Critique of judgment. 1790* (Trans. W. S. Pluhar, p. 212). Hackett.

47 For a summary of research into the correlational relationship between violent pornography and psychopathic behaviors, see Malamuth, N. M. (Ed.). (2014). *Pornography and sexual aggression.* Elsevier. Psychopaths may well be drawn to violent depictions of sex (and other behaviors), but this does not make such depictions causal in nature. Psychopathy also correlates with a preference for religious texts, without anyone attributing causality. The inability to distinguish fantasy from reality, and a lack of moral emotions, would seem to be the key predictors in psychopathic behavior.

48 It is psychologically plausible that fantasy can function as an outlet for otherwise unacceptable desire. Steven Pinker has exhaustively documented how violence—including sexual violence—has been decreasing across the board. (Pinker, S. (2011). *The better angels of our nature: The decline of violence in history and its causes.* Penguin). Has consumption of violent fantasy decreased? Quite the reverse. Violent video is everywhere, but violent crime has gone down. Now, as we all dutifully teach our stats students, "correlation is not causation." This is true, but strong and persistent correlations demand explanation. We should think long and hard about what the consequences of trying to ban fantasy might be. There are some arguments that those already inclined towards psychopathic behavior are attracted to violent pornography. This is likely true, but the supposed force of such an argument is lessened considerably when one reflects that the same could be said about attraction to religious literature, which is always incredibly popular among violent prisoners. This point is separate from a recognition that pornography makes a poor substitute for good sex education. See http://makelovenotporn.com/, Cindi Gallup Make Love Not Porn campaign, for some practical details of the differences and consequences of leaving the sex education of the next generation up to purveyors of fantasy.

49 Salmon, C., & Diamond, A. (2012). Evolutionary perspectives on the content analysis of heterosexual and homosexual pornography. *Journal of Social, Evolutionary, and Cultural Psychology, 6*(2), 193; and Salmon, C. (2012). The pop culture of sex: An evolutionary window on the worlds of pornography and romance. *Review of General Psychology, 16*(2), 152.

50 James, E. L. (2012). *Fifty shades of grey.* Arrow. (Original work published 2011). The author's real name is Erika Mitchell.

51 For more on slash fiction, see Salmon, C., & Symons, D. (2001). *Warrior lovers: Erotic fiction, evolution and female sexuality.* Yale University Press. It is still an unsolved puzzled as to why there are no women in slash stories with whom the female readership can identify. Salmon herself (personal communication, October 2017) suggests that this doubles the chance of a protagonist that the audience can eroticize, but I wonder if there is not something more at work here. For one thing, such concerns do not enter into more mainstream romances (like Austen) or more pulpy fantasies. Male transvestites (and auto-gynephiliacs, perhaps?) might possibly be understood as people who have fallen in love with a model of the opposite sex so completely, that they project themselves into it. Could slash fiction represent a related pattern in women? Maybe the projection into an idealized model of a male version of friendship turned into sexual love is easier to project oneself into? Perhaps the presence of an overtly female character in these highly sexualized scenarios would trigger jealous rival responses that would spoil the mood? For more on auto-gynephilia, see (e.g.) Blanchard, R. (1989). The concept of auto-gynephilia and the typology of male gender dysphoria. *Journal of Nervous and Mental Disease.*

52 Cited in Salmon, C., & Symons, D. (2003). *Warrior lovers: Erotic fiction, evolution and female sexuality*. Yale University Press.

53 Hazen, H. (1983). *Endless rapture rape, romance, and the female imagination*. Charles Scribbener. This book memorably drew attention to themes of dominance in female-driven erotica. Should we distinguish erotica and pornography? Moral judgments make for bad behavioral science, although such judgments themselves can provide interesting data for behavioral scientists. I am not about to try to distinguish erotica from pornography. The best expression of my sentiments would be captured by Ellen Willis, when she said "In practice, attempts to sort out good erotica from bad porn inevitably comes down to 'What turns me on is erotic; what turns you on is pornographic.'" Willis, E. (1981). *Beginning to see the light: Pieces of a decade*. Random House. http://heinonline.org/HOL/Page?handle=hein.journals/nyls38&div=26&g_sent=1&casa_token=&collection=journals.

54 King, R., & Belsky, J. (2012). A typological approach to testing the evolutionary functions of human female orgasm. *Archives of Sexual Behavior, 41*(5), 1145–1160.

55 Greer, G. (2019). TV violence and women. Article for *Radio Times* [Magazine]. www.radiotimes.com/news/tv/2018-05-05/tv-violence-women-the-bridge-germaine-greer/ last accessed 09/10/2019. The article finishes with the words: "The display of female victimhood in entertainment media is not the result of a conspiracy between wicked men to objectify, reify and sexualise women but a straightforward capitulation to market forces. Female victimisation sells. What should disturb us is that it sells to women." It need not disturb anyone if it remains fantastical and consensual. I will have more to say about this in later chapters.

56 Shresta, A., Dempsey, M., Tuohy-Hamil, S., & King, R. (2022). What does she see in him? Hybristophiles and spree killers. *Journal of Police and Criminal Psychology*, 1–13.

57 Jacqueline Gold, CEO of Anne Summers, is interviewed about the so-called "Fifty Shades of Grey" effect in *Cosmopolitan* magazine, available here: Briggs, Z. (2012). Fifty shades of grey increases sex toy sales. Interview with Jacqueline Gold, CEO of Anne Summers. *Cosmopolitan Magazine*. www.cosmopolitan.com/uk/entertainment/a16634/fifty-shades-of-grey-increases-sex-toy-sales/ last accessed 28/04/2018.

58 BDSM refers to bondage, dominance, sadism, and masochism and has become an umbrella term for arousal caused by erotic power exchange.

59 Nin, A. (1977). *Delta of Venus*. Bantam Books. (Original work published 1940)

60 Reage, P. (1954). *The story of O*. Pauvet Press. The author's real name was Anne Desclos, who kept quiet about her authorship for forty years. My copy of *L'histoire D'O* has a comment on the back about the book detailing "A woman being subjected to the full range of male sadistic desires." Said critic seemed painfully unaware that the book was a woman's fantasy.

61 Arsan, E., & Kovič, B. (1967). *Emmanuelle* (Vol. 1). le Terrain vague. Emmanuelle Arsan, also known as Marayat Rollet-Andriane, apparently based her novel Emmanuelle on *L'Histoire D'O*. She also published anonymously.

62 McNeil, E. (1986). *9 ½ Weeks*. Signet.

63 Gaitskill, M. (2012). *Bad behavior: Stories*. Simon & Schuster. One of the short stories is "Secretary," which inspired the movie of the same name.

64 Ellis, B. J., & Symons, D. (1990). Sex differences in sexual fantasy: An evolutionary psychological approach. *Journal of Sex Research, 27*(4), 527–555. For a more recent survey of the prevalence of "rough sex fantasies" in both men and women, see Lehmiller, J. J. (2018). *Tell me what you want: The science of sexual desire and how it can help you improve your sex life*. Hachette.

65 Zahavi, A. (1975). Mate selection—A selection for a handicap. *Journal of Theoretical Biology, 53*(1), 205–214.

66 For more on hornbills, see, for example, Poulsen, H. (1970). Nesting behaviour of the Black-Casqued Hornbill *Ceratogymna atrata (Temm.)* and the Great Hornbill *Buceros bicornis L. Ornis Scandinavica, 1*(1), 11–15.

67 For example, by post-modernists such as Foucault, M. (2012). *Discipline and punish: The birth of the prison*. Vintage.

68 Men also have submissive fantasies, of course. Humans are nothing if not a mutually sexually selecting species, but here I am discussing female fantasy and how it might relate to biology. Female submissive/male dominant fantasies do seem to outnumber the opposite pattern in our species, but the outnumbering is by no means total. See, for example, Jozifkova, E., & Flegr, J. (2006). Dominance, submissivity (and homosexuality) in general population. Testing of evolutionary hypothesis of sadomasochism by internet-trap-method. *Neuroendocrinology Letters, 27*(6), 711–718; Jozifkova, E., & Konvicka, M. (2009). Sexual arousal by higher-and lower-ranking partner: Manifestation of a mating strategy? *The Journal of Sexual Medicine, 6*(12), 3327–3334.

69 The differences between these categories are interesting, but I do not have space to go into them here. For an approachable discussion about the rise of open polyamory, see Geoffrey Miller's popular article in *Quillete*, an online journal: Miller, G. (2019). Polyamory is growing and we need to get serious about it. *Quillette* (Online article in popular science journal). https://quillette.com/2019/10/29/polyamory-is-growing-and-we-need-to-get-serious-about-it/ last accessed 07/11/2019.

70 Fernandes, E. M. (2009). *The swinging paradigm: An evaluation of the marital and sexual satisfaction of swingers*. Union Institute and University.

71 Kimberly, C., & Hans, J. D. (2017). From fantasy to reality: A grounded theory of experiences in the swinging lifestyle. *Archives of Sexual Behavior, 46*(3), 789–799.

72 Buss, D. M., Goetz, C., Duntley, J. D., Asao, K., & Conroy-Beam, D. (2017). The mate switching hypothesis. *Personality and Individual Differences, 104*, 143–149.

73 On the subject of jealousy see Buss, D. M., & Haselton, M. (2005). The evolution of jealousy. *Trends in Cognitive Sciences, 9*(11), 506. On sex-based differences in jealousy, see, for example, Buss, D. M., Larsen, R. J., Westen, D., & Semmelroth, J. (1992). Sex differences in jealousy: Evolution, physiology, and psychology. *Psychological Science, 3*(4), 251–256.

74 Findings are explored in Fernandes, E. M. (2009). *The swinging paradigm: An evaluation of the marital and sexual satisfaction of swingers*. Union Institute and University.

75 Dan Savage's advice column can be found here: www.thestranger.com/authors/259/dan-savage.

76 I'm grateful to my friend and Irish language scholar Karen Barry for bringing this saying to my attention.

77 See especially Eibl-Eibesfeldt, I. (1971). *Love and hate: On the natural history of basic behaviour patterns* (Transl. from the German by Geoffrey Strachan). Methuen. This quote comes from page 50. For a silent moving rendition (gif) of the "coy glance," see the insert at www.psychologytoday.com/blog/hive-mind/201406/passing-out last accessed 12/04/2018.

78 For a discussion of modularity, see especially Barkow, J. H., Cosmides, L., & Tooby, J. (Eds.). (1995). *The adapted mind: Evolutionary psychology and the generation of culture*. Oxford University Press. There is considerable debate as to the degree and meaning of modularity. For example, to what extent can the brain be conceptually divided into functional units specialized to do certain jobs? The details of this are interesting, but the crucial point is that no one in cognitive science disagrees on the fundamentals of functionalism—which is to say the computational model—as being the only game in town. For a discussion of this, see King, R. (2016). I Can't Get No (Boolean) Satisfaction: A Reply to Barrett et al. (2015). *Frontiers in Psychology, 7*, 1880.

79 Kobayashi, H., & Kohshima, S. (1997). Unique morphology of the human eye. *Nature, 387*(6635), 767. Interestingly, some (domesticated) breeds of dogs also have white sclera and seem to be able to follow human attentional gaze. See Gácsi, M., Miklósi, Á., Varga, O., Topál, J., & Csányi, V. (2004). Are readers of our face readers of our minds? Dogs (Canis familiaris) show situation-dependent recognition of human's attention. *Animal Cognition, 7*(3), 144–153.

80 Brooks, A. M. M. R. (2017). Eyes wide shut: The importance of eyes in infant gaze-following and understanding other minds. In *Gaze-following* (pp. 229–254). Psychology Press. For a review of infant gaze-following behaviors.

81 Moore, M. M. (1985). Nonverbal courtship patterns in women: Context and consequences. *Ethology and Sociobiology, 6*(4), 237–247.
    Moore, M. M., & Butler, D. L. (1989). Predictive aspects of nonverbal courtship behavior in women. *Semiotica, 76*(3–4), 205–216.

82 Birdwhistell, R. L. (1970). Masculinity and femininity as display. In *Kinesics and context: Essays on body motion* (pp. 39–46).

83 Lippa, R. A. (2007). The preferred traits of mates in a cross-national study of heterosexual and homosexual men and women: An examination of biological and cultural influences. *Archives of Sexual Behavior, 36*, 193–208.

84 Gangestad, S. W., Haselton, M. G., & Buss, D. M. (2006). Evolutionary foundations of cultural variation: Evoked culture and mate preferences. *Psychological Inquiry, 17*(2), 75–95.

85 For an advertising-based approach to this (advertisers need to be keen observers of actual human behavior or they will go broke), see Taflinger, R. (1996). Taking advantage. You and me, babe: Sex and advertising. *Taking Advantage You and Me, Babe: Sex and Advertisement, 26*.

86 Jerry Seinfeld makes this observation here: www.youtube.com/watch?v=bts28rz0lJ0 last accessed 12/04/2018.

87 Gersick, A., & Kurzban, R. (2014). Covert sexual signaling: Human flirtation and implications for other social species. *Evolutionary Psychology, 12*(3).

88 Pinker, S. (2005). *The blank slate* (p. 200). Southern Utah University.

3 Sex
The Gene's-Eye View

An unlearned carpenter of my acquaintance once said in my hearing: "There is very little difference between one man and another; but what little there is, *is very important*."

—William James (1890)[1]

This whole book is founded on the insight that humans are a mutually sexually selecting species.
Later, I will argue that previously neglected female choice mechanisms predominate in these selection processes. This should not come as a surprise, as the pattern is the same across all other primate
species. The ladies, be they macaques, chimps, or humans, always do the bulk of the choosing of
who gets to make it to the next generation. Much of this selection is obvious and public, but some
of it is cryptic and private. In humans, female orgasm—which partakes of both cryptic and public
elements—is a crucial part of these choosiness mechanisms. But, before I get to that part, I need to
lay some groundwork.

For a species to be mutually sexually selecting, there have to be two sexes, and they have to be
different from one another. I acknowledged in the introduction that asserting this fact as regards
humans—technically called sexual dimorphism—has become contentious, or at the very least somewhat confused, in some public quarters.[2]

In the introduction, I made the observation that many critics of the whole idea of *essentialist* human
sex differences might be less inclined to deny said differences, if they appreciated the Enlightenment
advance in science. Namely, that science has already replaced the study of *essences* with that of *processes*. It is now time to make good on my earlier promissory note and explore what sex differences
really mean, at least, what they mean to biologists. This is a scary topic—especially for a man to talk
about—but I take heart from the fact that a number of prominent feminist scholars—including some
previously hostile to the study of sex differences—have started to openly state that, for us to build a
better society, we first need to acknowledge that sex differences are real.[3] Because this is an area fraught
with moral and political implications (which can provoke reflexive defensive responses, rather than
considered ones), I hope I can be forgiven for starting discussing the topic in a slightly unusual way.

3.1 THE DIFFERENCE BETWEEN "VERY SMALL" AND "FAR AWAY"

I vividly remember being shown a roughly ten-minute video called *The Powers of Ten* while at
school.[4] I do not know if it is still shown in science class, but I hope so, because its effect on me—
and, I think, others—was profound. It started with an aerial view of a picnic (10 meters raised to the
power of 0, in other words, one meter per side of the frame) and then pulled out, gradually accelerating through all the scales of magnitude. Ten to the 0 meters is (roughly) human scale, 10 to the third
power in meters is a crowd, 10 to the sixth power, in meters, is roughly the size of the country I am
writing this in (Ireland), 10 to the eight meters is roughly the planet's diameter, and so on. Every ten
seconds the camera pulled back while the solar system, galaxies, and so on were all dwarfed, until
the film reached the limit of the observable universe (which was then roughly 10 to the 27th power
in meters). Then the direction reversed at dizzying speed until we were back at the picnic, and the
view proceeded to go inside one of the humans.

To divide by ten is to raise a number to a negative power. This makes 10^{-1} to be 1/10, 10^{-2} is
1/100 and so on. Ten to the minus third power is about the limit of human visual perception—about

DOI: 10.1201/9781003372356-3

the size of an LCD pixel, just visible if you are reading this on a screen. That is about the size of a human egg too. A nerve cell is about 10 to the minus 4 meters, while a chromosome exists at 10 to the minus 5 meters. The film moved through 10 to the minus 7 (the size of DNA, the building block of all life), but it went still deeper, down through atoms (10 to the minus 10 meters) and through the building blocks of atoms like quarks (10 to the minus 23) and further down through levels where human concepts, such as causation, cease to have the meanings we think we have such a good grasp on at our normal levels of existence.

Finally, the scale went down to the level that we think is the ultimate granularity of matter—the Planck constant—at 10 to the minus 35 meters. What became obvious to me, on reflecting on this was that a lot of the everyday concepts and perspectives that I had simply assumed were universal (such as *color* or even *causality*) were not merely *not* universal, but that, far worse, they were highly local and partial, barely existing at all as far as the rest of the universe was concerned. What I want to try to convey next is a similar sense of wonder about what might seem like a simple matter, sex. However, what it means depends crucially on the level of magnification that one views it under. I firmly believe that much of the dispute over the meaning of sex as it pertains to humans is a case of simply talking past one another.

3.2   LITTLE FLEAS DO NOT HAVE LESSER FLEAS, *AD INFINITUM*

Insufficient attention has been paid to the central importance of the discovery of glass to a human understanding of our place in the universe—to both science and philosophy.[5] Science requires measurement, and this requires the ability to (conceptually and/or physically) separate interesting bits of the universe that we want to study into observable, but differently treated, compartments, which can then be measured independently and compared.

Glass has the important (and rare) properties of being able to seal hermetically and to be transparent, while also controlling light in important ways. Without glass, our ability to separate out chosen sections of the universe, for example, into isolated containers or into temperature gauges where volume changes can be seen, would simply not exist. This is not an issue of mere convenience; the implications are profound. For example, before Louis Pasteur's controlled experiments on meat in sealed jars, it was widely believed that meat spontaneously devolved into maggots. Just take a second to reflect on the world view that would result from thinking that you and your fellow humans' bodies (not to mention your dinner) were all maggots-in-waiting. After his experiments, Pasteur could say, without hyperbole, that "Life is a germ, and a germ is Life. Never will the doctrine of spontaneous generation recover from the mortal blow of this simple experiment."[6]

It was not just the scientific paraphernalia of tests tubes and retorts that were important in changing our view of our place in the universe. Without lenses we would never have discovered the world of the very small (microscopes) or of the very large (telescopes). The implications of this perspective shift were devastating to those pillars of ancient knowledge—tradition, authority, and intuition. The Enlightenment showed all of these to be misleading at best.[7] The scales of magnitude that we humans are comfortable with are simply not the scales that the universe mostly exists on.

Following on from this, an awful lot of things that looked (and still look) like deep conceptual issues turn out to be, at least in part, issues of the scale of magnification one has observed them at. Take one of the first philosophical questions a smart child asks: "Is color real?" They sometimes ask this as "Is *my* blue the same as *yours*?" or "Is color out there, or just in my head?" which are related questions. How could we answer such a curious child?

I have a colleague who showed me some gold atoms the other day, in an electron microscope. She is a modern-day (reverse) alchemist who can literally turn gold into other things—by moving pieces of atoms at tiny scales. This is remarkable in and of itself. But, looking at the gold atoms allowed me to reflect on that question of is the color "gold" real? And there is a real definite answer to that question, "No." At least, the answer is "No" at that level of magnification. At that level, the philosopher's question "Is color in the object or in the observer?" becomes pretty much meaningless. An atom is of the order of 0.1 to 0.5 nanometers (about a million of these would be as thick as a human hair). Visible light exists at around 400 to 700 nanometers. Color, as we understand it, at our preferred level of existence, simply does not exist at the scale of atoms. It follows that gold atoms are not themselves gold colored (or any other color). It turns out that the (seemingly) conceptual question about the reality of color has a very complex empirical answer, namely: Certain emergent properties of matter interacting with evolved properties of our brains—such as the adaptively selected need to discriminate some features of the world—create the sensation of color (and there are huge complexities here) at greater scales of magnification. Color is a real thing, but what that means depends very much at which scale you explore it.

So much for color. What about sex—as in male and female—is that real? The answer, I hope to demonstrate, also depends very much on the scale of magnification one uses to look at the processes involved. I said in the introduction that, since the Enlightenment, science has advanced by considering processes, not essences. Pronouncements that there is no such thing as a female or male essence can be acknowledged, if such a victory can be said to be satisfying.[8] There never was an essence of one sex or the other, and no biologists would have argued that there ever was.

That said, these processes are not randomly distributed throughout the population. It may seem that calling human sex a *spectrum* is required to be socially progressive, until one appreciates that it logically commits one to denying the often-painful lived experience of many transsexual people. If there is nowhere for someone to transition from, and nowhere for them to transition to, then (one might be tempted to ask) why are these people in pain? Gender expression may be much more of a spectral affair than biological sex (see following section), but a spectrum implies something approaching equal representation of the population across all the gradations. This is simply not true of sex. It is more appropriate to view the pattern of sexual dimorphism as two large mountain peaks, with many small, but significant, foothills in between them. Exceptions to a naïve conception of a simple sexual binary can always be found, and they are interesting and important, but it would be misleading in the extreme to imply that there were not patterns of sex differences, or that these patterns do not make biological sense.

3.2.1 Sex Is Not Simplistically Binary, but It Is Not a Spectrum Either

A very helpful analogy is that sex is a set of dials rather than switches.[9] Viewed this way, sex—in the sense of male and female—is not just a simple binary event but a series of processes that get tuned by a variety of interconnecting processes—many of them involving environmental input—and at increasing levels of complexity. As the scale pulls back, we can look at anatomical structures, physiological processes, behavioral biases, and eventually, the sort of things that could be appropriately described as social constructions. Examples of the latter might include assumptions, social roles, and stereotypes, which derive at least some of their existence from a shared intentionality.[10]

This sort of negotiated space, such as gender identity, is often the ground for fierce and acrimonious dispute. These things are processes too and, in ethological terms, proximate mechanisms. I will have some things to say about this level of magnification later in the book, but mainly to clearly delineate the sort of questions that an ethologist can and cannot legitimately hope to answer. To forestall possible misunderstandings, ethologists can say things about the aggregate measurements of measured population-level differences that have environmentally relevant fitness-related effects. In other words, about personality as it exists at the population level.

Ethologists do not have anything much to say about social policy, law, or ethics. That admitted, I believe that engaging with the scientific detail is likely to disappoint those who wish to recruit facts as ammunition in all brands of ideological culture war. Traditionalists who want biological warrant for sexist attitudes will find little comfort. At the same time, idealists who think that humans float free of the natural world will be similarly dismayed. Humans are complex creatures, not to be summarized by easy slogans or intuitively appealing political sensibilities.

The most obvious place to start the exploration is at the most fundamental level of biologically meaningful magnification, the most basic of processes underlying sexual difference: differential parental investment.

3.3 DIFFERENTIAL PARENTAL INVESTMENT

It had been an exhaustingly long weekend of lovemaking in the balmy Brazilian summer, but the final result would all make it worthwhile. Many were the enthusiastic courtship attempts of many attractive members of the opposite sex (all keen to display their large stiff penises to Lyndsay). Despite all this, Lyndsay was not about to let their precious sexual resources go to just anyone, and resisted all their efforts for a considerable time. But, finally, Lyndsay had found someone to their liking. Love blossomed in their private cave and soon, Lyndsay would be a father.

Lyndsay, the male cave louse (*Neotrogla*), is a comparative rarity in nature—a male that is more careful about picking and choosing his mate than the female of the species.[11] But Lyndsay illustrates an important underlying process. Namely, that whichever sex that is obligated to bring the most to the (joint) reproductive party will be more successful if their behavior is relatively choosy, while the sex that can[12] bring the least has the option (not always taken) of being less picky (but more aggressive) about with whom it shares genetic material.

The technical term for what I have just described is "minimum potential investment," and it is important to appreciate that this minimum provides an option—or suite of options—not available to the opposite sex. In Lyndsay's case, this means that the females have penises (or *intromissal organs*, if you want to be pedantic) and compete ruthlessly for the males, who bring large, nutritious parcels of bat guano to the reproductive party.

While it might have seemed like a bit of a cheat of mine to call the female's intromissal organ a "penis," no one but the most ardent entomologist would have balked if I had done this in the case of a male arthropod. The male cave louse lacks an external sexual organ, and the females have evolved an erectile gynosome with which they mate aggressively. Females with penises, mating aggressively. Does this throw the whole logic of evolutionary biology into a loop? Far from it. It confirms one of its most fundamental functional precepts—differential parental investment.

Across the living world you will find organisms that change sex when it suits them,[13] species with 500 (or more) different sexes,[14] females with penises and phalluses, males who get pregnant, species where the males dominate a hierarchy, ones where the females do,[15] or ones where it seems roughly even-handed.[16]

How to make sense of all this variety? Here is one, utterly foundational, way. Throughout nature—in two-sex species—the sex whose minimum potential investment in offspring is the highest is going to be strongly selected to be the pickiest about those with whom it shares its gametes. In turn, the sex with the lowest minimum potential investment can sometimes (I stress *sometimes*) afford to be less choosy. Furthermore, this sex will tend to compete for the other sex. This superficially simple difference turns out to have a lot of important consequences.

The logic of this insight is laid out in Robert Trivers' key 1972 paper.[17] In it he does what all biological scientists dream of and predicts that species yet undiscovered will fit his explanatory pattern. History has proved him right, again and again. In many species the minimum potential investor is the male. It makes sense that the strategy built around small, fast-moving gametes—sperm—might, most often, be the one that can, on occasion, leave the other sex literally holding the baby.

However, there are plenty of species where the male's minimum potential investment is higher than the female's and, in *each and every single one of these species*, Trivers' prediction has been borne out. For instance, in Mormon crickets,[18] Jacanas,[19] seahorses,[20] and the recently discovered cave louse, that I discussed at the outset, the male brings so much extra to the reproductive party that the females can often be observed being pushy and competing for him.

In the case of our female cave louse, "compete for the males" is something of an understatement. Her huge spiny penis holds the struggling male in place and extracts his nourishing spermatophore, along with his sperm, in copulations lasting several days. Lyndsay is a male because, like all other males, he has small, fast-moving sex cells—sperm. The rest of his anatomy, physiology, and behavior derives from this. It does this in patterned ways, but human intuitions can be a poor guide to them. What, after all, did I mean by saying that being *male* amounts to nothing more than having small, fast-moving gametes? As we pull back our view from the fundamental underlying process, we are presented with the next level of magnification, where the crucial element is the size of the sex cells—the gametes.

3.3.1 ANISOGAMY

Sex, from the perspective of genes, is a strategy for getting genetic material into the next generation. Gametes—egg and sperm—are the organism's strategic payload. If these gametes were identical in size, they would be called "isogamous." When they are of different sizes—as they are in us— they are called "anisogamous." Over phylogenetic time, once any differences in size (and therefore optimal strategic options) take hold, then selection tends to exaggerate these differences as time goes on.[21] Once an ancestral gamete started down the "large, and energy-rich" route, then selection tended to keep on it, and the cells got larger over evolved time. These cells are what we call "eggs." Once something started down the "small, fast-moving route," it tended to stay on *that*. These are "sperm." Bodies—you and me and every other sexually reproducing species—are built around these two strategic options.

It did not have to be this way. Unicellular organisms—prokaryotes, like amoeba—divide themselves, and bdelloid rotifers (elegant translucent creatures less than a millimeter long) are the one complex animal (technically "eukaryote") that have gone back to cloning for their reproduction.[22] They are thus all females. No one is completely sure why they have done this, but among other all other known complex organisms, sexual reproduction is the rule.

Taking a gene's-eye perspective,[23] this can present something of a puzzle, or as Maynard-Smith put it, the "two-fold costs of sex."[24] By reproducing sexually, you are halving your genetic material in each generation, and you might not find a mate. Exactly *why* splitting yourself in half (genetically speaking) and then combining with another lump of genetic material is a successful way of doing this is an interesting story itself.

To your possible surprise, the answer is not entirely settled to everyone's satisfaction, but the most promising research looks at the strengthening of the immune systems of the resultant offspring.[25] For the moment, I need to note, simply, that sexual reproduction happens, and furthermore that egg and sperm have various strategies (this is what bodies and behavioral options *are* in ethological terms) built around them that can be understood as trying to maximize the efficiency of this process.

At this level of magnification, words like "male" and "female" are, for biologists, convenient shorthand for a large (but finite) set of strategic options. They are not essences—the critics are right about this—but then, no biologist ever thought in those terms. These strategic options are, however, limited and constrained by that initial split, in non-random ways. Trivers' insight about minimum potential investment is an argument about allocation of resources—about economics in its broadest sense. The things that might seem, to the casual observer, like exceptions to it, are in fact triumphant vindications of the principles at work.

Leaving aside some of the interesting strategic variety that has occurred across taxa, in humans, the sex cells—the gametes—are of very different sizes. I started the book by pointing out that eggs are costly, and sperm are cheap, but it would be remiss of me not to acknowledge that there has been some debate about whether the costliness of eggs and the cheapness of sperm, are really the drivers here. One good reason for thinking that this debate is in its closing stages is that this cost/benefit argument can be reframed in other terms, with identical predictions and findings. For example, if anisogamy had no downstream effects, we would expect to see random distributions of parental care across species, and this we most certainly do not observe. In addition, there are good abstract mathematical models for showing why this is the case. When this sort of thing happens in other branches of science, such as physics, we are usually justified in thinking that the problem is as solved as it is likely to be unless striking new evidence emerges.[26]

In any case, the relative sizes of sperm and egg are hardly the only crucial differences in our species. We are not turtles, who go to great lengths to lay eggs and then just abandon them to their fates.[27] Humans are what are known as "obligate investors." It is not just human eggs that are costly, but also human babies. They take an inordinately long time to grow, all the while reducing the food-gathering effectiveness of the woman carrying them and, until recently, they required some years of costly breast-feeding. But these investment requirements do not need to concern us yet.

Moving on again through the levels of magnification of genes, chromosomes, and the endocrine system, what sort of differences are there between the sexes? Can it all really come down to a few genes? Yes and no.

3.4   THE SRY GENE

At this level of magnification, sex differences boil down (more or less) to one gene. Does this admission contradict my earlier assertion that humans are too complex to admit of easy sloganizing? Only if the daily experience of human individuality is also incompatible with the biological reality that we are all homo sapiens. And it is not. As noted sex researcher Milton Diamond dryly points out, "Nature loves variety, unfortunately, society hates it."[28] But, before I get to those fascinating varieties, let me start with some of the basics.

Differential parental investment is the organizing principle behind sexual dimorphism, but you cannot see differential parental investment under a microscope. How are these processes made physically manifest? At this level, it is in the form of chromosomes. From high school biology, you probably recall that we humans inherit twenty-three chromosomes from each parent, and that these split and are then lined up against each other to reform into the unique individual. Furthermore, you may recall from school that one set of these chromosomes bucked the usual mirroring pattern, in that they could be XX or XY. The sex chromosomes.

If you stopped at high school biology, you may be forgiven for thinking that this pattern—with XX (*homogametic*—meaning similar gametes) leading to female offspring and XY (*heterogametic*, meaning different gametes) leading to male was universal across nature. It is not. The number of ways that sex determination can be realized in different species is considerable.[29] Some species (some *teleostei* fish, for example) are true hermaphrodites, while others (*hymenoptera*, such as honeybees) have a system called haploid-diploidy, where the males come from unfertilized eggs, while the (sterile) females come from fertilized ones. The birds in your garden may determine the sex of their offspring via ZO, ZW, or complex ZW chromosomal arrangements, where the males are homogametic—the opposite way to mammals. This bewildering complexity of sex determination is important, because it strongly implies that sexual reproduction (as opposed to its alternative) must be incredibly useful, given the huge numbers of different ways organisms have arrived at it. It is no mere accident.

People spend entire careers studying small aspects of these systems, but I am just going to talk about how mammals realize sex differences. With mammals (leaving aside some complications for the moment) we have a system where, when sperm meets egg, if it is carrying a Y chromosome, this will result (all other things being equal) in a male, and if carrying an X, a female. The initial finding of the role of X and Y chromosomes was done by Nettie Stevens, in collaboration with Edmund Wilson, and it was just lucky that their experimental species (mealworms) just so happens to have the same sex determination mechanism as we do.[30]

The Y chromosome is unique, in that it is not paired with any others. This means that any copying errors in reproduction are not corrected by the usual processes that clean up such errors. And, given that making an organism into a human male requires the coordination of anatomical structures (like penises), physiological functions (like testosterone production), and control mechanisms (like the sensorimotor sections of the brain corresponding to each of these things), it is frankly shocking that this is all set in motion by one gene.

Yet, this appears to be the case. The fascinating detective work leading to this discovery is explored in Siddhartha Mukherjee's recent work *The Gene: A Recent History*. A summary of these processes would mention that the crucial observational and experimental insight came from two sources.[31] It had been noticed that anatomically and physiologically normal young females with Swyers' syndrome failed to mature into normal adult females. When examined, their chromosomes were found to be XY, not the expected XX. Something had interfered to stop their cells responding to the Y chromosome. This finding prompted researchers to insert copies of likely candidate genes into experimental mice to see if they could artificially make them switch sexes. It turned out that they could. When extra copies of the SRY gene were inserted into XX female mice, they developed the anatomy, physiology, and behavior of typical males. This sex determination component of the Y chromosome turned out to be universal in mammals.[32]

3.4.1 A NOTE ABOUT INTERSEX CONDITIONS AND SEXUAL BINARY[33]

So, is sex simply binary in humans? Not so fast. For one thing, as I have already mentioned, when we pull back the scale of magnification and look at humans as individuals, or social groups, humans can do things like reflect on their biological sex, their sexuality, their sexual expression, and their gender roles, all in collection with other humans. And they may or may not be happy after these reflections. I have already mentioned people with Swyers' syndrome, which is a type of *androgen* (masculine sex hormone) insensitivity.

Is it obvious which sex these people should be identified as? It is only recently that we have started to ask the people in question rather than assuming or imposing.[34] They are not alone. It has been claimed that somewhere between 1 in 4,000, and 1 in 60, of all humans may have some sort of intersex condition, depending on exactly how one defines these, although that higher figure runs together some much more malleable gender non-conformity traits, and gender non-conformity— which is very common—should not be confused with sex.[35]

When we study tens of thousands of newborns, 99.79% are either XY or XX in line with their observed sex, and of the few with chromosomal abnormalities, those who are not sex typical number in single figures. These people need care and attention, but they do not need, deserve, or wish to be collateral damage in wider societal-level battles of the sexes. External genitalia are, in almost all cases, a reliable indicator of chromosomes.[36] However, the people who are in the tiny minority are of scientific interest as they allow us to test various hypotheses about how sex patterns manifest. One of the *most* interesting groups, for our purposes of understanding sex determination, are the people with 5-α reductase deficiency.

In the usual course of events, a baby, destined to be male, gets three doses of virilizing hormones—testosterone—to prepare its body and brain. The first is in the womb, the second just after birth, and the third at puberty. In people with 5-α reductase deficiency, the first two doses are somehow not seen by the body, and the people in question develop as anatomically and physiologically typical females. In the cultures that they grow up in (the Dominican Republic has the most famous family, the Batistas), they adopt female-typical interests and expectations. With puberty, the onset of masculinity, long delayed, takes place with sudden force, and in the space of a few months their testicles descend, vulvas (nearly) close up, and they develop into anatomical and physiological males. Most interestingly, despite a lifetime of being reared—and therefore socialized—as girls, they adjust to male-typical behavior and interests very rapidly.[37]

Or consider people with congenital adrenal hyperplasia (CAH). CAH is an umbrella term for a range of conditions, stemming from an autosomal recessive gene, that affect the production of sex steroids from the adrenal glands. This can result in several sorts of signs and symptoms, but of particular interest are two opposite states. The first is XX, i.e., chromosomally *female* individuals, with enlarged clitorises, or even somewhat functional penises, due to excessive androgen production. The second is XY, i.e., chromosomally *male* individuals, with apparently female external genitalia. In each case the expressed sexual identity, and sex-typical interests, tend to go opposite to what family and society have traditionally tried to impose.[38]

So, does this mean that so-called intersex (perhaps more properly termed people with disorders of sexual development) people fall somewhere in-between male and female? Once again, this thought, while tempting, would be highly misleading. To start with, *all* humans produce androgens. An androgen-insensitive individual—remember, someone who appears phenotypically female but who possesses undescended testes—will have a brain that has been *less* exposed to these androgens than a typical female, despite their having *actual testicles*. So, such a person will appear anatomically female, and have a mixed physiology, but have a brain that has been exposed to even fewer androgens than a typical female. These sort of mosaic effects are quite common in the study of human sex and sexuality, defying simplistic explanations.

All of this fascinating variety has led some people to think that expression of sex differences is a matter of simple choice, whim, or social learning. Some of these examples should undermine such notions. But, in case there was a lingering doubt, the underlying empirical claim, that humans are—in terms of brains at least—hermaphroditic has, somewhat tragically, been put to a direct test.

3.4.2 Psycho-Sexual Hermaphroditism? A Case Study

In place of a theory of instinctive masculinity or femininity which is innate, the evidence of hermaphroditism lends support to a conception that, psychologically, sexuality is undifferentiated at birth and that it becomes differentiated as masculine or feminine in the course of the various experiences of growing up.

Thus spoke John Money in 1955.[39] Coming at a time of increasing acceptance of previously ignored or, even worse, actively persecuted bodies and behavior (such as homosexuality or intersex conditions), Money's pronouncement seemed humane. Furthermore, it seemed to have scientific warrant.

His attitude certainly impressed the parents of David Reimer, an unfortunate boy who—at only eight months old—had lost his penis in a circumcision accident. When Money appeared on television, with a successfully sexually transitioned person, discussing the malleability of human sexual identity, Reimer's parents contacted him for help. David was given surgery to remove all trace of male gonads. His parents were then told to rename him and to bring him up as a girl, "Brenda."

However, and despite their best loving efforts to raise him as a girl, David never felt comfortable around the other girls and their preferred activities. David suffered from suicidal depression. At one point he threatened to kill himself, unless he could live as a boy. Things came to such a head that at age fourteen his parents finally told him what had happened to him, at which point he experienced some relief. The story does not end happily, alas. Despite attempts to live as a man, by having penis reconstruction surgery and getting married to a woman, David committed suicide at the age of thirty-eight.

This tragic case was still being misreported and misunderstood years later. At the time, Money reported that the boy who had lost his penis had transformed successfully into a young woman—a triumphant vindication of his theories that sexual identity was almost purely the result of socialization. Those very close to the case, such as Milton Diamond, have been consistently misrepresented—often by those who think that sex differences are all a matter of convention—as arguing for a purely hormonal basis for sexual identity, but this is something Diamond most emphatically does not support.[40]

Diamond supports (and ably documents) a "biased-interaction theory of psychosexual development," in which he gives weight to genetic, hormonal, and uterine factors. But additionally, Diamond documents several psychological processes that occur in the individual, such as comparing themselves in terms of *same/different* to particular peers, or role models, as they are presented in the wider culture.

Diamond argues that this sort of processing is quite general in humans and, it should go without saying, that he (as do the majority of those involved in sex research) supports the full range of mature gender identities that humans wish to express. In David Reimer's case, this assessment of difference manifested itself in his realizations that, for example, he did not urinate or fantasize in the same way as the girls he was expected by a wider society to be like. As Diamond points out, rather poignantly, "David Reimer's concept of masculinity was not based on the presence of a penis; he had none."[41]

3.4.3   Sex on and in the Brain

What is the most important sexual organ? There are no prizes for guessing that this is the brain, rather than the genitals. I hope I have given some sense of the complexities involved in sex determination, as we move up through the scales of magnification, from underlying evolutionary forces, through genes and chromosomes, to endocrine systems, anatomy, and lastly, physiology—which is anatomy in motion.

It would be, frankly, miraculous if all of this complexity did not manifest itself somehow in human brains, where something like half of our DNA is expressed in some form.[42] David Reimer's case highlights the interesting point that sexual identity is not simply rooted in genitalia.[43] Even if it were the case that genitalia were the sole foundation of sexual identity—and that is not true—then we would expect to find a corresponding set of brain regions that developed and had evolved to control and regulate these organs.[44] This we do find, of course, and it is interesting to note that this area, called the *somatosensory cortex*, is measurably and predictably different in women than in men. I gave some sense of this in the previous chapter, where I documented how misleading it has been in behavioral science when the male body and brain have been taken as the basic pattern, with females seen as deviant from this. There will be more on this later. But, before I get there, I also need to say something now about how the "male equals normal" assumption is interfering with general scientific progress.

The error in question has a technical name in psychology, "androcentrism." This is the error of assuming that males are standard and that females are therefore deviations from this norm. One might have thought that fifty-plus years of feminist critique of behavioral science had put an end to this sort of thing. Not so. Not only has continuing androcentrism made scientists mistake features for bugs when it comes to sexuality, but androcentrism has been, and still is, all too common in neuroscience and psychology as well.

For a variety of reasons, some good, some demonstrably less good, there are people who have thought it necessary to downplay or deny sex differences in human neuroscience. Some of the poor reasons have included the relative ease of obtaining male experimental participants, or of simply not realizing that differences might matter. Making one measurable thing a variable is not tantamount to saying that this measurable thing causes *all* the variance. Sex is not the only variable under the spotlight in brain research. One eminent neuroscientist put her finger on it nicely when she said, "underestimation or exaggeration of possible effects [of sex differences in the brain] retards progress in the field."[45] There are lots of reasons that androcentrism is a bad idea, but I want focus on only one for the moment:

> The X chromosome has played a critical role in the development of sexually selected characteristics for over 300 million years, and during that time, it has accumulated a disproportionate number of genes concerned with mental functions. . . Dosage differences in the expression of such genes (which constitute at least 15% of the total) are likely to play an important role in male–female neural differentiation, and in cognitive deficits and behavioral characteristics, particularly in the realm of social communication.[46]

When it comes to brains, neuroscientists can measure things at several, distinct levels of magnification. There exists the systems level (which can differ in both size and complexity), neural circuits (e.g., the reward circuits),[47] or the synaptic level (the nuts and bolts of how neurons work). Causally, we can look at the action of hormones and at the unequal doses of genes to be found on the X or Y chromosomes.

At all these levels, sex-based differences in human brains are found, and even small differences have been shown to be robust across multiple studies. For those who wish to follow this up, I have supplied a partial list of some of the more recent summaries and meta-analyses of these in the note section (it is rather long).[48] Note that many of these studies are summaries, meta-analyses, and reviews of dozens, sometimes hundreds, of pieces of work in the field. I encourage interested (or skeptical) readers to take the time to explore them and the back-up literature in considerable detail. They are a snapshot of how, when it comes to brains, complementarity between the sexes has been more successful as a research strategy than seeing women as being reflections of men, or to show so-called male superiority. Sex differences in functional connectivity are even measurable in the brains of fetuses.[49]

However, you do not have to be a neuroscientist to work out that male and female brains are, on average, readily distinguishable. Consider the unremarkable observation that heterosexuality is not a unisex, or unidimensional, trait. A human being is not just *heterosexual* in the way that they might be *tall*, or *someone who likes chili*. They are a person of a particular sex, who finds people with other typical secondary sexual characteristics desirable. Therefore, heterosexuality must manifest itself differently in male brains—whose owners tend to find *female*-typical bodies attractive, and female brains—whose owners tend to find *male*-typical bodies attractive. If you do not believe that finding people attractive comes from brain activity, then you have bigger problems to solve than the possibly vexed question of sexed brains. You have to explain why, uniquely among all creatures on the earth, humans are endowed with non-physical reproductive organs.

More worrying even than that, failure to include sex as one of many variables in neuroscience is medically irresponsible. Many of the studies mentioned in the notes above have clear medical implications, and failure to take sex (*not* gender, which is irrelevant here) into account increases the chances of misdiagnosis, dosage irregularity, and other dangers. It is also no accident that the increased attention to differences between men and women in medical research correlates with increasing participation of women in said research. For example, using a sample of 1.5 million medical research papers Nielsen et al. (2017) found unequivocal support for the fact that the greater representation of women in the research led to greater likelihood of sex and gender analysis within said research. Androcentrism is an anachronism.[50]

3.5 SUMMARY

Genes can be thought of as a set of instructions. They are not blueprints, for blueprint information is bi-directional. A blueprint could be recreated from the completed object, and this is not true of genes. The information only travels one way.[51] A better way is to think of genes as instructions, such as the instructions for folding an origami object. Except that, instead of folding a paper swan, the task is to fold an *actual* swan, and out of proteins, rather than paper. That is quite a set of instructions. Whenever I try to fold a paper swan, it ends up looking more like a disgruntled duck. But genes are much better at providing instructions than I am at following them—and most of the time a pretty good swan (or rose, or bee, or human) gets folded out of the proteins.

Change a few of the instructions in the genes, and you are folding a taller swan, or a fatter swan, or if you are unlucky, a lopsided swan. Given that variation in a single gene can cause profound differences in the physical attributes of the resulting organism, it should not be a surprise that altering an entire *chromosome* (which, depending on how you measure it, has about 800 genes on the X and 70 on the Y) can have profound differences in the resultant organism. Consider what you are being asked to believe if the sex difference deniers are right: That millions of years of evolution, which ruthlessly excises mutations that do not contribute to the organism's fitness, just allowed a whole host of differences to accumulate that have absolutely no effects whatsoever. The sheer number differential between X and Y should alert us to the fallacy of considering male brains (or bodies, for that matter) as somehow more basic than female ones. Sexual dimorphism at the level of hormones, genes, endocrine systems, and brains is just as much a fact, as it is about our more obviously visible bodies.

With sexual dimorphism now, I trust, firmly established as a basic fact about humans, how do we go about studying a trait—such as human sexual behavior—as an ethologist would approach it? That is the subject of the next chapter.

NOTES

1 James, W. (1890). The importance of individuals. *The Open Court, 4*(154), 24–37; reprinted in *The will to believe and other essays in popular philosophy* (1897; reprinted, New York: Dover, 1956), 255–262, pp. 256–257.

2 The claim that the acknowledgement of sex differences has become unnecessarily difficult in human behavioral science may seem like hyperbole, but I would be remiss if I failed to point out that there has been a slew of popular books that have received undue acclaim, for saying exactly this. For example, Rippon, G. (2019). *Gender and our brains: How new neuroscience explodes the myths of the male and female minds.* Pantheon; received a favorable review in *Nature*, of all places (Eliot, L. (2019). The gendered brain: The new neuroscience that shatters the myth of the female brain. *Nature, 566*(7745), 453–454.

  And another recent popular science book won a Royal Society award for, somewhat surprisingly, arguing that testosterone has no significantly measurable effect on human brains and that humans are therefore (cerebrally) hermaphrodites. Fine, C. (2017). *Testosterone rex: Unmaking the myths of our gendered minds.* For a detailed and accessible account of how fears of sexism are stifling research, see Claire Lehmann and Debra Soh's excellent summary: Soh, D., & Lehmann, C. (2017, April 17). *The rhetorical trap at the heart of the 'neurosexism' debate.* Published (and accessed) Online in *Quillette.* http://quillette.com/2017/04/17/rhetorical-trap-heart-neurosexism-debate/.

3 For example, Eagly, A. H. (2018). The shaping of science by ideology: How feminism inspired, led, and constrained scientific understanding of sex and gender. *Journal of Social Issues, 74*(4), 871–888.

4 EAMES Productions. (1977). *Powers of ten* [Film]. www.youtube.com/watch?v=0fKBhvDjuy0 last accessed 20/12/2022.

  The original had, I think 40 orders of magnitude. We have gone deeper since then.

5 Dartnell, L. (2014). *The knowledge: How to rebuild our world from scratch.* Random House.

6 Pasteur, L. (1861). *Sur les corpuscles organisés qui existent dans l'atmosphère: Examen de la doctrine des générations spontanées.* Leçon Professée a la Société Chimique de Paris, le 19 Mai.

7 There is an extended and insightful discussion of this in Pinker, S. (2019). *Enlightenment now: The case for reason, science, humanism, and progress.* Penguin Books.

8 E.g., Scientific American, *Visualizing sex as a Spectrum.* https://blogs.scientificamerican.com/sa-visual/visualizing-sex-as-a-spectrum/ last accessed 24/10/2018.

9 Schmitt, D. P. (2017). Sexual dials (not switches) theory: An evolutionary perspective on sex and gender complexity. *Evolutionary Studies in Imaginative Culture, 1*(1), 93–102.

10 Searle, J. R. (1995). *The construction of social reality.* Simon & Schuster. This is a foundational book that attempts to set out the logic of social facts, things that are (roughly) true by virtue of shared beliefs. Many things that seem to be of pressing interest (fashion for instance) fall under this description. But many things do not. At what point of magnification does what Searle would call "brute facts" turn into "social facts"? It is hard to draw a line, and no principled one may exist. This much is true—fashions about hair length would not exist once humans have gone extinct (unless replaced by a similar creature). Differential parental investment would still exist. To be contrasted with Berger, P. L., & Luckmann, T. (1991). *The social construction of reality: A treatise in the sociology of knowledge* (No. 10). Penguin UK, a book that sometimes argues as if all reality exists by social consensus. This (latter) point would confuse things that are ontologically dependent on shared intentionality (like money) with things that are epistemologically dependent on shared intentionality (like consensus views in a scientific field) and confuse both these with things that exist independently of humans (like most of the universe). A full discussion of this lies outside the scope of this book.

11 Yoshizawa, K., Ferreira, R. L., Kamimura, Y., & Lienhard, C. (2014). Female penis, male vagina, and their correlated evolution in a cave insect. *Current Biology, 24*(9), 1006–1010.

12 I only said "can." No implication that they "must" or (god forbid) "should" do this.

13 For example, clownfish (the star of the film *Finding Nemo*) would be more likely to do this than go off on a search for the missing parent. Buston, P. (2003). Social hierarchies: Size and growth modification in clownfish. *Nature*, *424*(6945), 145.

14 For species with bewilderingly large number of sexes, see: Erika Kothe, E. (1996). Tetrapolar fungal mating types: Sexes by the thousands. *FEMS Microbiology Review*, *18*(1), 65–87.

15 For females with social dominance, phalluses, and high testosterone, see Kruuk, H., & Kruuk, H. (1972). *The spotted hyena: A study of predation and social behavior* (p. 335). University of Chicago Press.

16 For more on the range of sexual strategies across taxa, see the witty and informed: Judson, O. (2002). *Dr. Tatiana's sex advice to all creation: The definitive guide to the evolutionary biology of sex*. Macmillan.

17 Trivers, R. (1972). *Sexual selection and the descent of man 1871–1971*. Aldine Press.

18 Gwynne, D. T. (1981). Sexual difference theory: Mormon crickets show role reversal in mate choice. *Science*, *213*, 14.

19 Jacanas are a Panamanian shorebird. See Emlen, S. T., & Wrege, P. H. (2004). Size dimorphism, intra-sexual competition, and sexual section in Wattled Jacana (*Jacana jacana*), a sex-role-reversed shorebird in Panama. *The Auk*, 391–403.

20 Berglund, A., Rosenqvist, G., & Svensson, I. (1986). Reversed sex roles and parental energy investment in zygotes of two pipefish (*Syngnathidae*) species. *Marine Ecology Progress Series*, *29*, 209–215.

21 For an extended discussion of this, see Cronin, H. (1993). *The ant and the peacock: Altruism and sexual selection from Darwin to today*. Cambridge University Press.

22 For more on what Maynard Smith (1978) called the "evolutionary scandal" of bdelloid rotifers, see especially; Judson, O. P., & Normark, B. B. (1996). Ancient asexual scandals. *Trends in Ecology & Evolution*, *11*(2), 41–46.

23 Dawkins, R. (2006). *The Selfish Gene:—with a new Introduction by the Author*. This is an excellent introduction to the gene's-eye perspective. Of course, as Dawkins would be the first to point out, genes are not literally "selfish." They have no motives. However, just as humans can be modeled as rational agents in economics (even though we fall far short of rational, alas), genes can be modelled as "selfish," even though they are not.

24 Smith, J. M. (1978). *The evolution of sex*. CUP Archive.

25 See, for example, Ebert, D., & Hamilton, W. D. (1996). Sex against virulence: The coevolution of parasitic diseases. *Trends in Ecology & Evolution*, *11*(2), 79–82.

26 Anisogamy and parental investment. A key researcher here is Hannah Kokko, who has produced a large and impressive body of sophisticated modeling of anisogamy and its effects at various levels—from population to single gene—and linking these levels via modeling of relevant behaviors such as parental care. A key insight is that sperm could evolve to have energy provision—like eggs do—but that costs would always outweigh benefits because it cannot be predicted which sperm will succeed. Thus, there are two routes to generating two sexes. When gametes find each other easily, then competition can generate a differential selection path for male and female gametes, but when there is limited gamete availability, this can also lead to anisogamy. For those who wish to pursue the details, central papers would include:

   Lehtonen, J., & Kokko, H. (2011). Two roads to two sexes: Unifying gamete competition and gamete limitation in a single model of anisogamy evolution. *Behavioral Ecology and Sociobiology*, *65*(3), 445–459.

   Jennions, M. D., & Kokko, H. (2010). Sexual selection. *Evolutionary Behavioral Ecology*, 343–364.

   Lehtonen, J., & Kokko, H. (2011). Two roads to two sexes: Unifying gamete competition and gamete limitation in a single model of anisogamy evolution. *Behavioral Ecology and Sociobiology*, *65*(3), 445–459.

   For those who want to dispute the role of anisogamy, this is a good place to start:

   Ah-King, M. (2013). On anisogamy and the evolution of "sex-roles." *Trends in Ecology & Evolution*, *28*, 1–2.

   And this is a solid reply to the objections raised:

   Kokko, H., Booksmythe, I., & Jennions, M. D. (2013). Causality and sex roles: Prejudice against patterns? A reply to Ah-King. *Trends in Ecology & Evolution*, *28*(1), 2–4.

   One, seemingly unanswerable (or at least, unanswered) point is that, if critics like Ah-King are right, and anisogamy has no effects, we would predict random distributions of parental care across species. This has not been observed. For a purely abstract argument in favor of why this is the way it is, see Queller, D. C. (1997). Why do females care more than males? *Proceedings of the Royal Society of London B: Biological Sciences*, *264*(1388), 1555–1557. However, it is important, once again, to emphasize that at root this is an economic argument about resource allocation and the large, but finite, number of fitness paths that can be traversed. It is not about male or female essences or anything likely to be confused with that.

27 Carr, A. F., & Ogren, L. H. (1960). The ecology and migrations of sea turtles. 4, The green turtle in the Caribbean Sea. *Bulletin of the AMNH, 121*, Article 1.

28 Milton Diamond's quote can be found here: Diamond, M. (n.d.). *Online comment.* www.hawaii.edu/PCSS/ last accessed 24/10/2018.

29 Bachtrog, D., Mank, J. E., Peichel, C. L., Kirkpatrick, M., Otto, S. P., Ashman, T. L., . . . Perrin, N. (2014). Sex determination: Why so many ways of doing it? *PLoS Biology, 12*(7), e1001899.

30 Ogilvie, M. B., & Choquette, C. J. (1981). Nettie Maria Stevens (1861–1912): Her life and contributions to cytogenetics. *Proceedings of the American Philosophical Society, 125*(4), 292–311.

31 Mukherjee, S. (2017). *The gene: An intimate history.* Simon & Schuster.

32 Koopman, P., Gubbay, J., Vivian, N., Goodfellow, P., & Lovell-Badge, R. (1991). Male development of chromosomally female mice transgenic for Sry. *Nature, 351*(6322), 117.

33 Some have called for the word "intersex" to be replaced by DSD (disorders of sexual development) to remove stigma. I am neutral about this and note only that constantly changing the nomenclature does little to remove stigma in other cases, but on the other hand, the very word *inter* being attached to *sex* may have led people to think in terms of a spectrum between egg and sperm, and there is not one. See Merrick, T. (2019). From 'intersex' to 'DSD': A case of epistemic injustice. *Synthese, 196*(11), 4429–4447.

34 Alice Dreger, in particular, has been a vocal and well-informed proponent of not forcing people, especially not forcing children, into categories without their ability to make informed choices. An increasing number of people wish to be identified as "third sexers," for example, and reject the idea of early so-called corrective surgery. See Dreger, A. D. (1998). *Hermaphrodites and the medical invention of sex.* Harvard University Press; Dreger, A. D. (1999). *Intersex in the age of ethics*; and Dreger, A. D. (1998). "Ambiguous sex"—or ambivalent medicine? Ethical issues in the treatment of intersexuality. *Hastings Center Report, 28*(3), 24–35. Whether we should treat people as a third sex is irrelevant to whether there actually *is* a third sex.

35 Defining intersex conditions: Fausto-Sterling, A. (2000). *Sexing the body: Gender politics and the construction of sexuality.* Basic Books.

36 Hamerton, J. L., Canning, N., Ray, M., & Smith, S. (1975). A cytogenetic survey of 14,069 newborn infants: I. Incidence of chromosome abnormalities. *Clinical Genetics, 8*(4), 223–243; Ohno, M., Maeda, T., & Matsunobu, A. (1991). A cytogenetic study of spontaneous abortions with direct analysis of chorionic villi. *Obstetrics & Gynecology, 77*(3), 394–398; Jacobs, P. A., Melville, M., Ratcliffe, S., Keay, A. J., & Syme, J. (1974). A cytogenetic survey of 11,680 newborn infants. *Annals of Human Genetics, 37*(4), 359–376. I am grateful to Emma Hilton, who tweets as @FondOfBeetles for drawing my attention to these studies and several others.

37 Imperato-McGinley, J., Peterson, R. E., Gautier, T., & Sturla, E. (1979). Androgens and the evolution of male-gender identity among male pseudohermaphrodites with 5α-reductase deficiency. *New England Journal of Medicine, 300*(22), 1233–1237.

38 For a review of clinical presentations and developmental sequelae of CAH, see Speiser, P. W., & White, P. C. (2003). Congenital adrenal hyperplasia. *New England Journal of Medicine, 349*(8), 776–788.

39 Money, J., Hampson, J. G., & Hampson, J. L. (1955). Hermaphroditism: Recommendations concerning assignment of sex, change of sex and psychologic management. *Bulletin of the Johns Hopkins Hospital, 97*(4), 284–300.

40 On misreporting Reimer, see for example: Butler, J. (2004). *Undoing gender.* Routledge. For reasons that are not quite clear, she argues for psycho-sexual hermaphroditism and misrepresents and misquotes Milton Diamond to boot.

41 Diamond, M. (2006). Biased-interaction theory of psychosexual development: "How does one know if one is male or female?" *Sex Roles, 55*(9–10), 589–600. The quote is from p. 594; Diamond, M., & Sigmundson, H. K. (1997). Sex reassignment at birth: Long-term review and clinical implications. *Archives of Pediatrics & Adolescent Medicine, 151*(3), 298–304; Diamond, M. (1965). A critical evaluation of the ontogeny of human sexual behavior. *The Quarterly Review of Biology, 40*(2), 147–175.

42 See Bae, B. I., Jayaraman, D., & Walsh, C. A. (2015). Genetic changes shaping the human brain. *Developmental Cell, 32*(4), 423–434, for a recent review.

43 If you just look for one thing at a time (technically called univariate analysis), then you get different results than with multivariate analysis. One interesting way that this has been shown is by comparing the brains of cis and transgendered people. When this is done it is found that the brains of transgendered

people do seem to converge on a single-sex continuum, as shown by differences in gray matter (GM). As one researcher put it, "[S]ex has a major effect on GM irrespective of the self-perception of being a woman or a man." Baldinger-Melich, P., Urquijo Castro, M., Seiger, R., Ruef, A., Dwyer, D., Kranz, G., Klöbl, M., et al. (2019). Sex matters: A multivariate pattern analysis of sex-and gender-related neuro-anatomical differences in cis-and transgender individuals using structural magnetic resonance imaging. *Cerebral Cortex.* https://doi.org/10.1093/cercor/bhz170. The quote comes from p. 10.

44 Penfield, W., & Boldrey, E. (1937). Somatic motor and sensory representation in the cerebral cortex of man as studied by electrical stimulation. *Brain, 60*(4), 389–443. I will have much more to say about this "homunculus" and its female counterpart in the next chapter.

45 The quote comes from page 236 of Pavlova, M. A. (2017). Sex and gender affect the social brain: Beyond simplicity. *Journal of Neuroscience Research, 95*(1–2), 235–250.

46 The quote comes from page 311 of Printzlau, F., Wolstencroft, J., & Skuse, D. H. (2017). Cognitive, behavioral, and neural consequences of sex chromosome aneuploidy. *Journal of Neuroscience Research, 95*(1–2), 311–319.

47 See, for example, Cahill, L. (2014). Fundamental sex difference in human brain architecture. *Proceedings of the National Academy of Sciences, 111*(2), 577–578, for a brief review of research showing that male brains lack a degree of connectivity found in female brains. One study cited had around 1,000 participants, which is unusually high for this sort of research.

48 A (partial) list of reliable and robust sex-based brain differences with clinical significance, *all* showcased in the January 2017 *Journal of Neuroscience Research,* published online November 2016.

Responses to experimental stroke (Dotson & Offler, 2017) and ischemic stroke (Spychala et al., 2017). Sex-dependent gene expression of activity-dependent neuroprotective protein (important in the response to Alzheimer's, autism, and schizophrenia and is evident in humans, mice, and birds; Gozes, 2017). Sex-specific hormones affect huge numbers of brain functions (for instance, estrogen affects functions in the hippocampus, prefrontal cortex, cerebral cortex, nigro-striatial pathway, mesolimbic dopamine pathways, paraventricular nucleus, brainstem, and periaqueductal grey matter and cerebellum with functional differences detectable in, among other things, learning and memory, excitability and seizures, neuroprotection, addiction, cardiovascular regulation, pain sensitivity, fine motor skills, coordination, and reaction time (McEwen & Milner, 2017). Decision making in rodents (Orsini & Setlow, 2017). Stress and emotional learning (Merz & Wolf, 2017). Various stress responses and social defeat mechanisms, important in mental health (Laman-Maharg & Trainor, 2017). Face processing (Proverbio, 2017). Relative corpus callosum size (greater in females and implicated in sex-typed patterns of social behavior; Holloway, 2017). Gustatory function (Martin & Sollars, 2017). Sex-dependent brain regions, and the actions of oxytocin or vasopressin (Rubin et al., 2017). Sex differences in number, morphology, and signaling of immunocompetent brain cells (Nelson & Lenz, 2017). Cortical area, brain volume, and white and gray matter volumes in adolescents, which were all assessed as being of large effect size, and more different in terms of effect size than all other measures of body changes at adolescence, apart from voice differentiation (Paus et al., 2017). Detectable differences in the development and etiology of Alzheimer's disease (Pike, 2017). Responses to hypoxia-induced brain damage (Netto et al., 2017). Sex-differentiated effects of decision-making following damage to the ventro-medial prefrontal cortex (Reber & Tranel, 2017). Sex-based differences in spatial-tasks (males typically higher) and verbal tasks (females typically superior) (Gur & Gur, 2017). Chronic pain responses (Rosen et al., 2017) and brain alterations across pain conditions (Gupta et al., 2017); responses to particular drugs that limit brain damage following injury in males, but not females (Cahill & Hall, 2017). Sex differences at the level of synapses (Dachtler & Fox, 2017). Cortisol responses to social stressors (Reschke-Hernández et al., 2017). Underlying neural differences in stress-responding structures (Seo et al., 2017). Glial cells in their opioid-receptor response (Doyle & Murphy, 2017). Type and susceptibility to various forms of epilepsy (Samba Reddy, 2017) and strokes (Sohrabji et al., 2017). Functional cerebral asymmetries (Hausmann, 2017). Nicotine preference (Pogun et al., 2017), including differences on the developing brain (Cross et al., 2017). Responses to parenting (Sundström Poromaa et al., 2017). Sensitivity to corticotrophin-releasing factor, implicated in cocaine addiction (McRae-Clark et al., 2017) and other forms of addiction (Becker et al., 2017). In other primates, sex differences in physical and social development, play, grooming, nursing, foraging, and the manipulation of objects (Lonsdorf, 2017). Brain-derived neurotropic factor (Chan & Ye, 2017; Wei et al., 2017). Hippocampal structure and function—especially important in memory (Koss & Frick, 2017). Neuro-immune modulation of memory (Tronson & Collette, 2017). Attentional bias towards negative stimuli—important in predicting mood disorders (Victor et al., 2017). Social brain differences and their complex interactions with gendered expectations (Pavlova, 2017); the interaction of hormones on psychiatric diseases across the range of behavior (Gobinath et al., 2017).

Specifically epigenetic processes underlying depression (Hodes et al., 2017). Rumination differences in depression (Shors et al., 2017). The visual and auditory processing of data when mediated by fearful or peaceful moods (Yang & Lin, 2017). The processing of visual information at all levels (Vanston & Strother, 2017); the mediation of voluntary physical activity—conserved mechanisms across humans and rodents (Rosenfeld, 2017). Differences in Broca's region—specifically the local grey-matter regions of Brodmann areas 44 and 45, important in verbal processing (Kurth et al., 2017). Migraine, which is significantly more common in women than men (Pavlovic et al., 2017). Psychiatric disorders, which are significant sex-linked, such as autism with males and depression with females (Jahanshad & Thompson, 2017); even where there are atypical sufferers the underlying neural structures can be different (Lai et al., 2017). Finally, how each sex forgets words (Kerschbaum et al., 2017).

49 Wheelock, M. D., Hect, J. L., Hernandez-Andrade, E., Hassan, S. S., Romero, R., Eggebrecht, A. T., & Thomason, M. E. (2019). Sex differences in functional connectivity during fetal brain development. *Developmental Cognitive Neuroscience*, 100632.

50 Nielsen, M. W., Andersen, J. P., Schiebinger, L., & Schneider, J. W. (2017). One and a half million medical papers reveal a link between author gender and attention to gender and sex analysis. *Nature Human Behaviour*, *1*(11), 791.

51 Crick, F. (1970). Central dogma of molecular biology. *Nature*, *227*(5258), 561.

4 Ethology
How to Understand Any Trait

It has been said that, in its haste to step into the twentieth century and to become a respectable science, psychology skipped the preliminary descriptive stage other natural sciences had gone through, and was soon losing touch with the natural phenomena.

(Tinbergen, 1963)[1]

How do we explain the nature and function of species-typical traits, such as female orgasm? What does it actually mean, when you get right down to it, to explain the nature and function of anything biological? As the great biologist D'Arcy Thompson asserted, "Everything is the way it is because it got that way."[2] This is a profound insight, but biologists have a variety of meanings behind that deceptively simple phrase, "got that way." Let me take a few pages to discuss how we answer questions like this in the most general terms, before turning to the specifics of female choice mechanisms.

Ethologists draw attention to the importance of studying precisely defined behaviors in natural settings, and they can answer four distinct kinds of questions about them. This is important because, for example, many of the alleged disputes in the science of studying female orgasm—in behavioral science in general too—result from not appreciating that different approaches are often answering different, but complementary, questions, rather than being genuinely competing answers.

Ethology—of whom Tinbergen is one of the founders—is the study of behavior in natural settings. To get an everyday sense of the value of this endeavor, consider the following: If you are my (middle) age, you have seen zoos change enormously in less than one lifetime. That change is an ethologically inspired one. There were many notable exceptions, but some of the older zoos were simply shocking places of cruelty. They often featured animals in tiny cages, rocking back and forth with boredom, or even self-harming in despair.

Some species—elephants and killer whales appear to be especially clear examples—seldom adjust to a life of captivity. As a striking example, consider the curious fact that, despite our obvious resemblance to prey species, there have never been documented attacks of orcas on humans in the wild, only in captivity.[3] Whatever the rights and wrongs of zoos in general, no one who has been paying attention can fail to have noticed that some animals are much happier in more modern wildlife parks, where they have natural settings, places to hide, lots more room, others of their kind to mix with, if and when they choose, and ingenious features of novelty, stimulation, and challenge. Every visitor who takes the time to observe can see for themselves that the animals' behavior is obviously different from that of those in restricted, impoverished enclosures. Behavior occurring in more natural settings can be said to be more *ecologically valid*, and the zoo example gives a sense of the importance of this. That is the first important feature of ethology to consider.

Second, ethologists do not draw very hard lines between behavioral and physical traits. They are all features of an organism about which the same four basic questions, that Tinbergen highlighted in 1963, can be meaningfully asked. What appear to be heated, principled, disputes in behavioral science can often be helpfully seen as confusions between which of these questions is being asked. Even when that is not the case, it is usually true that clarity about which question is being addressed by a particular study is hugely important, and I have found that the distinctions are surprisingly seldom followed.

Tinbergen's four questions are (respectively) as to mechanism, development, function, and evolution. I am going to have a lot to say about each of these types of explanation in this book, so I hope

DOI: 10.1201/9781003372356-4

that I can be permitted to take a few pages to describe their meanings in some detail and go back over these examples from time to time. If you already know all about the principles of ethology, and wish to skip to the next chapter, directly addressing sex research, then feel free.

4.1   HOW AND WHY? EXAMPLES OF THE TYPES OF BIOLOGICAL QUESTIONS WE CAN MEANINGFULLY ASK

A bright child asks us how our eyes work. What sort of things can we say to them? A *mechanistic* explanation about eyes might talk about how light-sensitive cells—rods and cones—worked. These are the nuts and bolts of what makes an eye a light-sensitive organ. We might show the child one of those enlarged plastic models that can be taken apart into pieces and point out all the different (named) parts, what those parts do, how they connect to the optic nerve, and so on. If they (or we) are not too squeamish, we might help the child to dissect an eye and see for themselves how the individual parts are fitted together. Some of the mechanistic questions are answered by looking at what happens when particular parts—such as rods or cones—fail, so we could look at the symptoms of certain diseases, or talk to the sufferers of those diseases, and get their perspective.

But that is not the only type of question that might be asked of eyes and vision. Another typical question that a child might ask is "Do we learn how to see?" The child might be interested in whether colors can be discriminated from birth, or how much learning is involved in vision. If they are a really smart child they might ask, "How do I know that when I see blue, you don't see red?" Strictly speaking this is a philosophical question, but details of the developmental story help us to answer it.

For instance, we can spot people who do not respond to the same color inputs as others do with color-blindness tests. Some of these types of questions are tough to answer, but with a bit of ingenuity we can derive tests that bear on them. It is hard to ask babies questions, but we can follow their attention when colors change in their environment and therefore discover that babies can discriminate a large color palette (and a lot of other things) pretty much from birth.[4] We also have the examples of people blind from birth (or nearly so) who have their sight restored by operation. It turns out that a blind person, who has their sight restored later in life, can integrate learning from one field into the visual one. For instance, such a person could take his experience of feeling the hands of a clock without a face and rapidly learn to tell the time with his restored sight. Questions about these sorts of things are developmental (technically *ontogenic*) questions.[5]

A moment's reflection will reveal that these two types of question I have just discussed are about *how* the overall (visual) system works, but they are still very different questions from one another. These two *types* of question (mechanism and development) are often called *proximate* questions, and here "proximate" means "close."[6] We could, in principle, answer each of them by looking only at one normal individual (although in practice that is pretty unlikely). In a similar vein, extraordinary individuals—case studies—can often tell us a lot about these sorts of questions, such as the few people who had their sight restored after long blindness. *How* questions look within the organism for answers.

The proximate level of explanation is not the be-all and end-all to understanding a trait. Ethology teaches us that we can also ask *ultimate* questions. While proximate questions are "how" questions, the ultimate ones are "why" questions. There are two types of why question as well—*function* and *phylogeny*, and I will detail each of these in turn. When we ask about a trait's evolutionary function, we mean something very specific. We do *not* mean "how it works" (that is the mechanistic explanation). By "function" ethologists mean "how does (or did) that trait contribute to fitness?"

And, by "fitness" they mean something very specific too: "How did it contribute to the chances of that organism's reproducing successfully?" These distinctions are crucial. I have read many examples of sincere non-ethology specialists say things like "female orgasm has (or even can have) *no* adaptive function because it does not (appear to) do *this particular* mechanistic thing we looked for."

Well, maybe it does not do that *specific* thing, but that does not mean that it does not contribute to fitness in some other way. To fail to find a particular proposed mechanism, and thereby conclude that something appeared by accident, would be to give up a bit too easily. Imagine if you took your car to the shop to find out the source of the strange knocking sound when you accelerate. The mechanic looks at the spark plugs, sees that they are not the issue, and then sadly hands back the keys saying, "Sorry, can't help you, it's one of them *phenomena*." You would not be satisfied at this point. You might insist that *something* was happening and that you would like him to keep searching.

Back to eyes: what is the adaptive function of eyes? There are numerous ones, but at least this much—eyes assist in building a useful model of the organism's environment (its *umwelt*) by gathering and collecting electromagnetic radiation in a particular spectrum.[7] Beings without eyes (or with malfunctioning ones) do less well than others with eyes in those environments, and we now have numerous examples of eyes evolving in separate species, so useful is this trait.[8] Eyes are costly items to build and maintain though (having written much of this book with a detached retina, I can attest to this, on a personal basis). This cost is easy to see in species where light is no longer part of the environment. Creatures that have switched to living in caves retain only vestigial eyes, or eyes covered over with skin. Nature does not give free lunches.[9]

The final type of question, also a "why" question that we can meaningfully ask, is called the *phylogenetic* one. Phylogenetic questions trace the evolution of a trait through the paths of speciation over potentially huge tracts of time. It is tough to answer this sort of question when we are looking at traits that do not fossilize. Tough, but not impossible. Behavior does not turn conveniently into rocks for us to study, but rocks are not the only things that persist through time. Genes do as well. And genes underlie behavioral traits every bit as much as physiological ones.

For example, all personality traits that we have measured show some degree of heritability—the amount of the variance that can be attributed to genetic factors. It is rarely less than 50%.[10] Given that personality is linked with reproduction (something to be discussed in detail in later chapters), and it is considerably heritable, it is something that evolution by natural selection can see just as clearly as a bright tail feather, or a faster turn of speed. When it comes to the mechanisms underlying female orgasm, it turns out that comparison with other mammalian species is very illuminating. But, for the moment, back to eyes again. As well as looking at how eyes work mechanistically, seeing how they develop in normal individuals, and considering how they contribute (or contributed) to fitness, we can also compare our eyes with those of other eyed creatures. Estimates vary, but eyes have emerged and evolved maybe fifty separate times on this planet, so useful are they.[11]

Comparing these separate phylogenetic pathways can answer otherwise puzzling questions. For instance, you probably remember from those high school biology lessons that humans have a blind spot. I remember when I first discovered this, and the sense of astonishment that there is a huge blob of nothing in my field of vision—that I am consciously unaware of—has never quite gone away. The phylogenetic explanation of this weird fact is that evolution works with what it has—producing often non-optimal fixes for traits. Humans have a blind spot but octopodes—whose eyes are otherwise remarkably similar to ours—do not. Why is this? Human eyes evolved down a pathway that developed eyes from the inside out. Millions of years ago (long before we were human), nerve cells extended beyond the central nervous system to the outside world. This meant that a rather ugly ad-hoc fix (a *kludge*)[12] was necessary—the optic nerve goes through the light-sensitive retina and then is wired backwards through it. This means that a large chunk of retina has no light-sensitive cells—a blind spot. Octopuses' ancestors, on the other hand, evolved light-sensitive cells outside on their skins, and these gradually turned into eyes over time, connecting inwardly to a developing inner central nervous system. So, no blind spot for them.[13] Looking at our evolutionary history often tells us important things about the constraints on the raw materials that evolution had to work with.

Taken together, functional, and phylogenetic, questions are usefully called *ultimate* questions, to distinguish them from the proximate ones. As I suggested earlier, a very helpful way to think of these differences is to think of proximate questions as pertaining to *how* a trait works, whereas ultimate ones are about *why* it got that way. It is very important to appreciate that these are not rival explanations, in competition with one another. A complete explanation of a trait will answer all four questions.

I realize that I might be accused of over-egging this particular pudding, but I have my reasons. Highly specialized scientists, not to mention those who report on them, can easily get these different questions confused, and my evidence for this is that there are all-too-frequent examples of totally different types of these four explanations being offered as rivals to one another, which they are emphatically not. Let me give a recent example of exactly this sort of confusion relevant to this book.

4.1.1 "THE FEMALE ORGASM HAS BEEN SOLVED!"

A fairly recent paper on female orgasm raised the question of the so-called true meaning (or *function*, or *origin*, or *purpose*) of female orgasm. I saw headlines—and not only in non-scientific newspapers—saying each of these things. The original paper in question, by Pavlicev and Wagner, helped move the science forward by looking at other species' sexual behaviors and proposing the interesting idea that female orgasm arose as a sort of phylogenetic holdout—a relic of a past where ovulation was not spontaneous (as it is in humans) but generated by external stimuli (such as male presence, as it is in some other species, like cats).[14]

What Pavlicev and Wagner had actually done was to trace some likely evolved pathways of ovulation in mammalian and pre-mammalian species, and to then speculate about some likely historical pathways that led to human spontaneous ovulation. This is intriguing because it might have something to do with some of the hormonal correlates of both orgasm and ovulation. That is all very interesting and a potentially useful contribution to the field, but it did not warrant the sort of talk that accompanied it. Some representative headlines over their orgasm paper said things like "Mystery of the Female Orgasm revealed" (*Irish Times*),[15] "Female Orgasm . . . [S]olved" (*Guardian*),[16] or "It has been shown that all [sic] female orgasms are is an . . . [E]volutionary Leftover" (*Science*).[17]

Implicit in all these headlines—and their accompanying pieces—was the idea that if we could find out the (phylogenetic) origin of the female orgasm, then there would be nothing more to know about it. I hope I have made this clear by now—this is simply wrong. We need to know (at least) four somewhat independent sorts of things before we understand a trait, and just one of them on its own is no more the one true answer than any of the others. All are needed.

When a persistent error like this one occurs over and over again, the thought arises that simple ignorance is not to blame. As a psychologist, my suspicions immediately turn to one of the multitude of cognitive glitches that human reasoning is subject to. Simply rehearsing the details of Tinbergen's four questions, and their relative independence, is not always enough to counteract such a glitch. In this case, I think the confusion is what logicians call a genetic fallacy—the belief that the origin of something is its true nature. The history of behavioral science is littered with examples of people confusing origins, purposes, and essences, and I will have more to say about that in subsequent chapters. But, for the moment, I just want to share a way I have come across that helps to slay this particular glitch of the genetic fallacy. If you find yourself immune to this glitch, then please feel free to pass over this digression and skip on to the next chapter.

4.1.2 DIGRESSION: ETYMOLOGY AND ESSENCE

What is the *true* meaning of a word or a phrase? Is it just a case of looking it up in the dictionary? But dictionaries can lag behind everyday usage, and such textbook definitions can be technically correct but decline in usage, so the dictionary cannot be the whole story. A personal case in point will establish this principle, I hope. Example: As much as I dislike the fact that the phrase "begging the question" is changing from its useful and distinct meaning, of "assuming the conclusion in the premises," to the banal (and already defined) usage of "prompts the question," I have to accept defeat. I am outnumbered, and the new meaning is winning. *Merriam-Webster*—who track usage more closely than most dictionaries—has me cornered on this one.[18]

But the true meaning of a phrase is not just its common usage either. No matter how often my greengrocer puts in that extra apostrophe, to indicate that those "potatoe's" belong to someone, he has not won that one. Not yet, anyhow. But it is worse even than that when we try to pin down the one true meaning of a word or phrase.

The term of logician's art "connotation" is used to refer to what a word conjures up in the listener's mind, while the term "denotation" is used to refer to the unique set that the word picks out.[19] Woe betide the person who innocently (or not so innocently) says "But all I meant by using *that* word was . . ." They often get themselves into trouble, especially on social media, such as Twitter.[20]

This exact confusion be-devilled the popular science educator Neil de Grasse Tyson recently. Tyson is an obvious candidate for atheism (although he never himself asserts it), but he said "God-speed" in a Tweet about a space shuttle launch.[21] "But . . . but . . . we thought you were an atheist" came the howls of protest from his Twitter fans. Tyson pointed out that "Godspeed" was a phrase traditionally used in space launches. The headlines ran with "Godspeed John Glenn" back with the pioneering 1962 launch, after all. Tyson's fans were still outraged.[22]

Tyson then pointed out (not unreasonably) that his Twitter followers all happily used the word "Goodbye." "Well, what of it?," his fans replied. "Goodbye," Tyson pointed out, is a contraction of "God be with ye." The (virtual) room went quiet. Oh. They had learned something of the etymology of the phrase. So, is the etymology, the origin of the phrase, its one true meaning? No, but it contributes something to our understanding.

What I hope this example shows is that the attempt to nail down a single true meaning of a phrase is not as simple as it might first appear. But that lack of simplicity does not mean that "anything goes." Far from it. The true meaning of a phrase includes its connotation, its denotation, but I also hope I've at least sown a seed that implies that its etymological root—the story of its origins—tells you *something* about the expression. Something useful. It is part of the complete picture. It does not replace the other elements; it enhances them. Now, this is the important bit:

The ultimate perspective is to a trait, what the etymology is to a phrase.

4.1.3   END OF DIGRESSION

With the four distinct types of question that ethology can answer firmly in mind, I now need to address how female orgasm has been studied. It will turn out that questions of ultimate and proximate matter hugely to what we have found.

NOTES

1 Tinbergen, N. (1963). On aims and methods of ethology. *Zeitschrift für Tierpsychologie, 20*(4), 410–433.

2 This phrase definitely typifies D'Arcy Thompson's work—and especially his monumental (1917) *On growth and form*. However, it is tough to pin down to an actual citation.

3 For more on elephants in captivity, see Rizzolo, J. B., & Bradshaw, G. A. (2016). Prevalence and patterns of complex PTSD in Asian elephants (*Elephas maximus*). There is also the amazing work of Lek in Thailand, as detailed in https://theecologist.org/2005/nov/12/elephant-whisperer last accessed 26/04/2018. See also Mason, G. J. (2010). Species differences in responses to captivity: Stress, welfare and the comparative method. *Trends in Ecology and Evolution, 25*(12), 713–721. *Science Direct.* Web. 27 Nov. 2013.

   For a popular take on orcas in captivity, see Cowperthwaite, G. (2013). *Blackfish* [TV Documentary]. Magnolia Home Entertainment.

   There have been some recent attacks of orcas on yachts, even coordinated ones. Yet, even here, no humans have been directly attacked. See, for example, the NPR article, Killer whales are 'attacking' sailboats near Europe's coast. Scientists don't know why. www.npr.org/2022/08/20/1117993583/orcas-attacks-spain-portugal-killer-whales last accessed 10/08/2023.

4   Ricci, D., Cesarini, L., Groppo, M., De Carli, A., Gallini, F., Serrao, F., . . . Mosca, F. (2008). Early assessment of visual function in full term newborns. *Early Human Development, 84*(2), 107–113.

5   Richard Gregory goes into considerable detail about some specific cases (as well as many similar issues) in Gregory, R. L. (2015). *Eye and brain: The psychology of seeing* (Vol. 38). Princeton University Press.

6   The first modern discussion of this crucial distinction is in Mayr, E. (1961). Cause and effect in biology: Kinds of causes, predictability, and teleology are viewed by a practicing biologist. *Science, 134*(3489), 1501–1506. Students of philosophy will notice that there are striking resemblances to Aristotle's discussion of different types of causation (*Aristotle's physics: Books I and II*. Oxford University Press, 1983). This is not an accident, as Aristotle is responsible for many of the foundations of modern science, largely through his admirable practice of being utterly clear, so that subsequent disagreements with him were equally progressive. Mayr's paper, among other things, brings out the implications of the modern synthesis of Darwinian evolution, with Mendelian genetics. See Dickins, T. E., & Barton, R. A. (2013). Reciprocal causation and the proximate–ultimate distinction. *Biology & Philosophy, 28*, 747–756, for a fuller discussion of this.

7   The "umwelt" is a very useful term, meaning "ecology" in German, co-opted by Jakob von Uexküll to describe the different environments experienced by an organism. This may happen even if organisms occupy (e.g.) the same space, because they are not alert to the same things, and also for many other reasons. von Uexküll, J. (1931). *Der Organismus und die Umwelt*.

8   For example, see Jacob, F. (1977). Evolution and tinkering. *Science, 196*(4295), 1161–1166.

9   For a discussion of vestigial organs in salamanders and other creatures, see Fong, D. W., Kane, T. C., & Culver, D. C. (1995). Vestigialization and loss of nonfunctional characters. *Annual Review of Ecology and Systematics, 26*(1), 249–268.

10 See, for example, Plomin, R., DeFries, J. C., & McClearn, G. E. (2008). *Behavioral genetics*. Macmillan. To a first approximation, behavioral geneticists measure human differences, while evolutionary psychologists measure human similarities. Genetics indisputably underlies both disciplines.

11 Gehring, W. J. (1996). The master control gene for morphogenesis and evolution of the eye. *Genes to Cells, 1*(1), 11–15.

12 Kludge: Whether one considers an evolutionary design solution a masterpiece of elegance or a cobbled-together piece of haphazard jury-rigging largely depends on the scale of magnification one applies. A billiard ball looks like the moon surface under an electron microscope.

13 For a discussion of the eyes of octopuses, see Ogura, A., Ikeo, K., & Gojobori, T. (2004). Comparative analysis of gene expression for convergent evolution of camera eye between octopus and human. *Genome Research, 14*(8), 1555–1561. For the record, the plural of octopus is not "octopi." It is a Greek word and has a Greek plural ("octopodes," if one wishes to be pedantic), not a Latin one.

14 Pavlicev, M., & Wagner, G. (2016). The evolutionary origin of female orgasm. *Journal of Experimental Zoology Part B. Molecular and Developmental Evolution, 326*(6), 326–337. For cat estrous see Wildt, D. E. (1991). Fertilization in cats. In *A comparative overview of mammalian fertilization* (pp. 299–328).

15 The *Irish Times* reports on Pavlicev and Wagner (2016). www.irishtimes.com/life-and-style/health-family/mystery-of-the-female-orgasm-revealed-1.2742453 "Mystery revealed."

16 Davis, N. (2016). Mystery of the female orgasm may be solved . . . Scientists believe they can explain the evolutionary reason for women's sexual response. *The Guardian*. www.theguardian.com/society/2016/aug/01/mystery-of-the-female-orgasm-may-be-solved last accessed 25/08/2023.

17 *Science Mag* reports on reports on Pavlicev and Wagner (2016). www.sciencemag.org/news/2016/08/new-theory-suggests-female-orgasms-are-an-evolutionary-leftover "Female orgasms are an evolutionary holdover . . . having no evolutionary function"

18 "Begging the question," the formal fallacy of *petitio principii* used to (and according to some obstinate historians, still *should*) mean "assuming the conclusion." Literally, asking the question to support itself. In other words, it is a term denoting a circular argument. Common usage is killing off this meaning, to be replaced with the banal "This prompts the question," for which we already have adequate terms.

19 For more on connotation and denotation, see Russell, B. (1905). On denoting. *Mind, 14*(56), 479–493.

20 Now called X, although this usage has not caught up with everyone yet.

21 Neil de Grasse Tyson on the Twitter debacle: https://twitter.com/neiltyson/statuses/40892790736949249 wishing the Shuttle astronauts godspeed.

22 The original "Godspeed John Glenn" headlines: www.pennlive.com/midstate/index.ssf/2013/10/scott_carpenter_delivered_the.html.

5 The Two Traditions of Female Orgasm Research

There is a notion that scientists start by defining their terms. This is rubbish . . . They start with a rough hunch and start digging. It is notoriously difficult to define the word *living*.

(Francis Crick, att.)

I hope you will excuse me from following that stern advice given to student essayists: First define your terms. I want to delay setting out exactly what constitutes the female orgasm, for the moment at least. Instead, I want to talk about the ways female orgasms have been studied up until now.

I have my reasons. As we saw in chapter four, biological science can ask both *how* questions and *why* questions.[1] It turns out that the ways we have of studying the (proximate) questions of *how* an orgasm works have had profound consequences for our investigations of the (ultimate) questions of *why* it exists at all. This relationship goes both ways. Assumptions about ultimate (*why*) functions of female orgasms have conditioned the ways that we have studied proximate (*how*) mechanisms. At the core of the scientific method is the control of variables. But what variables can we control so we can study sex properly, while retaining its character as an authentically recognizable human activity? To put this another way: Can we recreate everything we need for the study of sex under laboratory conditions?

As I noted earlier, many other animals often do not mate well in captivity. In this chapter, I argue that humans are not so different from those other mammals. Studying the full range of human sexual behavior necessitates (sometimes) leaving the laboratory, but some key studies into female orgasm never did this. That is fine, as far as it goes, if one does not thereby conclude that the phenomenon has been studied to its fullest. Humans, like many other animals, prefer privacy to allow themselves the vulnerability of sex, and I will argue that this is not some optional add-on to the behavior. On the contrary, it allows us to do one of the key things that we use sex for: To make an accurate assessment of our partner's qualities. And, further, it is only in this state that the full expression of sex, including female orgasm, is at its fullest.[2]

5.1 THREE TYPES OF ANSWER TO THE QUESTION, "WHY DO WOMEN HAVE ORGASMS"?

Why—in that ultimate sense of *why*, discussed in the previous chapter—do women have orgasms?[3] There are three broad possibilities: pair-bonding, sperm selection, and by-product. Recall, from chapter four, that evolution by natural selection concerns the contribution of a trait to fitness, and fitness means differential reproduction. In what ways might female orgasm contribute to this?

The first possibility is that maybe female orgasm increases fitness by cementing pair bonds between sexual partners, and hence increasing the chances of offspring survival. This suggestion was made famous by Desmond Morris, in his popular science work, *The Naked Ape*. His rather inelegantly titled "poleaxe" hypothesis, featuring the idea that orgasmic females are erotically knocked out by orgasm (thereby persuading the males to hang around and protect them), does not have much empirical support. Indeed, it does not even really fit with much of the known phenomenology of the orgasms of both sexes. If anyone is liable to be "knocked out" after orgasm it is the men—who get a dose of post-orgasmic prolactin—and their attendant sleepiness, which can leave their partners frustrated.[4]

However, the noted scholar of Russian literature Daniel Rancour-Laferriere produced a more sophisticated version—or really a set of interconnected versions—of the pair-bonding (also called "mate-choice") hypothesis. His version makes some interesting and testable predictions, and I will have more to say on this later.[5]

A second type of adaptationist proposal has been that female orgasm somehow increased fertility by directly improving chances of pregnancy with that partner, usually by some mechanism promoting the movement of sperm into the reproductive tract, or by retaining it in usable form while there. This idea (sometimes also called the "sire choice hypothesis") goes back to the ancient scholars such as Aristotle and has certainly been the most extensively researched. Mechanisms by which it might happen have, however, been tricky to investigate.

The word "tricky" hardly covers the difficulties. For example, the proposals of the team of Robin Baker and Mark Bellis—that female orgasm could be timed to increase sperm intake and instigate so-called sperm wars between competing male partners—has failed to replicate.[6] Many scholars in the field are familiar with this failure of replication, but Baker and Bellis' proposal is not the only, or even the best, way to investigate differential sperm selection.[7] This is a large topic, and I will talk about it more thoroughly across several chapters. For the moment let us acknowledge that differential sperm selection is a tough phenomenon to study, and that this has led to some false starts, all of which is perfectly normal in science.

Standing apart from those two types of proposal, the seeming intractability of understanding female orgasm from an adaptationist perspective has led some to call for giving up trying to find a female-centric function for orgasm at all. Thus, we get the third major type of answer to the "why" question: That female orgasm has no function of its own—existing solely as a by-product of selection on male anatomy. An analogy is drawn by these by-product advocates between the clitoris and functionless male nipples. This idea was first put forward by Donald Symons in his ground-breaking *The Evolution of Human Sexuality* in 1979, but after that it was enthusiastically championed by noted paleontologist Stephen Jay Gould and then Elizabeth Lloyd in her highly influential 2005 book, *The Case of the Female Orgasm: Bias in the Science of Evolution.*[8]

5.2   HOW HAVE WE STUDIED ORGASMS?

Zoologists, Russian scholars, evolutionary biologists, anatomists, entomologists, and paleontologists—all with differing views of the possible adaptive function (or lack thereof) for the female orgasm? And that selection does not cover even half of the disciplines that have offered an opinion. This bewildering breadth of suggestions speaks to the knottiness of the problem. And it gets worse. It turns out that these proposals of ultimate function have prompted rather different ways to measure proximate mechanisms. Furthermore, the distinctions have implications for one another and require quite a bit of disentangling. The obvious place to start the disentangling process might seem to be by considering the proximate questions first: For example, what is happening during female orgasm?

However, even *this* deceptively simple question turns out to be rather more complicated to answer than it first appears. We have already had a parade of different disciplines offering perspectives, and it is about to get more complex still. The noted physiologist Roy Levin put it thusly: "To characterize the phenomenon of orgasm in both men and women obviously needs the combined skills of the physiologist, psychologist, endocrinologist, brain imager and the subject but even then we have some obvious difficulties."[9] Levin found thirteen separate definitions of "orgasm" in 1981,[10] and Mah and Binik documented twenty-six, some two decades later.[11] This sort of complexity is likely to bewilder all but the hardiest.

I think there is a way through this maze. To try to make sense of what is going on, I propose to start by exploring two separate strands of behavioral science that have both generated productive research. While both strands have been fruitful, one is especially consistent with what we find from studying other species, collecting behavioral and phenomenological reports in our own species, performing neuroscience, and examining endocrinology. But, before I get to those details, I need to suggest that a particular tradition of studying behavior—including, but not limited to sexual behavior—has been unfairly overlooked.

5.3 ETHOLOGY AND THE TWO TRADITIONS OF FEMALE ORGASM RESEARCH

In the previous chapter I drew attention to the importance of studying behaviors in as natural a setting as possible. This is especially important when it comes to sexual behavior. For example, it is rather unfair to blame panda infertility on their own proclivities (or supposed lack of them) rather than their response to confinement. It might surprise you to discover that pandas have rich and complex sex lives, when they are allowed to have them in private.[12] Humans are not really so different in that respect, and they are not that unusual among other mammals either, in their preferring of intimacy at reproductive times. As I mentioned in the preface, my nearby zoo—really a wildlife park, which prides itself on providing lots of space for animals to hide from prying eyes—has just had a year where seemingly all the big cats have had kittens.[13] This is a good sign of animals that feel comfortable and able to express a fuller range of behavior.

I want to draw your attention to this otherwise obvious set of facts, because there will still exist a sneaking suspicion in some minds that only behaviors that take place in *unnatural* settings can be considered as valid for scientific study. Put that way, the fallacy is obvious. But, if you say something like, "only laboratories offer the degree of control necessary to isolate variables and so do proper science," your pronouncement will be joined by an (imaginary) background of serried ranks of white-coated supporters nodding sagely.

I do lab-based studies myself, they are invaluable, and lots of this book will require standing on the shoulders of some giants in the field who performed them. But, and it is a serious "but," laboratories are part of the scientists' toolkit—they are not the be-all and end-all of science. For anyone who thinks that no proper science can exist outside of labs, then I have one word for them: astronomy. We cannot force a nebula into our controlled conditions. No one thinks that the comparative lack of control invalidates the science of astronomy.

This matter, of what we technically call *ecological validity*, is acute for sex research because (unless one has some very unusual preferences) laboratories are not places of intimacy, passion, and eroticism. This is a problem for human sexual science because things like intimacy, passion, and eroticism are not weird add-ons to human sexual behavior. They are central to our being able to make sense of it at all, or even for certain behaviors to manifest themselves in their fullest expression. Therefore, lab-based studies must be used with caution. To be even more specific, I am claiming that the fact that a sexual behavior has not been observed in a lab is not, of itself, sufficient reason for arguing that it does not exist, has no role, or has been explored to the fullest. As the archeologists are fond of reminding us, absence of evidence is not evidence of absence.

From an early stage it was recognized by the pioneers in ethology that behavior in natural settings could be radically different from that in captivity. Early ethologists studied things like foraging and territoriality, which, almost by definition, cannot really exist in labs at all. Sometimes key insights were taken from the field back to the lab and recreated there (like Konrad Lorenz's brilliant imprinting studies).[14] However, I should emphasize that this is not an all-or-nothing affair. No one has to take sides or (God forbid) take part in one of those unedifying "my science is bigger than yours" contests. Science is a big tent and can accommodate variety. That said, ecological validity matters.

With thoughts of ecological validity (and sexual passion, perhaps) firmly in mind, I want to turn back to those two key modern traditions of sex research pertaining to female orgasm. The first is laboratory based, and the second brings in elements of ecological validity—namely, sex in a natural setting. Obviously, we need to build on both. If I seem overly critical of that first strand, it is not because that tradition has not been an indispensable part of the science. It is only because some elements of methodology have occasionally been taken to be the last word on the topic—and they are not. Far from it. The two strands in question, exploring the nature and function of female orgasm, are those of the American sex researchers William Masters and Virginia Johnson, and the British team of Cyril and Beatrice Fox.

5.4 MASTERS AND JOHNSON

Masters and Johnson are the sex researchers that most non-specialists will have heard of—and understandably so.[15] Their investigations, through the 1950s, 1960s, and into the 1970s, built upon Kinsey and Pomeroy's[16] impressive and (then) startling interview and observational analyses of human sexual behavior, and opened up modern public discourse about sex.[17] Using new investigative techniques, to be detailed presently, Masters and Johnson attempted nothing less than a comprehensive analysis of both male and female sexual response, although I will focus on the female element here.[18]

5.4.1 WHAT HAPPENS DURING SEXUAL AROUSAL?

To aid in the investigation of the cycle of human sexual response in general, Masters had already divided it up into four, somewhat arbitrary, phases. These phases were, in order:

1) Excitement
2) Plateau
3) Orgasm
4) Resolution

How does this model apply to female orgasm? During the *excitement* phase, somewhere between a few seconds and half a minute, after the start of sexual stimulation—be that physical or psychological—vaginal lubrication begins. Blood moves to the vaginal walls, a process technically called "vasocongestion." The glans of the clitoris (that external part) enlarges, and the internal parts of the clitoris (the shaft) also fill with blood. Breasts and nipples usually become more aroused at this stage, as blood flows there as well. The vagina expands and the cervix and uterus move upwards. These extra blood flow patterns can often be visibly detected (as a "sex flush") in the area just above the stomach (this is higher up the body than most people assume, and is technically called the "epigastrium"). That is quite a lot of action already, and things have barely started. Assuming that sexual activity continues . . .

Plateau is now reached. The heartbeat goes up. Sometimes way up—from somewhere between 100 beats a minute to 175, nearly three times the normal resting rate. The body is readying itself for action. All that blood must go somewhere, and one obvious place is the labia—the outer genitalia—which can darken considerably. That "sex flush" I mentioned earlier can now sometimes be visible in other areas across the body. At the same time, this continued flow of extra blood to the genital region causes the tissues of the outer vagina to expand even more. Simultaneously, the uterus lengthens, and the glans of the clitoris (usually) becomes hidden by the clitoral hood. Blood pressure goes up too. Tension is mounting.

This plateau stage can be maintained for a considerable time. Compared to men, women can maintain sexual excitement, without orgasm, for longer periods without discomfort. In addition, their refraction period—the time when they can recover and orgasm again—is faster. That said, there is usually some (often verbally expressed) desire for the orgasm to occur to relieve that tension just described. That relief of tension is the *orgasm* phase, as classically described by Masters and Johnson.

Thus far, what has been said is uncontroversial to sex researchers. Masters and Johnson do acknowledge that women report their orgasms differently, but they maintain that "an orgasm is an orgasm." We will have occasion to revisit this assertion but, for the moment, let us accept that so far everyone agrees: Following effective stimulation, the female partner orgasms—a highly pleasurable sensation accompanied by muscular contractions of the uterus, the outer third of the vagina, and the anal sphincter.

During the final phase, *resolution*, the clitoris returns to its normal size and its glans moves out from under its protective hood. The visible sex flush dissipates. As stated previously, women often have a comparatively short refraction period, and may be ready for sexual activity again soon. This is usually, alas, in stark contrast to men.

This EPOR (excitement, plateau, orgasm, resolution) model allowed scientific attention to focus on conceptually distinct phases in the arousal pattern of both sexes and, with some important reservations, is still in use by many sex researchers.[19] Masters and Johnson have to be congratulated for opening up a field of inquiry in the teeth of cultural opposition, which at that time could be fierce.

5.4.2 CAVEATS TO MASTERS AND JOHNSON

What about limitations in their methods? It needs to be noted from the outset that when it came to studying female orgasm, what Masters and Johnson were directly studying was not sexual intercourse per se; it was an attempt to recreate what they felt to be its key features. To emphasize, *none* of the roughly 10,000 cycles of human sexual response recorded, in either the Department of Obstetrics and Gynecology at Washington University in St. Louis or their St. Louis Reproductive Biology Research Foundation included even a *single instance* of actual penetrative sexual intercourse.[20] Here is how they described their work themselves:

> With the introduction of *artificial* coital techniques, the reactions of the vagina during coition became available to direct observation and repeatedly have been recorded through the medium of cinematography.[21]

When it came to the detailed examination of the internal physiological effects of female orgasm (or supposed lack thereof), the essence of their technique was to insert a rigid glass artificial penis—nicknamed *Ulysses*—containing recording equipment into the vaginas of female participants. These participants—and there were only six repetitions of this part of the study—then masturbated themselves to orgasm, through stimulation of the mons and clitoral glans—the external part of the clitoris.

Key among Masters and Johnson's conclusions were the assertions that only the external parts of the clitoris, or the area close to it, were sexually sensitive, and that there was no difference between so-called clitorally based or vaginally based orgasms. They were particularly emphatic that this was fatal to the idea of a possible role of female orgasm in creating any uterine sucking effect that others had reported and could be a vehicle for sperm selection. I explain this in much greater detail later. Those were their findings. What are some of the implications?

5.4.3 IMPLICATIONS AND ISSUES ARISING FROM MASTERS AND JOHNSON'S RESEARCH

If Masters and Johnson were correct in their assertion that only the clitoris and its immediate surrounding area are capable of producing sexual stimulation in the female, then it would follow that only those sexual positions that allow direct or, more probably, indirect stimulation of the outer part, or glans, of the clitoris would lead to female orgasm. This is, indeed, what they do assert:

> Only the female superior and lateral coital positions allow direct or primary stimulation of the clitoris to be achieved with ease . . . in the knee-chest coital position no direct stimulation of the clitoris is possible . . . stimulation of the clitoris (receptor organ) developing during active coition is the secondary or indirect result of penile traction on the minor labial hood.[22]

One set of predictions of these claims is that female orgasm from coition in certain positions (e.g., what is today commonly referred to as "doggy-style") and orgasm after certain invasive operations (such as clitoridectomy) should be as near to impossible as makes little difference. These particular predictions have been tested. I will have a lot more to say about them in the next chapter and in chapter eight. To anticipate somewhat, these predictions have not been supported, but I want to take their other assertion on board first. This is their central and crucial claim—that only what they (incorrectly) viewed as *clitoral* orgasms could exist:

> From an anatomic point of view, there is absolutely no difference in the responses of pelvic viscera to effective sexual stimulation, regardless of whether the stimulation occurs as a result of clitoral-body or mons area manipulation, natural or artificial coition, or, for that matter, specific stimulation of any other erogenous area of the female body.[23]

Masters and Johnson failed to detect any of the internal pulsing sensations that some researchers and participants had previously noted during female orgasm.[24] In retrospect, this might have provided a clue to them that their research procedure was not capturing the full range of female orgasm properties but, in the event, it did not. On the contrary, they reported that the lack of such reported sensations in their research put paid to any idea of orgasm having an insuck function. This claim is a pretty bold piece of conceptual architecture. It might surprise people to learn just how slender are the stilts upon which it rests.

What was their methodology? They got six women to have a cervical cap inserted and then to masturbate to orgasm. No internal sensations were reported by anyone involved. In addition, Masters and Johnson made it plain that they could see nothing happening inside through the medium of *Ulysses*' camera. "In none of the six individuals was there evidence of the slightest sucking effect on the media in the artificial seminal pool."[25] But their conclusions could be seen as premature. As I will detail in chapters six and nine, penises are importantly different from glass tubes and show every sign of having been selected by females over evolved time to provide internal stimulation.[26] In addition, the cap that Masters and Johnson inserted was placed over the mouth of the cervix, and would have effectively blocked any potential insuck mechanism, were it to have occurred.[27]

5.5   LEAPIN' FOR THE CLITORIS?[28]

Before we get to considering the alternative—and ecologically richer—approach to studying female orgasm, there is a rather pressing anatomical issue with Masters and Johnson's account. What they called the "clitoris" is only the *external* part—the glans—of this complex, sensitive, largely internal, and functionally integrated organ.

The full details of the importance and implications of this oversight deserve an entire chapter of their own, and it is the one following this one. For now, I just want to note that mistaking the glans of the clitoris for the whole thing has led some commentators—typically those who espouse the by-product view—to view the whole mechanism as having a "Rube Goldberg" haphazardness about it.[29]

But, if they are right, then the situation is far worse than this: If it was actually true that only the external part of the female urogenital system was sensitive to sexual stimulation, then calling the whole arrangement a "Rube Goldberg" device would be an understatement. It would seem more as if a cruel and capricious god had deliberately set up women to be aroused, but almost guaranteed to be unsatisfied, by sexual intercourse.

Are women really the recipients of some cruel cosmic joke, on the part of a malevolent Mother Nature? There are some reasons for thinking not. For one thing, there are good reasons for thinking that public masturbation does not capture the be-all and end-all of women's sexual experience. Therefore, there are good prima facie reasons for thinking that investigation restricted to using that particular methodology will give an incomplete picture. At least one of Masters and Johnson's potential participants thought the same thing at the time and said so. In a letter to them she wrote:

> I just know that if someone would watch me copulate with a partner, the best I could do would be a little outer clitoral climax, as fast as possible to get the silly situation over with. I do not call that an orgasm.[30]

The writer raises important points about privacy and orgasmic variability, but these are not the only points at issue. At the time when the research was being conducted, a number of scientists were strongly of the opinion that the so-called orgasmic platform (whose activities Masters and Johnson described) was, while important, not the only site of female sexual sensitivity and response. Specifically, a number of contemporary researchers—notably the husband-and-wife team of Josephine and Irving Singer—drew attention to the uterine contractions that could accompany those of the orgasmic platform.[31]

Other contemporary scientific commentators expressed dismay that Masters and Johnson had rushed to publication. For example, a letter to the *British Medical Journal* in 1967 criticized the Masters and Johnson findings for failing to account for the role of prostaglandins in possibly generating uterine contractions. I will have a more extensive discussion of this set of mechanisms later, but for the moment, I just want to note the conclusion to this letter, which by no means expressed a lone scientific voice at the time:

> As Drs. Cyril and Beatrice Fox imply, it is a pity that Masters and Johnson's findings were published in the form of a popular book, lacking the experimental detail that would have been demanded in a more formal publication.[32]

A number of sex therapists have noted that pathologizing women's oft-reported low sexual desire often comes about from assuming that men's and women's sexual cycles are more or less identical. This is a mistake. Consider, for example, the work of sexual medicine pioneer Rosemary Basson on this issue:

> Both Leiblum (1998) and Tiefer (1991) stressed that the focus on genital responses and traditional indicators of desire, including sexual fantasies and a need to self-stimulate, ignores major components of women's sexual satisfaction: trust, intimacy, the ability to be vulnerable, respect, communication, affection, and pleasure from sensual touching.

I could not put it better. Masters and Johnson's research eliminates practically all of these features from the sexual cycle.[33]

Some historical perspective may help at this point. Scientists are meant to be objective searchers for truth, but it would be naïve to pretend that personalities do not sometimes intrude. It is often important to know whom a particular scientist stands in opposition to, in order to grasp their particular emphases and biases. In this case, the villain of Masters and Johnson's world was undoubtedly Sigmund Freud.[34] Freud, as many people know, had argued that female orgasms were of two kinds—clitoral and vaginal. Masters and Johnson were keen to overturn this (and much else that Freud had said). As we shall see, Freud had underestimated the nature of the clitoris, but so too did Masters and Johnson. As I detail in the next chapter, the clitoris is a much larger and more complex organ than either set of scientists were aware of. But one does not have to (nor should one) accept Freud's idea that some orgasms were superior to others, or more mature than others, in order to realize that female orgasmic complexity is greater than Masters and Johnson could find with their techniques.

Masters and Johnson were also very keen to overturn a separate tradition of exploring the nature and function of female orgasm that associated it with internal and reproductively important effects. These included sensations of pulsing, which might be a sign of an internal mechanism at work. These internal sensations, which some thought might be connected to sperm transport, had been reported (but not objectively and directly measured) in humans. On the other hand, actual sperm transport *had* been directly measured (but not, of course, reported) in a host of other mammals since the 1930s. Species studied included dogs, rats, horses, and cows. Later findings expanded this menagerie to include pigs and a range of primate species.[35] However, at the time, internal effects resulting from female orgasm had not been measured in humans. The husband-and-wife team of physiologists Drs. Cyril and Beatrice Fox, the same ones mentioned in that earlier letter, decided to try to bring these two lines of research together using more direct methods of investigating sex and female orgasm.

5.6   FOX AND FOX

At roughly the same time that Masters and Johnson were finding that, perhaps surprisingly, female orgasm did not appear to do anything very biologically significant, another team was studying female orgasm in the UK. These scientists used methods that, while still appropriately controlled, had much more in common with what most people would recognize as actual sexual intercourse. Specifically, they were using a couple, who knew each other well and were having sex in a natural setting: their own marital bedroom.

The pioneering work of the husband-and-wife team of Cyril and Beatrice Fox in the late 1960s through to the 1970s was the first to find solid empirical evidence of a potential functional property of female orgasm, via the uterine pressure changes that could occur during it. By using inserted radio-telemetry devices during real coitus, they found that there was evidence for an insuck function to female orgasm, which could have fertility implications. In other words, female orgasm was shown by them to generate some sort of internal action (with details to be specified) that might influence fertility. This (proximate) mechanism was—if it existed—something of a kind that could then be under (ultimate) selection. If pressure changes privileged some sperm intake over others—and therefore some sexual partnerings over others—or increased fertility chances (and all this happened in patterned ways), then this would be something that natural selection could "see."[36]

Why were the Fox team even looking for this? Claims as to the possible fertility implications of the female orgasm go back at least as far as Aristotle, according to early pioneers of sexuality studies, such as Havelock Ellis. As to what the actual mechanism might be, Ellis had speculated that "in women as in mares, bitches, and other animals, the uterus becomes shorter, broader, and softer during the orgasm, at the same time descending lower into the pelvis, and its mouth opens intermittently."[37]

A lot of women had reported, and still report to this day, feeling sensations consistent with internal pressure changes during some orgasms—*uterine peristalsis* is the technical term—which might bring about rapid transport of sperm or some other mechanism. This would be important because it might privilege some sexual encounters over others, and thereby increase the odds of fertility. *This* mattered because these peristaltic effects would unify human research with what science had shown happening in a whole host of mammals, in research going back to the 1930s, as detailed earlier.

Given all this comparative ethology background, if humans did *not* have sperm transport mechanisms associated with uterine peristalsis, then they might start to look like something of an exception among mammals, but not so fast. Orgasm is not quite the same thing as uterine contractions, and a mechanism that could increase fertility had not yet been demonstrated in humans. That is where the Fox team came in. With human participants, they could correlate the self-report of orgasm during natural coition, with objective measures of changes in internal states. And, in a series of classic pioneering papers, that is precisely what they proceeded to do. To do this, they used an inserted pressure change reader (which recorded pressure changes using H_2O, in ways similar to a barometer) and an associated radio-telemetry device. They then proceeded to show that the pressure changes—intra-vaginal *and* intra-uterine—during female orgasm were remarkable:

> "There was an increase in intra-uterine pressure as the female climax was reached, with two peaks of +48cm H_2O in the first uterine experiment and +40cm H_2O in the second uterine experiment." Following these peaks, there were then sharp falls, after which "regular contractions returned."[38]

The Fox team had not demonstrated insuck of sperm per se, because all they had attempted to show was a mechanism (pressure change) consistent with it.[39] But that, pace certain critics, is exactly what they *did* show. This was the result of sex between a couple who knew each other well, were comfortable during the sexual encounter, and were having said sex in their own bed at home. It does not get much more ecologically valid than that. These differences between their methods and those of Masters and Johnson in generating and measuring female orgasm cannot be lightly set aside. It is the difference between natural sex and sex in captivity. It has stimulated a rich and varied line of research over the next fifty years, showing associated mechanisms of sperm insuck, including our own team's more recent research demonstrating orgasmic sperm retention.[40]

However, the Fox research drew some criticism. Specifically, Elizabeth Lloyd, while championing the by-product account of orgasm, decided to take on the Fox research team's findings in her widely read 2005 book.[41] Her criticisms relied on (1) asserting that masturbatory female orgasms (as used in Masters and Johnsons research) capture all the relevant properties of female sexual response; (2) claiming that no movement of sperm (or sperm-like material) was shown by the Fox studies; (3) proposing that any effects shown were trivial; and (4) reporting on the small sample size.

Obviously, Lloyd's claims require detailed responses. The responses to her criticisms, and the research built on those responses, will form a large part of the rest of this book. The brief overview of these is that: (1) can be shown to be false—all female orgasms are not the same; (2) differential sperm selection was not being searched for in the initial experiments of the Fox team, but has been shown subsequently; (3) appears to be a confusion over the reporting of the readings; and (4) follow-up studies have built on the Fox studies, replicating key elements (such as pressure changes and insuck demonstrations) even if their precise methodology has not been used again.

Before I tackle those issues, I need to prepare the ground, by saying more about the fascinating and oft-misrepresented organ at the heart of the puzzle of female orgasm—the clitoris. That is the focus of the next chapter.

5.7 SUMMARY

Masters and Johnson are giants in the field of sex research. However, even giants do not have the last word on a subject, and these particular giants do not occupy the whole of the field. At the same time that a line of inquiry into female orgasm was being explored by Masters and Johnsons' lab-based methods, a more naturalistic line was being pursued by another team—the Foxes. Their work deserves to be more widely known, particularly as it has generated a fruitful scientific program, detailing the nature and function of female orgasm. It is not hard to appreciate why female orgasm is hard to study scientifically: It is challenging to replicate its key features under conditions of appropriate scientific control. Thus, scientists have to be especially cunning, putting together a jigsaw of different approaches utilizing a range of methods.

NOTES

1 Students of philosophy might see echoes of Aristotle's four causes in Tinbergen's typology of causes. There is considerable overlap, but they are not identical, although the insight that we can (and should) ask different kinds of causal questions is common to both. Some have argued that Aristotle's fourth cause—teleological, or end-driven—ascribes psychological factors to nature, but this has been disputed. See Sorabji, R. (1980). *Necessity, cause and blame.* Cornell University Press, for a fuller discussion.

2 There are exceptions, of course. Someone could reasonably ask: What about exhibitionists and voyeurs? These phenomena are real enough and an important part of human sexuality. However, the fact that humans can play with the normal parameters of sex to add elements such as danger shows the boundaries of those normal parameters. The implications of various human kinks, fetishes, and so on are important enough, but will have to wait for now.

3 A trait could have been under previous selection but no longer be under such. However, rather than constantly say "does or could have contributed to" each time, I am going to stick to one tense to avoid clumsy locutions.

4 Alwaal, A., Breyer, B. N., & Lue, T. F. (2015). Normal male sexual function: Emphasis on orgasm and ejaculation. *Fertility and Sterility, 104*(5), 1051–1060.

5   Morris, D. (1967). *The naked ape*. Jonathan Cape. See also Rancour-Laferriere, D. (1983). Four adaptive aspects of the female orgasm. *Journal of Social and Biological Structures, 6*, 319–333.

6   Baker, R., & Bellis, M. A. (1993). Human sperm competition: Ejaculate manipulation by females and a function for the female orgasm. *Animal Behaviour, 46*, 887–909.

7   Birkhead, T. R., & Møller, A. P. (1998). *Sperm competition and sexual selection*. Academic Press., failed to replicate Baker and Bellis' findings. Following this, Tim Birkhead devoted some pages to how "his idol was tarnished." (See Birkhead, T. R. (2000). *Promiscuity: An evolutionary history of sperm competition*. Harvard University Press., for details.)

8   The by-product view was initially proposed by Symons, D. (1979). *The evolution of human sexuality*. New York: Oxford University Press. It was then taken up by Gould, S. J. (1987). Freudian slip. *Natural History, 96*, 14–21, and then Lloyd, E. (2005). *The case of the female orgasm: Bias in the science of evolution*. Harvard University Press.

9   Levin, R. J. (2004). An orgasm is . . . who defines what an orgasm is? *Sexual and Relationship Therapy, 19*(1), 101–107. The quote is from page 107.

10 Levin, R. J. (1981). The female orgasm—a current appraisal. *Journal of Psychosomatic Research, 25*, 119–133.

11 Mah, K., & Binik, Y. M. (2001). The nature of the human orgasm: A critical review of major trends. *Clinical Psychology Review, 21*, 823–856.

12 Martin-Wintle, M. S., Shepherdson, D., Zhang, G., Zhang, H., Li, D., Zhou, X., . . . Swaisgood, R. R. (2015). Free mate choice enhances conservation breeding in the endangered giant panda. *Nature Communications, 6*, 10125.

13 Fota Wildlife Park has had an impressive 200 cheetahs born in captivity since 1985. For examples of some of the research into making them feel comfortable around humans and other animals, see O'Donovan, D., Hindle, J. E., McKeown, S., & O'Donovan, S. (1993). Effect of visitors on the behaviour of female Cheetahs *Acinonyx jubatus* and cubs. *International Zoo Yearbook, 32*(1), 238–244.

14 For example, Lorenz, K. (1937). Imprinting. *Auk, 54*(1), 245–273.

15 Most recently in Maier, T. (2013). *Masters of sex: The life and times of William Masters and Virginia Johnson, the couple who taught America how to love*. Basic Books.
    This was made into a TV series for Showtime: Ashford, M., Timberman, S., Beverly, C., Lipman, A., & Verno, J. (2013) *Masters of sex* [TV Show]. Distributed by CBS Television, produced by Round Two and Timberman/Beverly productions.

16 Kinsey, A. C., Pomeroy, W. B., Martin, C. E., & Sloan, S. (1948). *Sexual behavior in the human male* (Vol. 1). Saunders.

17 Two key texts are Masters, W. H., & Johnson, V. E. (1965). The sexual response cycles of the human male and female: Comparative anatomy and physiology. In F. A. Beach, (Ed.), *Sex and behavior* (pp. 512–534). Wiley; and Masters, W. H., & Johnson, V. E. (1966). *Human sexual response*. Churchill.

18 Masters, W. H. (1959). The sexual response cycle of the human female: Vaginal lubrication. *Annals of the New York Academy of Sciences, 83*(1), 301–317.

19 For reservations about the EPOR model, see e.g., Levin, R. (1998). Sex and the human female reproductive tract-what really happens during and after coitus. *International Journal of Impotence Research, 10*(Suppl.1), S14–S21.

20 The St. Louis Reproductive Biology Research Foundation was renamed the Masters and Johnson Institute in 1978.

21 The artificial coition techniques quote is from Masters and Johnson (1966, pp. 66–67) (emphasis added).

22 The sexual positions quote is from Masters and Johnson (1966, pp. 59–60).

23 The "all orgasms are clitoral" quote is from Masters and Johnson (1966, p. 66).

24 Those internal pulsing sensations include (but are not limited to) those of the vagina—which could themselves stimulate the penis to ejaculation—and deeper inside. These were noted by several researchers as well as reports from sexual actors. See especially Fox, C. A., & Fox, B. (1971). A comparative study of coital physiology, with special reference to the sexual climax. *Journal of Reproduction and Fertility, 24*, 319–336.

25 It sometimes surprises people to learn that Masters and Johnson carried out their insuck experiment on only six participants, but this is true. An awful lot of sex research rests on a small (and brave) sample of participants. The quote about the negative results concerning sucking effects comes from Masters and Johnson (1966, p. 123).

26 Schultz, W. W., van Andel, P., Sabelis, I., & Mooyaart, E. (1999). Magnetic resonance imaging of male and female genitals during coitus and female sexual arousal. *British Medical Journal, 319*, 1596–1600.

27 Similar remarks pertain to the Kinsey team (Kinsey et al., 1948), who also used insertions of metal specula that would tend to cover up crucial areas of internal sensitivity.

28 Jones, T. (1983). *Monty Python's the meaning of life* [Motion picture]. Celandine Films.

29 Rube Goldberg is the North American equivalent of Heath Robinson. Both invented fantastical and elaborate machines, in cartoon form, that (just about) performed the functions laid out for them. For the first to apply this model to female orgasm, see Symons, D. (1979). *The evolution of human sexuality.* Oxford University Press. The idea was then taken up enthusiastically by Gould, S. J. (1987). Freudian slip. *Natural History, 96,* 14–21, and then Lloyd, E. (2005). *The case of the female orgasm: Bias in the science of evolution.* Harvard University Press.

30 The quote about external orgasms comes from Lowen, 1967, p. 238, quoted in Singer, J., & Singer, I. (1972). Types of female orgasm. *Journal of Sex Research, 8,* 255–267.

31 Singer, J., & Singer, I. (1972). Types of female orgasm. *Journal of Sex Research, 8,* 255–267.

32 Pickles, V. R. (1967). Uterine suction during orgasm. Letter to the *British Medical Journal, 1*(5537), 427.

33 Basson, R. (2000). The female sexual response: A different model. *Journal of Sex &Marital Therapy, 26*(1), 51–65. The quote comes from page 52.

34 Freud, S. (1932). Female sexuality (Trans. J. Riviere.) *International Journal of Psycho-Analysis, 13,* 281–297.

35 For work on the effects of uterine peristalsis on rapid sperm transport in early experiments on humans and other animals, see Ammersbach, R. (1930). Sterilität und frigidität. *München. Medizinische Wochenschrift, 77,* 225–227, and Trapl, J. (1943). Neue Anschauungen über den Ei- und Samentransport in den inneren Geschlechtsteilen der Frau. *Zentralblatt für Gynäkologie, 67,* 547, for humans (although the details are sparse). For work on dogs, see Evans, E. I. (1933). The transport of spermatozoa in the dog. *American Journal of Physiology, 105,* 287–293. For work on rats (postulating a different mechanism), see Genell, S. (1939). Experimental investigations of the muscular functions of the vagina and the uterus in the rat. *Acta Obstetricia et Gynecologica Scandinavica, 19,* 113–175. For more on rats, see Hartman, C. G., & Ball, J. (1931). On the almost instantaneous transport of spermatozoa through the cervix and uterus of the rat. *Proceedings of the Society of Experimental Biological Medicine, 28,* 312–314, and Toner, J. P., & Adler, N. T. (1986). Influence of mating and vaginocervical stimulation on rat uterine activity. *Journal of Reproduction and Fertility, 78,* 239–249. For similar work on stump-tailed macaques, see Goldfoot, D. A., Westerborg-van Loon, H., Groeneveld, W., & Slob, A. K. (1980). Behavioral and physiological evidence of sexual climax in the female stump-tailed macaque (*Macaque arctoides*). *Science, N.Y., 208,* 1477–1479. For rapid sperm transport in rabbits, see Krehbiel, R. H., & Carstens, H. P. (1939). Roentgen studies of the mechanism involved in sperm transportation in the female rabbit. *American Journal of Physiology, 125,* 571–577. For the same mechanism in horses, see Millar, R. (1952). Forces observed during coitus in thoroughbreds. *Australian Ethology Journal, 28,* 127–128. For work exploring similar mechanisms in cows, see Van Demark, N. L., & Moeller, A. N. (1951). Speed of spermatozoan transport in the reproductive tract of the estrous cow. *American Journal of Physiology, 165,* 674–679. Finally, for more up-to-date work on rapid sperm transport in pigs, see Knox, R. V. (2001). *Artificial insemination of swine: Improving reproductive efficiency of the breeding herd.* www.gov.mb.ca/agriculture/livestock/pork/pdf/bab13s04.pdf last accessed 24/09/2010.

36 The classic papers by the Fox team include:

Fox, C. A. (1970). Reduction in the rise of systolic blood pressure during human coitus by the β-adrenergic blocking agent, propranolol. *Journal of Reproduction and Fertility, 22*(3), 587–590.

Fox, C. A., & Fox, B. (1969). Blood pressure and respiratory patterns during human coitus. *Journal of Reproduction and Fertility, 19*(3), 405–415.

Fox, C. A. (1976). Some aspects and implications of coital physiology. *Journal of Sex and Marital Therapy, 2,* 205–213. Fox, C. A., & Fox, B. (1971). A comparative study of coital physiology, with special reference to the sexual climax. *Journal of Reproduction and Fertility, 24,* 319–336.

Fox, C. A., Meldrum, S. J., & Watson, B. W. (1973). Continuous measurement by radio-telemetry of vaginal pH during human coitus. *Journal of Reproduction and Fertility, 33*(1), 69–75.

Fox, C. A., Wolff, H. S., & Baker, J. A. (1970). Measurement of intra-vaginal and intra-uterine pressures during human coitus by radio-telemetry. *Journal of Reproduction and Fertility, 22,* 243–251.

I include the (near) complete set because they give a flavor of the scope and detail of sexual response that the team were exploring. No one has challenged their findings on Ph values, blood pressure, or respiratory patterns on the grounds that the sample size was small. Perhaps those elements of physiology do not excite people's scientific curiosity in quite the same way as does a potential reproductive role for orgasm. The Fox team also noted that prolactin secretion might have important effects on the behavior of sperm in the reproductive tract. Their predictions turn out to have some support (e.g., Krüger, T. H. C.,

Haake, P., Hartmann, U., Schedlowski, M., & Exton, M. S. (2002). Orgasm induced prolactin secretion: Feedback control of sexual drive? *Neuroscience & Biobehavioral Reviews*, *26*, 31–44), and I will come back to the effects of these in chapter eight.

37 Ellis, H. D. (1933). *The psychology of sex: A manual for students*. Heinemann. The quote about ancient sources such as Aristotle is from page 21.

38 Fox et al. (1970, p. 247).

39 "Insuck" and "upsuck": For sources referring to "upsuck," see Baker, R., & Bellis, M. A. (1993b). Human sperm competition: Ejaculate manipulation by females and a function for the female orgasm. *Animal Behaviour*, *46*, 887–909., and Lloyd, E. (2005). *The case of the female orgasm: Bias in the science of evolution*. Harvard University Press. The original papers, e.g., Fox, C. A., Wolff, H. S., & Baker, J. A. (1970). Measurement of intra-vaginal and intra-uterine pressures during human coitus by radio-telemetry. *Journal of Reproduction and Fertility, 22*, 243–251, use the term "insuck." At some point the term has been changed with no one going back to source to see the error. But I prefer to use the expression from the primary sources.

40 This research forms a great deal of the rest of the book. To anticipate some highlights of it, this includes the role of oxytocin in female orgasm (e.g., Blaicher, W., Gruber, D., Bieglmayer, C., Blaicher, A. M., Knogler, W., & Huber J. C. (1999). The role of oxytocin in relation to female sexual arousal. *Gynaecologic and Obstetric Investigation, 47*, 125–126, Carmichael, M. S., Humbert, R., Dixen, J., Palmisano, G., Greenleaf, W., & Davidson, J. M. (1987). Plasma oxytocin increases in the human sexual response. *Journal of Clinical Endocrinology and Metabolism, 64*, 27–31, measurements of spermatogenic material in the reproductive tract by oxytocin action (Wildt, L., Kissler, S., Licht, P., & Becker, W. (1998). Sperm transport in the human female genital tract and its modulation by oxytocin as assessed by hysterosalpingoscintigraphy, hysterotonography, electrohysterography and Doppler sonography. *Human Reproduction Update, 4*, 655–666. Zervomanolakis, I., Ott, H. W., Hadziomerovic, D., Mattle, V., Seeber, B. E., Virgolini, I., . . . Wildt, L. (2007). Physiology of upward transport in the human female genital tract. *Annals of the New York Academy of Science, 1101*, 1–20); and measurements of sperm retention King, R., Dempsey, M., & Valentine, K. A. (2016). Measuring sperm backflow following female orgasm: A new method. *Socioaffective Neuroscience & Psychology, 6*(1), 31927.)

41 Elizabeth, L. (2005). *The case of the female orgasm. Bias in the science of evolution.*

6 "The Lady Vanishes"

6.1 THE CASE OF THE VANISHING CLITORIS

It's a mystery worthy of Miss Marple herself. The clitoris vanishes every few generations, and then needs to be rediscovered. Sometimes, this rescue from obscurity is to considerable fanfare. What on Earth is going on? In this chapter I explore the physiology of the human clitoris and aim to show that its complex and beautifully designed form are perfect for its central role in female sexual choice. However, how can it be that this detailed, complex structure is not more widely known? How is it that what many (perhaps most) people think of as the clitoris is just the tip of the iceberg?

First, do not panic. A vital and complex anatomical structure is not really vanishing; it just largely disappears from anatomical textbooks every so often. Second, do not just take my word for this, test it for yourself. Go into any decent bookshop and check out the medical anatomy section. Look up "clitoris" in the pictorial index. The last time I did this, roughly half of the medical textbooks—the exact same ones that are used to train doctors and surgeons—were in disagreement with the other half.[1] One half labeled, as if complete, a structure they called the "clitoris" as a wholly external organ, but this is really the clitoral *glans*—a small but highly sensitive part of the clitoris, but definitely not the whole thing.[2] The remaining textbooks detail a far more complex structure, about four inches in length, mostly internal, and with a variety of nerve pathways, complex mapping onto the sensory cortex of the brain, and a host of integrated organelles. If you want a life-size model of this organ, then you can use three-dimensional (3D) printing technology to create one out to have beside you as you read this chapter.[3] But, back to the textbooks. Are half the female population notably different from the other half? No, it is the half of the textbooks that fail to note these details that are at fault.

You might be forgiven for thinking that I am exaggerating, or that we live in an age of such sexual license that such matters can be freely discussed—especially in medical terms? Not so. As I was working on an early draft of this chapter, I learned that Twitter had banned the promotion of the scholarly work *The Vagina Bible* by the gynecologist Dr. Jen Gunter. The ban—based on the fact that the word "vagina" is automatically flagged as "dirty" on the platform—was eventually lifted. However, the author is no stranger to the misinformation that women are routinely given about their own anatomies, sometimes from ignorance and sometimes for sexual control.[4]

Why does this matter to the argument I am making, namely that female orgasm needs to be seen in the light of *female* sexual selection, not male? It might surprise people to learn that some of the most prominent theories about the nature and function of the female orgasm rest largely on an anatomical misunderstanding. I started to tease apart the various strands that led to this state of affairs in the previous chapter. One issue was (and still is) that female orgasms, like male ones, brought about by masturbation of the surface of the genitalia are somewhat easier to study under laboratory conditions. However, to mistake ease of study with comprehensiveness of study is akin to the old joke about the drunkard looking for his keys under a lamppost, and, when asked if that is where he dropped them, replying "No, but the light is better here." We need to be wary of making that sort of mistake in science, but that is only part of the story. I can set the stage for the next step of the vanishing clitoris mystery with a relevant comparison.

Imagine, for a second, that we knew next to nothing about the male erectile system for the human penis. In this imaginary world, penises rarely get erect at all, often needing artificial drugs to aid this process, resulting in many people seeing the whole system as obviously a by-product of the fact that basic fetal morphology is female.[5] In this imaginary world (as far as we know), men ejaculate without too much difficulty, but they do this when their prostate gland is stimulated—via the anus. As it happens, stimulating the prostate (in the real world now, not our imaginary one) will indeed

DOI: 10.1201/9781003372356-6

add something to male sexual pleasure. But imagine, if you will, that this "anal penetration only" method was the dominant way to bring about male orgasm in our world. Further, imagine that it just so happened that this is how all males achieved orgasm—at least, that was the conventional wisdom. If we wanted to study male ejaculate (say) we would stimulate the prostate under laboratory conditions, and then take relevant measurements and samples. This imaginary scenario is physically entirely possible in real life, incidentally. Not only that, but scholars note that these "emission-type reflexive orgasms," as they are known, can be obtained through prostatic massage, with much less recovery time to be ready to orgasm again (technically called "refraction") between them than is normal (again, in the real world) with male orgasm. In other words, anal stimulation would be the most efficient method for producing male orgasms in the laboratory, or elsewhere.[6]

Then, someone comes along in our imaginary world and says, "Hey, isn't this interesting? If you stimulate the penis *itself*, it gets engorged with blood . . . furthermore, if you carry on stimulating it, it causes an ejaculation—and an interesting and exciting one at that—all on its own. You don't have to stimulate the prostate at all (although it's still pleasant if you do). Isn't that something?"[7]

I submit that we are in something a bit like that situation with the (mislabeled) clitoris and the female orgasm. Like the penis, the clitoris is quite a large organ, with both erectile and sensitive tissue. A major difference is that the clitoris is mostly inside the body. There are other differences of course. Men and women do not become aroused in exactly the same ways, and the vagus nerves provide an important sex-specific arousal path in women only.[8] But, these two key factors—internal (and largely invisible) stimulation and female-specific arousal paths—have led some sex researchers, and also some well-meaning sex therapists, to the state where it seems logical and natural for them—with the best will in the world—to describe female sexual response as having little, or nothing, to do with intercourse. However, the only indisputable fact we have here is that there is an orgasm gap: Male orgasm (once you remove erectile impotence from the numbers, and it is not clear why that is legitimate) is easier to produce during intercourse than female orgasm.

This fact can be interpreted in various ways. Freud averred that it meant that many women were psychologically immature, having not transferred their arousal site from the clitoris to vagina. He was wrong about this. But Kinsey, who criticized Freud severely for upsetting generations of women by getting them to try to achieve the "biological impossibility" of intercourse-based orgasms, was wrong too. Both Kinsey and Freud shared the same concept of the clitoris as being a small, purely external, organ.[9] Kinsey wrote that "the vaginal walls are quite insensitive in the great majority of females," but this is definitely not what the majority of *aroused* females report. As an example of this (and many more are going to follow), a generation later than Kinsey, noted feminist scholar Germaine Greer retorted at Kinsey: "It is nonsense to say that a woman feels nothing when a man is moving his penis inside her vagina. The orgasm is qualitatively different when the vagina can undulate around the penis instead of a vacancy."[10]

As we shall see in this chapter, the clitoris—the site of female orgasm—is bigger than either Freud or Kinsey realized, and it can get aroused just as penises can, only mostly internally—and therefore invisibly. However, Kinsey's response to Freud (and after him, Masters and Johnson) inspired a line of thinking that resulted in the idea that women were simply not designed by natural selection to orgasm through intercourse at all—the by-product theory. Another perspective than either of these is now possible. Said realization lies at the heart of this book, and it is this: A complex, refined, exquisitely designed measuring instrument—the clitoris—should not be judged according to the yardstick of a comparatively crude and more obvious one—the penis. The clitoris has a nature and function all of its own.

6.2   A BIT OF HISTORY

For reasons still not fully understood by scholars, people in some time periods and cultures celebrate a much fuller expression of human sexuality—especially female sexuality—than do others. Matching this, the anatomical disappearance of the clitoris from learned textbooks appears to be somewhat cyclical in history, rather than linear. People from certain time periods seem to be happier with the true nature and function of the clitoris than do others. To take just a couple of historical examples, the ancient Romans knew perfectly well what rubbing the glans of the clitoris (the *landica*) would do, and they had many obscene graffiti describing exactly this process, to be found across the empire.[11] At roughly the same time as the Romans were sharing their thoughts on the clitoris, the Indians of the Chadela dynasty were carving some very specific and anatomically highly detailed suggestions of how to stimulate this organ into the statues of their Khajuraho temples in the region of Madhya Pradesh.[12] Why was this knowledge lost or suppressed in later years? No one is very sure, but it seems plausible that suppression of sexual knowledge may be of a piece with other forms of sexual repression and control such as genital mutilation. In the case of the clitoris, the ignorance of its large size and complex nature has, at times, extended to textbooks, peer-reviewed papers, and even learned works of anatomy and physiology.

The case of the vanishing clitoris goes back some time, in western medicine at least. Renaissance scholars drew upon the supposed wisdom of the ancients, but the ancients sometimes shared the modern uncertainty about the clitoris. For example, in the sixteenth century, Falloppio (after whom the tubes are named) described clitoral structure,[13] challenging the received wisdom of the eminent scholars Galen and then Vesalius, who had denied it existed as a distinct organ. Vesalius thought that female genitalia was an inverted version of the males'.[14] Eight hundred years after Galen (who wrote in the second century of the common era), the Persian anatomist Avicenna was referring to the clitoral glans as "el bathr," *the (woman's) penis.*[15] In modern medicine, the complex—and largely internal—structure of the clitoris was documented by Kobelt as early as 1851.[16]

Sigmund Freud, that alternately reviled and revered pioneer of sex research, added to the confusion by insisting that clitoral orgasms were immature versions of vaginal ones.[17] Of course, by "clitoral" he meant what we would now see as pertaining to the clitoral glans, not the whole internal structure. In addition, what he almost certainly meant by "immature" referred to orgasm brought about by external masturbation alone. Nowadays, we might find it odd to regard masturbation as immature, but it is perhaps clearer why Freud, presented as he was with sexually frustrated clients, divided the world of women's sexual pleasure in the way he did. However, his nomenclature (*vaginal* or *clitoral* orgasm) persists to this day.[18]

In their eagerness to overturn Freud, Masters and Johnson, those two giants of sex research I discussed in the previous chapter, did not look as closely as they might have at the anatomical details either, concluding that masturbation of the surface of the clitoral glans would generate all the sexually pertinent effects they wanted to measure.[19] They concluded that female orgasms performed no function. I argued in the last chapter (and will build on this in the next) that they were far too hasty in concluding this, and that another tradition of sex research has unearthed a rich and complex functionally integrated set of female orgasm properties consistent with cryptic—or semi-cryptic—female choice.

Moving forward a hundred years from Kobelt, clitoral complexity was re-described by Dickinson in 1949, showing a large organ that wrapped around a (potentially) inserted penis, but many textbooks were slow to catch up with research.[20] Advancing in time once more, in the early part of this century, Helen O'Connell's team produced beautifully rendered magnetic resonance imaging (MRI) scans showing the size and complexity of the clitoris in three dimensions, now examinable in states of actual arousal.[21] The complete clitoris has at least eighteen distinct anatomical sections. The part of the clitoris that is typically visible is homologous to the glans of the penis. However, the rest of the clitoris is internal and, after an inch or two, it divides and the remainder of its erectile tissues, typically known as the bulbs of the vestibule, but which O'Connell argues should be renamed the *bulbs* of the clitoris, run down four or five inches on either side of the vaginal opening, as illustrated in Figure 6.1.

I have regularly given talks to interested folk—some of whom are biologists—who tell me that this is the first they have heard of the clitoris' size and complexity. Now that I have a lovely 3D life-size (it is about the size of a large rose) model of the clitoris to hand around, this can just add to their

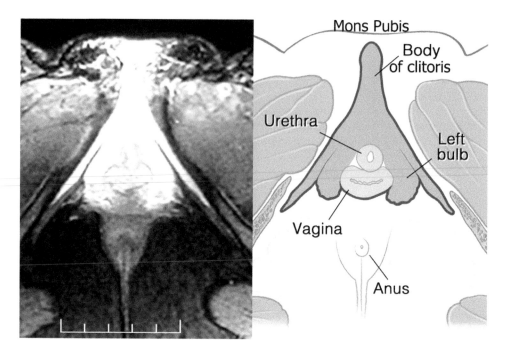

FIGURE 6.1   MRI of clitoris and components in axial plane. The bulbs, crura, and corpora are easy to see.

Source: Reprinted from O'Connell (2005), with the permission of the original publication

amazement. More often than should be necessary, I have read (and sometimes commented on in peer review) papers that still treat the clitoris as if it were a vestigial organ, like a male nipple.

The eminent physiologist Roy Levin once told me that the clitoris was an "iceberg organ, seven eighths of it is always hidden,"[22] but if you look at the above diagrams, this is actually something of an understatement. In addition, the largely internal nature of the organ seems insufficient to explain repeated human ignorance of its size and role. Why are these things not more widely known? It is worth considering that persistent human sexual ignorance might have been the result of some consistent cognitive or emotional glitch, rather than simple lack of access to knowledge. I believe that this may be the case here.

6.3   FLOWERS AT MIDNIGHT

For reasons that are important enough to demand a separate chapter, some men want (or think that they want) some kind of magic button that will make their sexual partners satisfied. By "satisfied" such men typically mean "orgasm during sex."[23] When female orgasm doesn't happen, there is a marked tendency to regard the woman as somehow broken or inferior. Sigmund Freud was responsible for some of this mischief with his notion that clitoral (glans) orgasms were somehow immature. Men almost always orgasm during sex, so it is argued, and women do not, therefore the search for some solution to this supposed malfunction begins.[24] If someone seems to have found a magic button to make women's sexuality work in the same way that men's does (or appears to), then it is not hard to see why people get excited. Sometimes this view of females as imperfect copies of males, in sexual terms, can extend to a general theory about women's sexuality. This is because what works sexually for men does not work for women—at least not every time. It's as if I found some flowers at midnight and concluded that they must be malfunctioning because the petals were closed. This "flowers at midnight" view finds its fullest expression in the by-product theory of female orgasm.

6.4 FEMALE ORGASMS AS BY-PRODUCTS OF MALE ORGASM?

As I argued in chapter four, the otherwise miraculous appearance of complex design in nature is explained by adaptation—the gradual fitting to the environment of traits that increased reproductive success. However, rather like teenagers who think that their clothes magically pick themselves up from the floor, and arrive fully washed and folded in their clothes drawers, some of us seem to have lost our sense of the scope and scale of the problems that this insight solves. Adaptation really is the only non-magical explanation for functional complexity in the living world. Some complex structures are the result of direct Darwinian selection—natural or sexual. Some others exist because they piggyback on other adaptations; these structures have no functions in their own right, and they make no contribution to fitness. The by-product theory of female orgasm argues that female orgasm exists only as a by-product of male orgasm, and supporters of this theory draw an analogy here with male nipples.[25] But, if this is the case, then something near miraculous would have happened: Namely, a set of anatomical structures and physiological systems would exist as the by-product of ones *less* complex than they are themselves.

The ontogenic warrant (e.g., the developmental story) for the by-product theory is now quite famous: It is a reasonably well-known factoid that we all start as anatomically (if not chromosomally) female embryos, with those destined to be male receiving a dose of testosterone at about four weeks after conception. This prepares the brain (usually) for being housed in a male body, and sends signals to relevant tissue that will grow it into a penis. At the same time, structures that will eventually grow into female breast tissue remain vestigial in males. However, it would be a serious mistake to assume that the clitoris is a vestigial structure, like a male nipple. As shown in the O'Connell pictures in Figure 6.1, it grows into a complex organ with multiple sites of sensitivity. Despite this, the by-product argument then proceeds along the line that because males (allegedly) needed sensitive penises to reward their sexual behavior (females apparently not needing this reward, for reasons unstated), women then get a free ride (as it were) with an enjoyable but functionless clitoris.

There is nothing inherently wrong with emphasizing the phylogenetic or ontogenic constraints—the raw materials—that evolution has to work with. In this vein, Stephen Jay Gould famously drew attention to the panda's thumb, a less-than-optimal kludge evolving from the sesamoid bone to enable the stripping of bamboo.[26] He pointed out that this so-called thumb was much less complex than it might be (and much less so than the panda's other digits) but that evolution could not roll back the phylogenetic tape and start over, in order to make something more optimal. This is true. But is it true that the female orgasm stands in relation to male orgasm, as male nipples do to female ones? This is the picture that Gould[27] and then later Elizabeth Lloyd[28] championed, following from an earlier proposal by Donald Symons.[29] Male nipples are small, functionless by-products of selection on female glandes,[30] which are large and richly innervated. Rarely do I find myself talking about the topic without someone bringing up this imagination-grabbing analogy, so I hope I will be excused for spending some time on it.

Not to put too fine a point on it, this orgasm/nipple analogy does not stand up to detailed scrutiny, informed by up-to-date anatomical knowledge. To begin with, what exactly *is* the analogy supposed to be here? Is it that *clitorises* are meant to be like male nipples, or that female *orgasms* are? I have already dealt with some of what is wrong with the first interpretation of this ambiguity, and I will come back it later. But let me reiterate: The clitoris is not a second-rate penis. Clitorises are large, complex organs, with their own attendant somatosensory array, not tiny vestigial structures like male nipples. Could the second option—that female orgasms are somehow vestigial—make more sense? I think not. The analogy loses a lot of its meaning once it is appreciated that a nipple is a structure, whereas an orgasm is an event. But this proposed reading of the by-product analogy is even less plausible once one recognizes that (often) a female orgasm is quite a dramatic event. The famed biologist Robert Trivers is reputed to have once quipped of female orgasms that "One has to wonder how often Steve [Gould] has witnessed this blessed event to regard it as a side-effect."[31] That comment may be a tad unkind. However, a relevant point was raised. As I showed in the previous chapter, and will show a great deal more of in the next chapter, the female orgasm can be an event with a large number of complex mechanisms, including, but not limited to, waves of pleasurable sensations (that do not occur randomly), pressure changes, and differential sperm retention.

Which of these two possible interpretations—structure or event—do the by-product champions favor? It is not always 100% clear, and sometimes writers equivocate, but I think that Gould meant the first interpretation, because he endorsed the view that there was no internal female sexual sensitivity. Gould says this in relation to female orgasm:

> In his study of genital anatomy, Kinsey reports that the female clitoris is as richly supplied with sensory nerves as the male penis—and therefore as capable of excitation. The walls of the vagina, on the other hand, "are devoid of end organs of touch and are quite insensitive when they are gently stroked or lightly pressed. For most individuals the insensitivity extends to every part of the vagina."[32]

There is no real room for maneuver here. By "clitoris" in this description, Gould can only mean the external part or glans, not the richly sensitive internal parts. Lloyd's view seems to mirror Gould's. For example, in follow-up papers to her 2005 book, she measures a structure that she labels as the clitoris, but is in fact merely its glans.[33] And, there is no getting away from this, this picture is simply not in line with current anatomical knowledge. What about Symons, whose proposal in *The Evolution of Human Sexuality* was the foundation for both Gould's and Lloyd's arguments? Symons also endorses the Masters and Johnson view that all intercourse could do would be to cause "mechanical traction . . . on both sides of the clitoral hood" and thus seems also to be endorsing the idea that clitorises are small external penises.[34] Perhaps Lloyd was just unfortunate in not having the most up-to-date anatomical data—such as the work of O'Connell—to hand when her book was written in 2005?

This lacuna as to the true size of the clitoris did persist for a few years after O'Connell's dramatic MRI publications, however. For example, in 2008, Kim Wallen and Elizabeth Lloyd, in a follow-up to Lloyd's (2005) book, analyzed various data sets to attempt to show that clitorises are significantly more variable in length than are penises.[35]

But this 2008 paper did not measure the full extent of the clitoris. All that had been measured was the length of a number of clitoral glandes. Wallen and Lloyd had intended to argue that, because some clitorises are farther away from point of contact with a penis during intercourse than others, then this was an argument in favor of the by-product hypothesis. As I have argued, this is anatomically incorrect. Furthermore, it turns out that there are measurement issues, even if granted assumptions about what constitutes a clitoris. When Lynch re-analyzed their data using volumes rather than lengths, he found that variability between the two organs was not significantly different anyway.[36] Not that this should have mattered for a by-product argument in any case, because variability is not the final answer when it comes to judging matters of selection.[37] It is true that the great genetic scholar Fisher argued that "the rate of increase in fitness of any organism at any time is equal to its genetic variance in fitness at that time."[38] But, it does not follow from this that there is little or no variation in fitness-related traits. Fisher was not talking about components of fitness. Traits that fall under strong sexual selection may show dramatic variation, as a glance at the fabled peacock's train (a mere feather of which was reputed to make Darwin feel sick) reminds everyone. This variation underscores sexual selection, often with females choosing the best, brightest, or boldest.

I have included some figures above that would be good to look at now if you have not already done so. But, we live in a digital age, and we can do better than mere pictures. If you happen to have access to a 3D printer, you can actually print out, for yourself, a model of the clitoris so you can have it besides you while we talk about it:

www.thingiverse.com/thing:1876288

(The resultant shape even looks a bit like a flower—a tulip, according to journalist Minna Salami, writing in the *Guardian*, and I can see her point.)

6.5   "NO MAN SHOULD MARRY UNTIL HE HAS STUDIED ANATOMY AND DISSECTED AT LEAST ONE WOMAN." HONORE DE BALZAC

As should be obvious from either the model or the pictures, the clitoris is not a tiny vestigial penis. It is larger than an un-erect penis, and is also a complex organ with at least eighteen distinct interacting functional parts, including muscular, erectile, and sensitive tissues. It wraps around the opening where the penis enters, enfolding it in highly innervated and vascular tissue. Blood carries oxygen to wherever the action is in the human body. Given a sufficiently sensitive magnetic imager, the iron in this blood can be detected, and the tool we have for this is called a functional magnetic resonance imager (fMRI). Using this tool to scan internal states is a skilled job, but using it to scan the behavior of two people at a time really deserves credit—as do the intrepid volunteers. In the late 1990s, the Schultz team managed to achieve this—scanning a couple having intercourse in one single scanner[39]—and the results were striking. It turns out that penises do not remain straight inside the vagina. Instead, they bend to conform to the contours of the clitoris and to interact with specific internal nerve centers. These nerve centers are not scattered about haphazardly. The nerves from the clitoris map to their own specific regions in the female brain's somatosensory cortex. Nothing like this can be said of male nipples. Clitorises are therefore not vestigial. They are active organs. How are they active? They are the seat of female orgasm. And, as such, the clitoris has its own dedicated brain regions that it is connected to, like a map of the terrain. I now need to explain how this mapping works.

6.5.1   THE HERMUNCULUS

As a child I visited Bourton-on-the-Water, a beautiful village in the Cotswolds. There, in the garden of the Old New Inn, is a model village, big enough to walk around, while feeling like a giant. Close inspection reveals this diorama to be a replica of the village of Bourton-on-the-Water itself. Obsessive that I am, I immediately rushed to look at the model version of the Old New Inn pub to see if there was a model of the model in that, and was delighted to find that there was. Was there also a model of the model of the model? Alas, no—that's where it ended. The little fleas only had lesser fleas upon their backs. It did not go on ad infinitum.

A similar sort of nested self-referential modeling applies to human brains. Brains exist to do at least this—generate predictions about the future of the organism they rest in. They often do this by building internal models of the external world. Like my Bourton-on-the-Water example illustrates, any sufficiently sophisticated model will have to contain a model of itself as well. Our brains are at least this sophisticated.

In humans this model of the modeler is technically called the somatosensory cortex. Our brain contains a model of our bodies. It is not the same shape as our bodies, and it is not in proportion to body size but rather, the importance of sensing from and moving that body part. You have likely seen one of these weirdly distorted models that relates brain region to body parts, with enlarged face, hands, genitals, and pipe-cleaner limbs. This is because the face, hands, and genitals take up a lot of brain real estate, and limbs a lot less. Pictures of this model are sometimes called a Penfield homunculus, after the pioneering neurosurgeon Wilder Penfield. He was the first to discover that you could open up obliging people's brains and play their bodies like a sophisticated piano, if you knew which keys to press.[40] For instance, stimulating one area of the brain causes music to be heard, and another, memories to be relived. Every part of the body is represented there. Watching this happen for the first time (there are many videos available online) is an unforgettable experience that starts to undermine that nagging sense that we all have that humans are, in essence, a ghost living in a machine. It turns out that we are all machine, but a very beautiful and sophisticated one.

Assumptions are the things we don't know we are making.[41] For years it was simply assumed that the female somatosensory cortex was like the male one. And, if this was true, it followed that there should be a female homunculus that corresponded exactly to the male one. But it transpires this is not true. Why should it be? It runs a very different system. Gillian Einstein has rather wittily termed this female system the *hermunculus*, and has argued convincingly that it is well past time for neglect of direct study of female sexual anatomy and brain matter to be turned around.[42] Drawing on recent research, she sketched out some key differences in male and female brains as they relate to the genital region. I will quote them directly but, before I do so, you possibly need to be made aware that rCBF refers to regional cerebral blood flow and terms like M1 or S2 refer to specific regions in the sensorimotor cortex. How does the female representation of orgasm in the brain differ from that of the male? For example (there are many others), Penfield noted that the penis mapped onto the "medial wall (or paracentral lobule) of the postcentral gyrus" while others have localized it at the "dorsal surface of the postcentral gyrus, lateral to the representation of the toe" (Kell et al. 2005).[43] However, clitoral stimulation (in comparison with both resting states and voluntary tensing of muscles in the genital region) produced

> significant rCBF activation bilaterally on the dorsal surface of the postcentral gyrus in S1 and laterally in the left postcentral gyrus within S2, or parietal operculum. Brain activity during orgasm showed peak rCBF in S1 with significant clusters of activity in dorsal M1 cortex, the paracentral lobule, and the dorsal aspect of the central sulcus.
>
> (Georgiadis et al., 2006)[44]

And it does not stop there: "[T]here is decreased rCBF in the neocortex during female orgasm but not male" (Georgiadis et al., 2009).[45] Therefore, Di Noto et al. (2013) go on to say that "from this, we can infer that the clitoral localizations suggested by this study are not consistent with Penfield's localization of the homologous structure, the penis." The authors go on to discuss the alternative mapping scheme suggested by Barry Komisaruk and Beverly Whipple.[46] This important research, often based on patients with spinal cord damage, is such an important topic to be discussed (as it implies some very separate sex-specific pathways unique to women) that it needs more space than this chapter can provide. For the moment, I just want to note that Komisaruk and Whipple have provided evidence for a "female-specific sensory map of the genitalia—which includes the clitoris, cervix, and vagina—on the homunculus."[47]

Brain tissue is the most expensive real estate in the human body. At the risk of being seen as over-emphasizing this point, that is why the male nipple does not have its own set of specialized dedicated brain regions (activated by this stimulation, in male-specific ways) to regulate functions, and to report sensations that are unique to males. The clitoris, and related structures, have exactly these things in relation to women.[48] To argue that this level of functionally integrated physiological and neurological complexity arose, and is maintained, as a functionless by-product strains credibility.

6.6   FORM FOLLOWS FUNCTION

There are not many laws in biology—not as many as there are in physics anyway—but if there is one upon which you can rely it is "mother nature will never pack you a free lunch." Animals, whose ancestors had sight but that now live in caves, will develop flaps over their now nerveless eyes and turn pale after millennia in the darkness. Primates that no longer use their tails to balance will end up with tiny vestigial tails—like the ones that you and I have. In general, if something is costly, then once the value of paying for its costs goes, then so do the structures and the mechanisms required to support them. Sometimes these traits may emerge again as curious scars of evolution, like vestiges and atavisms, but they do not retain their costly complexity.

Given the aforementioned integrated functional complexity of the female sexual system with its attendant brain hardware, ask yourself if it seems likely that female orgasms are a "wonderful bonus," rather than a system with a job (or jobs) to do?[49] Viewed in this light, the existence of sexually frustrated women takes on a rather different complexion—that of a system that is performing a complex gatekeeping job. Knowledge of the clitoris' true complex structure prompts this view, especially its nature when aroused—as depicted in three-dimensional sonography in 2009 by Foldes and Buisson[50] or the O'Connell team. Far from being a second-rate penis, the clitoris is a highly sensitive, beautifully designed measuring instrument. Rather than seeing the clitoris as a second-rate penis, that women are lucky to have been stimulated by sex at all, our attention should focus on the ways the clitoris and its attendant neural structures are designed to be stimulated.

6.7 SUMMARY

The (threatened) vanishing clitoris is revealed to be a complex organ, with many interlocking parts and its own unique set of control centers. The idea that all this integrated complexity is some sort of vestige, some functionless leftover from male sexuality, looks highly implausible. What are we to make now of the fact that penetrative intercourse is relatively inefficient at bringing about female orgasm? The simple answer to this is that it is no accident that efficiency is not typically taken to be a hallmark of good sex. Let us be clear about what is meant by "good" here. Enjoyment is one aspect, and this enjoyment may well co-vary with some functions to be explored in the next chapter. But (and it is surprising how little this fact is emphasized in sex research) good sex is sex you want to have again and with the same person. The comparatively low female orgasm (across the population) rate through penetrative sex alone says nothing about whether female orgasm is *facultative* adaptation—responsive to *particular* partner behaviors and features. This implies, in case we had forgotten it, that human sex is not meant to be easy; indeed, its very complexity and requirement to acquire general physical and interpersonal skills, and be responsive to specific partner requirements and preferences, are likely crucial elements. As the information technology crowd would have it, maybe the much-touted difficulty in bringing about orgasm in women is not a bug, but a feature. Put simply, women are choosy—here as they are in every other aspect of sex.

However, even if we have cleared up the event/structure and clitoris/mini-penis confusions, and we now have a clear picture of the beautiful and complex organ that lies at the seat of orgasm, does that clear up the matter as to why female orgasms exist? Not yet. But, we are now in a position to revisit that tradition of sex research informed by a functionalist lens that I introduced in the last chapter: work that builds on that of the Foxes.

NOTES

1 I do not think it would be fair to name just some of them and leave out others. However, I will hope that they update themselves soon.
2 The clitoral glans is an important site of sensitivity. A good example of this is its crucial role in genitoplasty. Schober, J. M., Meyer-Bahlburg, H. F. L., & Ransley, P. G. (2004). Self-assessment of genital anatomy, sexual sensitivity and function in women: Implications for genitoplasty. *British Journal of Urology*, 94, 589–594.
3 Fillod, O., & Richard, M. (2016). [Downloadable model of clitoris, life size for 3D printing]. www. thingiverse.com/thing:1876288 last accessed 25/08/2023. To print out a clitoris of your very own, with the modern magic of 3D printing, see the creative commons website Matilda detailed above: Many thanks to Lisa O'Connor for drawing my attention to this oneiric feature.
4 See, for example, www.theguardian.com/lifeandstyle/2019/sep/08/jennifer-gunter-gynaecologist-womens-health-bodies-myths-and-medicine last accessed 16/09/2019. See also Gunter, J. (2019). *The vagina bible: The vulva and the vagina: Separating the myth from the medicine*. Citadel Press.
5 Of course, no one would argue that the frequent failure of men to achieve erection was suggestive that the male urogenital system did not evolve to perform a reproductive function, yet male erectile impotence is sufficiently widespread to make companies that create drugs that treat it huge profits. An analysis is here: www.fool.com/investing/2017/02/28/pfizer-stock-history-how-the-drugmaker-became-an-i. aspx last accessed 10/02/2022.

   As a side note, the fact that fetuses are *morphologically* female does not mean that sex differentiation at the chromosomal level is not present prior to birth. It most assuredly is.
6 Perry, J. F. (1988). Do men have a G-spot? *Australian Forum*, 2, 37–41.

7 Hite, S. (1981). *The Hite report on male sexuality*. Ballantine Books reports on the extra pleasure obtained from prostatic massage along with penile stimulation.

8 For information on the vagus nerve, see Komisaruk, B. R., Bianca, R., Sansone, G., Go, L. E., Cueva-Rolo, R., Beyer, C., & Whipple, B. (1996). Brain-mediated responses to vaginocervical stimulation in spinal cord–transected rats: Role of the Vagus nerves. *Brain Research, 708*, 128–134. Komisaruk, B. R., & Sansone, G. (2003). Neural pathways mediating vaginal function: The Vagus nerves and spinal cord oxytocin. *Scandinavian Journal of Psychology, 44*, 241–250.

9 Kinsey, A. C., Pomeroy, W. B., Martin, C. E., & Gebhard, P. H. (1953). *Sexual behavior in the human female*. Saunders. The quotes come from chapter 14.

10 Greer, G. (1971). *The female eunuch*. MacGibbon and Kee. The quote is from page 48.

11 Varone, A. (2002). *Erotica Pompeiana. Love inscriptions on the walls of Pompei* (Trans. R. P. Berg). L'Erma di Bretschneider.

12 Gangoly, O. C. (1957). *The art of the Chandelas* (Vol. 1). Rupa.

13 Fallopius, G. (1561). *Observationes anatomicae, ad Petrum Mannam*. Venetiis (work).

14 Vesalius, A. (1564). *Observationum anatomicarum Gabrielis Fallopii examen* (p. 143). Francesco de'Franceschi da Siena, p. 1564.

15 For discussion of Avicenna, see Browne, E. G. (2001). *Islamic medicine. Fitzpatrick lectures delivered at the Royal College of Physicians in 1919–1920* (Reprint). Goodword Books.

16 Kobelt, G. L. (1851). *De l'appareil du sens génital des deux sexes dans l'espèce humaine et dans quelques mammifères, au point de vue anatomique et physiologique, traduit de l'allemand par Kaula H, Berger-Levrault et fils, Strasbourg et Paris (1851)* (1re éd). allemande (Original work published 1844).

17 Freud, S. (1932). Female sexuality. *International Journal of Psycho-Analysis, 13*, 281.

18 For example, the majority of our participants spontaneously described their orgasms this way without prompting by us. King, R., & Belsky, J. (2012). A typological approach to testing the evolutionary functions of human female orgasm. *Archives of Sexual Behavior, 41*(5), 1145–1160.

19 Masters, W. H., & Johnson, V. E. (1965). The sexual response cycles of the human male and female: Comparative anatomy and physiology. *Sex and Behavior*, 512–534; and Masters, W. H., & Johnson, V. E. (1966). *Human sexual response*. Little, Brown.

20 Dickinson, R. L. (1949). *Human sex anatomy: A topographical hand atlas*.

21 O'Connell, H. E., Sanjeevan, K. V., & Hutson, J. M. (2005). Anatomy of the clitoris. *The Journal of Urology, 174*(4), 1189–1195.

22 Roy Levin (personal communication at International Academy of Sex Research, Prague, 2010).

23 This can manifest itself in a seeing female orgasm as a male achievement, as explored in Chadwick, S. B., & van Anders, S. M. (2017). Do women's orgasms function as a masculinity achievement for men? *The Journal of Sex Research, 54*(9), 1141–1152. Another way of seeing this is that proximate mechanisms of male reward at female orgasms have been ultimately selected for as they enhance fitness. In other words, men have evolved to care about female orgasm because female orgasm has fitness consequences.

24 Of course, men do not always orgasm during sex either. For one thing, they frequently do not achieve erections—as the scale of the industry in drugs such as Viagra™ or Cialis™ attests. Large-scale surveys reveal that, in some regions, more than one-quarter of men have erectile difficulty, and yet this is never taken as evidence that the male erectile system was not formed by natural selection to facilitate intercourse. It is worth considering this discrepancy in detail, but it will have to wait for a later chapter. Laumann, E. O., Nicolosi, A., Glasser, D. B., Paik, A., Gingell, C., Moreira, E., & Wang, T. (2005). Sexual problems among women and men aged 40–80 y: Prevalence and correlates identified in the Global Study of Sexual Attitudes and Behaviors. *International Journal of Impotence Research, 17*(1), 39.

25 Technically this is known as *sexually antagonistic pleiotropy*, and it happens throughout nature. For instance, female leopard geckos have (functionless) paired hemipenes as a result of (strong) selection on male anatomy. But, as is common with such vestigial structures, they are small (one-twentieth the size of the male version) and with little to no associated support systems. None of this is true of the human clitoris. See Holmes, M. M., Putz, O., Crews, D., & Wade, J. (2005). Normally occurring intersexuality and testosterone induced plasticity in the copulatory system of adult leopard geckos. *Hormones and Behavior, 47*, 439–445, for details.

26 Gould, S. J. (2010). *The panda's thumb: More reflections in natural history*. W. W. Norton & Company.

27 Gould, S. J. (1991). Male nipples and clitoral ripples. *Bully for Brontosaurus*, 124–138. The quote given is from pp. 128–129. It is worth noting that, even at this stage, vaginal sensitivity was being reported in the sex research literature. See, for example, Alzate, H. (1985). Vaginal eroticism and female orgasm: A current appraisal. *Journal of Sex & Marital Therapy, 11*(4), 271–284.

28 Lloyd, E. A. (2005). *The case of the female orgasm: Bias in the science of evolution*. Harvard University Press.

29 Symons, D. (1979). *The evolution of human sexuality*.

30 *Glandes* is the plural of "glans."

31 The story is repeated here: www.theguardian.com/books/2005/aug/27/featuresreviews.guardianreview9 last accessed 17/10/2018.

32 The quote is from (Gould, 1991, p. 128).

33 E.g., in this paper, Wallen, K., & Lloyd, E. A. (2008). Clitoral variability compared with penile variability supports nonadaptation of female orgasm. *Evolution & Development, 10*(1), 1–2.

34 Symons, D. (1979). *The evolution of human sexuality*, 87. Here he is quoting Masters and Johnson (1966, p. 59).

35 Wallen, K., & Lloyd, E. A. (2008). Clitoral variability compared with penile variability supports nonadaptation of female orgasm. *Evolution & Development, 10*(1), 1–2.

36 Lynch, V. J. (2008). Clitoral and penile size variability are not significantly different: Lack of evidence for the by-product theory of the female orgasm. *Evolution & Development, 10*, 396–397.

37 Hosken, D. J. (2008). Clitoral variation says nothing about female orgasm. *Evolution & Development, 10*, 393–395. Their reply is here Wallen, K., & Lloyd, E. A. (2008b). Inappropriate comparisons and the weakness of cryptic choice: A reply to Vincent J. Lynch and D. J. Hosken. *Evolution & Development, 10*(4), 398–399. However, it does not address the central issue—that of misidentification of the clitoris.

38 "The rate of increase in fitness of any organism at any time is equal to its genetic variance in fitness at that time" is the fundamental theorem of natural selection. That quote comes from page 35 of Fisher, R. A. (1930). *The genetical theory of natural selection*. Clarendon Press. He later restated it with even greater clarity as "The rate of increase in the average fitness of a population is equal to the genetic variance of fitness of that population" in Fisher, R. A. (1941). Average excess and average effect of a gene substitution. *Annals of Eugenics, 11*(1), 53–63.

39 Schultz, W. W., van Andel, P., Sabelis, I., & Mooyaart, E. (1999). Magnetic resonance imaging of male and female genitals during coitus and female sexual arousal. *British Medical Journal, 319*, 1596–1600.

40 Schott, G. D. (1993). Penfield's homunculus: A note on cerebral cartography. *Journal of Neurology, Neurosurgery, and Psychiatry, 56*(4), 329.

41 I think it was Douglas Adams who first made this observation in Adams, D., & Carwardine, M. (2013). *Last chance to see*. Penguin Random House.

42 Di Noto, P. M., Newman, L., Wall, S., & Einstein, G. (2013). The hermunculus: What is known about the representation of the female body in the brain? *Cerebral Cortex, 23*(5), 1005–1013.

43 The quotes come from Di Noto et al. (2013, p. 1008). The cited piece is Kell, C. A., von Kriegstein, K., Rosler, A., Kleinschmidt, A., & Laufs, H. (2005). The sensory cortical representation of the human penis: Revisiting somatotopy in the male homunculus. *Journal of Neuroscience, 25*, 5984–5987.

44 The quotes come from Di Noto et al. (2013, p. 1008). The cited piece is Georgiadis, J. R., Kortekaas, R., Kuipers, R., Nieuwenburg, A., Pruim, J., Simone Reinders, A. A. T., & Holstege, G. (2006). Regional cerebral blood flow changes associated with clitorally induced orgasm in healthy women. *European Journal of Neuroscience, 24*, 3305–3316.

45 The quote comes from Di Noto et al. (2013, p. 1010). The cited piece is Georgiadis, J. R., Reinders, A. A. T., Paans, A. M. J., Renken, R., & Kortekaas, R. (2009). Men versus women on sexual brain function: Prominent differences during tactile genital stimulation, but not during orgasm. *Human Brain Mapping, 30*, 3089–3101.

46 The paper cited by Di Noto et al. (2013) at this point is Komisaruk, B., Wise, N., Frangos, E., Liu, W. C., Allen, K., & Brody, S. (2011). Women's clitoris, vagina, and cervix mapped on the sensory cortex: fMRI evidence. *Journal of Sexual Medicine, 8*, 2822–2830. This work is part of a much larger body of work which will be addressed in detail later in the book.

47 Di Noto et al. (2013, p. 1008).

48 The title alone of this paper explains a great deal of what is important here: Wise, N. J., Frangos, E., & Komisaruk, B. R. (2017). Brain activity unique to orgasm in women: An fMRI analysis. *The Journal of Sexual Medicine, 14*(11), 1380–1391. I will go into much more detail about this in chapter nine.

49 Lloyd, E. A. (2005). *The case of the female orgasm: Bias in the science of evolution*. Harvard University Press.

50 Foldes, P., & Buisson, O. (2009). REVIEWS: The clitoral complex: A dynamic sonographic study. *The Journal of Sexual Medicine, 6*(5), 1223–1231.

7 Standing on the Shoulders of Giants

If we are uncritical we shall always find what we want: we shall look for, and find, confirmations, and we shall look away from, and not see, whatever might be dangerous to our pet theories. In this way it is only too easy to obtain what appears to be overwhelming evidence in favor of a theory which, if approached critically, would have been refuted.

(Karl Popper)[1]

Popper's warnings are much to be heeded. Furthermore, as he would be quick to remind us, nothing in science is ever proven true beyond all doubt. Our best scientific models are those that have resisted the most determined attempts at refutation, and so we must always be doubly grateful to those who attempt such refutations, especially the refutations of pet theories. Each resistance to such refutation justifies a little more confidence in the theory in question. Especially good theories generate rich lines of research, which themselves propagate even more hypotheses for us to test. It is time to apply this principle to the idea that female orgasm is linked to a particular fitness-related effect—that of differential sperm retention.

In chapter five I introduced the research of the Fox team's ground-breaking naturalistic studies into pressure changes brought about by female orgasm. I noted that their work had come under intense criticism by Elizabeth Lloyd on the grounds that (1) masturbatory orgasms (as used in Masters and Johnsons research) are claimed to capture all the relevant properties of female sexual response; (2) claiming that no movement of sperm (or sperm-like material) was shown by the Fox studies, (3) claiming that any effects shown in the original studies are trivial; and (4) reporting on the small sample size. Later objections have focused on the lack of replication of the Fox team's studies, which amounts to a similar objection to (4). It is time to deal with these objections in some more detail. The resistance of these studies (and those that have developed from them) to attempted refutation entitles us to having a bit more of that confidence that Popper was talking about.

I have said a few things about (1) already, but it is such a large topic that it will have to wait for a chapter of its own (chapter eight). For the moment, I shall just note that female orgasms vary in location, intensity, and associated sensations, and that these are robust and replicated findings going back decades, using multiple methodologies, and across cultures. Female orgasms, unlike male ones, are not all the same. Putting that to one side for the moment, and with the details of the integrated functional complexity of the clitoris fresh in mind from the last chapter, it is now time to explore the other three objections to female orgasm having a sperm-selection function in a bit more detail. I have already mentioned the hormone oxytocin. Before I go further, I need to say a few more things about this important chemical.

7.1 THE ROLE OF OXYTOCIN

Oxytocin is a key mammalian chemical involved with breast-feeding,[2] pair-bonding,[3] feelings of calm and security,[4] as well as uterine contractions of all sorts,[5] but especially those associated with female arousal, and orgasm.[6] This means that the action of this hormone not only produces potentially important internal effects that correlate with female orgasm, but that many of these effects are consciously self-detectable—and therefore reportable—by humans.

Recall that Lloyd's second objection to the Fox studies into sperm insuck was that no sperm movement was shown by them, but it is important to note that none was being specifically searched

DOI: 10.1201/9781003372356-7

for *by* them. Studies carried out by the Wildt team, some three decades later, consciously built on the Fox team's findings and further explored them.[7] Using a sophisticated series of methods such as hysterosalpingoscintigraphy, the recording of intrauterine pressure, electrohysterography, and Doppler sonography of the fallopian tubes, Wildt's team discovered that the application of oxytocin to the vaginal fornix had dramatic effects.

This finding matters hugely, because it has been known for some time that increases in the hormone oxytocin are attendant on female orgasm. What happens when you apply this chemical artificially? The amplitude (but not the frequency) of uterine contractions increases by statistically significant amounts, roughly doubling. In addition, the changes in pressure gradient between the cervix and the fundus also increase by a significant amount. The Wildt team knew exactly why this might matter: Their findings could be considered to be of crucial significance in the treatment of certain kinds of infertility and in the explanation of the functional role of sperm transport in humans:

> Data from the present study, collected in a large number of women, unequivocally demonstrate that ipsilateral transport is not the consequence of tubal pathology or an artefact of the method, but a reflection of the physiological function of the uterus and the oviduct.[8]

The mention of *ipsilateral* transport is important, because this refers to the preferential transport to the ovary bearing the dominant follicle—where the egg is to be found. That this transport is mediated by oxytocin, and functionally connected to pregnancy, can therefore be considered somewhat important in helping to frame female orgasm in adaptationist terms. Wildt's team conclude that:

> [O]ur data demonstrate that the uterus and Fallopian tubes represent a functional unit with regard to transport of spermatozoa. The uterus seems to act as peristaltic pump that provides the pressure gradients necessary to transport spermatozoa from the vagina to the fallopian tubes . . . Oxytocin may play a critical role in the control of this process by activation of pump mechanisms via contraction of uterine smooth muscles.[9]

So, while it is true that the Fox team found no movement of sperm or sperm-like particles, subsequent investigation, consciously building on the work of the Fox team, found exactly those (predicted) things in response to administration of the hormone that mediates orgasm in women.

Objection (3) also takes some disentangling. Here is what Lloyd has to say on the issue:

> The only problem with this uterine-upsuck hypothesis of female orgasm is that, according to these studies, uterine upsuck would be expected to occur frequently and in the absence of orgasm, which makes orgasm unnecessary for any increase in fertilisation upsuck might provide. Fox and his colleagues admit this very point.[10]

Because Lloyd makes so much of this, it is worth considering in detail what the Fox team actually claim on this point. This is what they say: "Our hypothesis must therefore be considered to explain only one, albeit the most efficient, method of fertilisation."[11] The Fox team are clearly well aware (as is everyone) that fertilization can occur without female orgasm, and they were quite open-minded about the other mechanisms involved. Given that the Foxes say that "orgasm may play a role in fertilisation through the release of hormones into the blood stream,"[12] the Wildt team's later work is properly seen as supporting and extending their discoveries, detailing more proximate mechanisms such as vaginal distension and cervical contact, both of which increase the chance of oxytocin release and therefore orgasm. One of the hallmarks of good science is just such the generating of new, testable hypotheses that, step by step, generate a complete picture.

In a related objection to point (3), Lloyd is troubled that the Fox team document regular uterine contractions in the absence of orgasm, and that therefore female orgasm would not add anything of significance. Regular uterine contractions do indeed occur "at the rate of four per minute,"[13] according to the Fox team, and as Lloyd affirms, in addition have the regular value "+12cm H_2O pressure at their peak, and a small phase of negative pressure (-2 to -6 cm H_2O) at their trough." These are the baselines that Lloyd mentions. Surprisingly, what she does not then go on to mention is that this itself *is* the evidence for an orgasmic-based insuck mechanism.

The Fox team are quite explicit on this point, and given the central importance of it, I hope I can be excused the repetition: "There was an increase in intra-uterine pressure as the female climax was reached, with two peaks of +48cm H_2O in the first uterine experiment and +40cm H_2O in the second uterine experiment." Following these peaks, there were then sharp falls, after which "regular contractions returned."[14] The "regular contractions" cannot, therefore, be taken as somehow undermining the special behavior of the uterus during female orgasm; it is the huge departure from regularity that gives the Fox team the warrant for claiming to have found mechanisms for sperm insuck. Incidentally, the original papers do not use the term "upsuck" but "insuck." At some point there has been a vocabulary shift, and few people appear to have gone back to check the original usage.

It is, perhaps, possible that the source of the confusion is that the Foxes talk about the fact that female orgasm causes the abolition of the "regular uterine contractions, which occur irrespective of orgasm," and that this has been taken to imply that pressure changes have been abolished by orgasm. I hope it is plain, from the quotes given here, that orgasm abolishes the *regularity* of the contractions, not the *contractions* themselves. Furthermore, it is important to appreciate that both intra-vaginal *and* intra-uterine pressure changes occur. This pressure differential could cause insuck.[15] It is as if an engine is running on idle, and then the accelerator is pressed. No one would think of saying that the existence of a purring turnover of the engine prior to acceleration somehow invalidates the action of the accelerator pedal.

There may be a yet simpler explanation for the confusion. As this replication of one of the key Fox figures shows (Figure 7.1), the arrow of time goes in the opposite direction to that which is usual on graphs. Typically, time in graphs runs along the x-axis from left to right as you look on the page, but the Fox team ran it from right to left. Is it possible that critics have been misled by this into thinking that female orgasm does the opposite of what the original researchers found?[16]

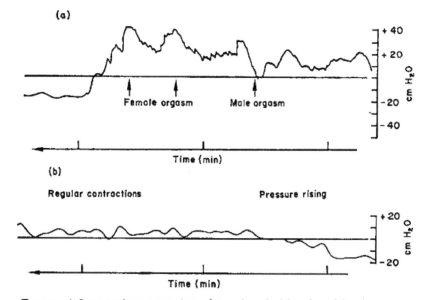

TEXT-FIG. 4. Intra-uterine pressure (second experiment): (a) male and female orgasm; (b) post-orgasmic (female) phase.

FIGURE 7.1   Intra-uterine pressure changes before, during, and post orgasm from the Fox studies.

As to objection (4), *small sample size*, it is true that the Foxes carried out dramatic studies, but using only one couple. Physiology studies often have much smaller sample sets than other types of study (recall that Masters and Johnson's key study into this had only six participants), and sometimes this is justified on the basis that a single case's very uniqueness is worth studying—as can happen with case studies. As noted neurologist Ramachandran pointed out, only one instance of a talking pig is needed to demonstrate that pigs can talk.[17] Human functional anatomy, and physiology, cannot vary hugely among individuals. But, in any case, we might ask whether there are independent reasons for thinking that an insuck mechanism might occur during some women's orgasms, and have these studies been carried out on larger sample sizes? In a word, yes.

7.2   A PERISTALTIC PUMP

As stated earlier, the key hormone of interest here is oxytocin, which rises sharply during female orgasm. What happens when we just administer oxytocin to women? Briefly, we get the same effects that the Fox team showed, but this time the scientists of the next generation were ready with the means to measure actual movement of appropriate material in the female reproductive system. As mentioned before, the Wildt team consciously built on the findings of the Fox team (who were well aware of the actions of oxytocin) and used direct methods to demonstrate actual moment of a sperm simulant within the reproductive system:

> It has been shown that oxytocin is released from the posterior lobe of the pituitary gland in response to vaginal distension, in response to cervical stimulation and during intercourse in response to tactile as well as emotional stimuli (Fox et al., 1970; Fox & Fox, 1971; Fox, 1973, 1976; Gilbert et al., 1991; Murphy et al., 1987). . . . [C]ontractions of the non-pregnant human uterus in response to oxytocin were studied systematically by Knaus who showed that injections of posterior pituitary extract containing oxytocin promptly induced contractions during the follicular phase, but not after ovulation (Knaus, 1950). Our observations confirm and extend these findings by demonstrating a striking effect of oxytocin on transport mechanisms.[18]

To summarize, Wildt's team were perfectly well aware of a continuing line of research, of which the Fox team were also well aware, into the various ways that sperm might be transported into the uterus. Oxytocin has been a well-known correlate of orgasm for decades,[19] and thus there is absolutely no reason to believe that increased contractions during oxytocin rise would undermine the orgasm insuck hypothesis. In addition, the Wildt team had over fifty participants in one of their trials, which should be sufficient to even out any individual anomalies. However, some—especially the noted physiologist Roy Levin—have objected that the use of oxytocin in this way is illegitimate because doses may exceed those occurring naturally.[20] This is a surprising objection. It seems logically equivalent to arguing that administering painkillers is illegitimate (and therefore tells us nothing about the nature of pain relief) because natural levels of endorphins are exceeded by such administration. Oxytocin generates the action predicted, and in the manner predicted, in women. In relation to the by-product idea, it should be noted that men do not even have oxytocin in the same way that women do, but its analogue, vasopressin.[21] The onus would therefore now be on the skeptic to explain how a complex integrated functional system can be activated in a predictable, patterned, functional way if it just came about as a by-product—and a by-product of a system that does completely different tasks and uses different mediating hormones to perform those tasks, at that.[22]

Women have been shown to have what the Wildt team describe as a "peristaltic pump" that is activated by a hormone—oxytocin—that is generated by orgasm. It is hard to see how this system exists as a side-effect of anything in the male system—a central requirement of the by-product account. Levin has maintained that cervical tenting occurs at female orgasm, elevating the cervix from the posterior vaginal wall and therefore taking it away from the seminal pool.[23] It should be noted that if he is correct, and this is the proximate mechanism by which female orgasms generate a fitness response—by increasing sperm viability—then its adaptive significance—increasing fertility—remains the same.[24]

In addition, it is important to note that women's refraction period—the time when they are ready to orgasm again—is much shorter than that of men.[25] Focusing on a single event—one orgasm—as the be-all and end-all of the sexual encounter unconsciously imports a very male-centric model of orgasm (yet again). Fuller discussion of this will have to wait until later but, for the moment, it is worth noting that the general raising of oxytocin level in women within a time frame of an hour (or more) is all going to contribute to the peristaltic pump action described. Once again, the pattern for women is utterly different from that of men, with much higher pituitary activation following orgasm for some time.[26] There is no naturalistic requirement for one single orgasm to do all the work here and, indeed, women's patterns of sexual desire and arousal would tend to argue against this too.

7.3   BACKFLOW STUDIES

The potential for differential sperm retention is the crux of the issue here, however it comes about. We can measure this indirectly—via looking at pressure changes, or the action of known correlates of orgasm such as oxytocin—but sperm retention can also be measured directly. This is because, in mammals, somewhere between five minutes to two hours after intercourse, the female expels a quantity of seminal fluid that is not taken up into the reproductive tract. Minimizing this backflow is a major issue in the farming industry, where artificial insemination is an expensive process. The administering of oxytocin to recipients of artificial insemination is a key part of this process.[27] In humans, this backflow is not under conscious control, but it is detectable by the women in question. It was backflow that Baker and Bellis were referring to in their controversial studies, but they used estimates of this rather than direct measures.[28]

The balance between high ecological validity and experimental control means that any study is a work of trade-offs. It is the business of honest science to balance those trade-offs and attempt to generate a complete picture from the necessarily fragmentary pieces that each study offers. In the case of female orgasm, one trade-off is between studying genuine penetrative intercourse that stimulates internal areas of female sensitivity in aroused partners (and possibly the action of prostaglandins) and controlled measurement of internal mechanisms that are obviously hard to access at this time. We found one way to solve part of this puzzle. Modern sex toys include the Magic Wand™ made by Hitachi. This device—using mains power—generates a truly colossal amount of RPM (6000) that can create deep tissue responses, even via stimulation of the external genitalia only. This makes it possible to stimulate internal sensitive clitoral tissue, without penile penetration.

That partially solved the internal sensitivity issue, but what about the male part of the equation? Men do not produce identical amounts of seminal material, and to measure backflow we needed controlled amounts for comparison. We overcame these factors by using a measured dose of a sperm simulant with similar viscosity properties to semen and a collecting device called a Mooncup™, which some women use instead of sanitary pads or tampons. Thus, we could generate an orgasm with internal sensations and measure the backflow of a controlled amount of sperm simulant, with each woman acting as her own control (i.e., in orgasm/non-orgasm conditions).[29] This technique is neutral about the proximate mechanisms occurring inside. It could be that female orgasm generates immediate insuck to the fallopian tubes, or it could be that the raised oxytocin levels just generally increase the efficiency of this system over the next half hour or so. Either way, the results, in terms of reduced backflow, are directly measurable.

This technique has only been piloted on half a dozen participants, but the results are encouraging: An orgasm, readily identified by participants as originating deep inside, and accompanied by internal pulsing sensations, reduced the amount of backflow by between 10% and 24%. The results obviously stand in need of replication, and I will go into more details about the implications of this in the final chapter, but here I want to note that this is not simply an area of academic or even recreational interest. The prominent fertility specialist Lord Robert Winston is on record as opining that persistent inorgasmia likely accounts for some of the variance in fertility for persistently infertile couples, once other variables have been controlled.[30]

7.4 SUMMARY

The Fox team opened up a line of research exploring what happens during female orgasms that occur in natural settings. For a variety of reasons, this involves stimulation of the full range of clitoral erectile and sensitive tissue, much of which is internal. It also involves the release of large amounts of oxytocin, which stimulates a peristaltic pump that produces measurable flow of spermatogenic material that can be measured either by looking directly at the mechanisms in question or by measuring backflow of suitable sperm-like material following female orgasm. This complex, integrated set of interlocking mechanisms does not, however, stand in isolation. They are locked into a set of control and decision-making mechanisms surrounding sexual behavior. We have not yet solved the mystery of the female orgasm. To take us closer to that, I first need to clear up a remarkably persistent myth—that women lack internal sexual sensitivity at all and only pretend to experience this to please male partners. The myth of the myth of the vaginal orgasm.

NOTES

1   Popper, K. (1957). *The poverty of historicism*. Routledge. Ch. 29.
2   Nishimori, K., Young, L. J., Guo, Q., Wang, Z., Insel, T. R., & Matzuk, M. M. (1996). Oxytocin is required for nursing but is not essential for parturition or reproductive behavior. *Proceedings of National Academy of Science U S A, 93*, 11699–11704.
3   Carter, C. S., Devries, A. C., & Getz, L. L. (1995). Physiological substrates of mammalian monogamy: The prairie vole model. *Neuroscience and Biobehavioral Reviews, 19*, 303–314; Carter, C. S., Williams, J. R., Witt, D. M., & Insel, T. R. (1992). Oxytocin and social bonding. *Annual New York Academy of Science, 652*, 204–211; Fisher, H. E., Aron, A., Mashek, D., Li, H., and Brown, L. L. (2002). Defining the brain systems of lust, romantic attraction, and attachment. *Archives of Sexual Behavior, 31*, 413–419.
4   Zak, P. J., Kurzban, R., & Matzner, W. T. (2005). Oxytocin is associated with human trustworthiness. *Hormones and Behaviour, 48*, 522–527. It is worth noting that simplistic ideas of oxytocin as a generalized "love hormone" are refuted by studies that show that lactating humans are demonstrably more aggressive to those seen as threats. E.g., Campbell, A. (2008). Attachment, aggression and affiliation: The role of oxytocin in female social behavior. *Biological Psychology, 77*(1), 1–10.
5   Ayinde, B. A., Onwukaeme, D. N., & Nworgu, Z. A. M. (2006). Oxytocic effects of the water extract of *Musanga cecropioides* R. Brown (Moraceae) stem bark. *African Journal of Biotechnology, 5*, 1350–1354; Russell, J. A., Leng, G., & Douglas, A. J. (2003). The magnocellular oxytocin system, the fount of maternity: Adaptations in pregnancy. *Frontiers in Neuroendocrinology, 24*, 27–61.
6   Blaicher, W., Gruber, D., Bieglmayer, C., Blaicher, A. M., Knogler, W., & Huber J. C. (1999). The role of oxytocin in relation to female sexual arousal. *Gynaecologic and Obstetric Investigation, 47*, 125–126; Carmichael, M. S., Humbert, R., Dixen, J., Palmisano, G., Greenleaf, W., & Davidson, J. M. (1987). Plasma oxytocin increases in the human sexual response. *Journal of Clinical Endocrinology and Metabolism, 64*, 27–31; Carmichael, M. S., Warburton, V. L., Dixen, J., & Davidson, J. M (1994). Relationships among cardio-vascular, muscular and oxytocin responses during human sexual activity. *Archives of Sexual Behavior, 23*, 59–79.
7   Wildt, L., Kissler, S., Licht, P., & Becker, W. (1998). Sperm transport in the human female genital tract and its modulation by oxytocin as assessed by hysterosalpingoscintigraphy, hysterotonography, electrohysterography and Doppler sonography. *Human Reproduction Update, 4*, 655–666; Zervomanolakis, I., Ott, H. W., Hadziomerovic, D., Mattle, V., Seeber, B. E., Virgolini, I., et al. (2007). Physiology of upward transport in the human female genital tract. *Annals of the New York Academy of Science, 1101*, 1–20.
8   Wildt et al. (1998, p. 663).
9   Wildt et al. (1998, p. 665).
10   Lloyd (2005, pp. 187–188).
11   Fox et al. (1970, p. 250).
12   Fox et al. (1970, p. 250).

13 Fox et al. (1970, p. 247).

14 Fox et al. (1970, p. 247).

15 "Regular uterine contractions, which occur irrespective of orgasm . . ." (Fox et al., 1970, p. 247). Note that, despite what Lloyd claims, this does not mean that "uterine upsuck would be expected to occur frequently and in the absence of orgasm, which makes orgasm unnecessary" (Lloyd, 2005, p. 188). Just glancing at the graph should show the differences in pressure that are occurring. There are *baseline* contractions but they are dwarfed by the orgasmic ones.

16 Kim Wallen (Wallen, K. (2006). Commentary on puts' (2006) review of the case of the female orgasm: Bias in the science of evolution. *Archives of Sexual Behavior, 35*(6), 633–636) claims that there are technical issues to do with the transducers in the Fox study, which could have been affected by temperature as well as pressure changes, and laments that "The data in the Fox study have never been replicated and, while heroic in their attempt, it seems uncertain that post-orgasmic negative intrauterine pressures were actually found in this study" (p. 635). I am surprised to find Wallen claiming this after looking at the figure(s) in question. Only truly huge internal temperature drops could otherwise explain the pressure changes (unless we invoke a freak weather condition), and other independent evidence is that heat, blood flow, and oxygen tension increases in the female genital region during intercourse (e.g., Sommer, F., Caspers, H. P., Esders, K., Klotz, T., & Engelmann, U. (2001). Measurement of vaginal and minor labial oxygen tension for the evaluation of female sexual function. *The Journal of Urology, 165*(4), 1181–1184.) rather than decreases as would have to happen for the pressure to change in the way that Wallen describes. The Fox team controlled for these factors in follow-up studies anyhow (see Fox & Fox, 1971 and Fox et al., 1970 for details). In any case, key elements of the Fox team's findings have been replicated, for example, by the Wildt team (Wildt et al., 1998; Zervomanolakis et al. (2007)) and ourselves (King, R., Dempsey, M., & Valentine, K. A. (2016). Measuring sperm backflow following female orgasm: A new method. *Socioaffective Neuroscience & Psychology, 6*(1), 31927).

17 Ramachandran, V. S., & Blakeslee, S. (1999). *Phantoms in the brain: Human nature and the architecture of the mind.* Fourth Estate Ltd.

18 "It has been shown that oxytocin . . ." comes from (Wildt et al., 1998, p. 664). The Knaus reference is Knaus, H. (1950). *The physiology of human reproduction* (Originally, Knaus, H. H. (1950). *Die Physiologie der Zeugung des Menschen.* Maudrich.)

19 Documented by, e.g., Blaicher, W., Gruber, D., Bieglmayer, C., Blaicher, A. M., Knogler, W., & Huber, J. C. (1999). The role of oxytocin in relation to female sexual arousal. *Gynecologic and Obstetric Investigation, 47*(2), 125–126.

20 See Levin, R. J. (2011). Can the controversy about the putative role of the human female orgasm in sperm transport be settled with our current physiological knowledge of coitus? *The Journal of Sexual Medicine, 8*(6), 1566–1578.

21 Heinrichs, M., & Domes, G. (2008). Neuropeptides and social behaviour: Effects of oxytocin and vasopressin in humans. *Progress in Brain Research, 170*, 337–350.

   More strictly speaking, the expression of vasopressin seems mainly dependent on androgens. Lu, Q., Lai, J., Du, Y., Huang, T., Prukpitikul, P., Xu, Y., & Hu, S. (2019). Sexual dimorphism of oxytocin and vasopressin in social cognition and behavior. *Psychology Research and Behavior Management*, 337–349.

22 This is what Wildt had to say on the matter: "We did not do a detailed dose-response study on the dose of Oxytocin, we just used a dose with which we saw an increase in contractions and peristalsis in ultrasonography, hysterosalpingoscintigraphy and in direct pressure monitoring within the physiological range. Thus of course, plasma levels of oxytocin may be well above those measured in peripheral blood during physiological basal conditions, but we have to take into account that the critical factor are local concentrations within the uterus which have not been measured so far. I do not think that the criticism of Levin is justified. There are contractions of the uterus during orgasm, this may be caused by central release of oxytocin or by oxytocin produced within the uterus, the release of intrauterine oxytocin may be caused by the action of prostaglandins. If oxytocin is administered peripherally, you will need much, you may well need to achieve high levels in peripheral blood in order to reach physiological levels within the uterus. In my view, orgasm is one phenomenon, sperm transport the other, which are interdependent. This should not be interpreted in such a way that orgasm is a prerequisite for sperm transport." (Wildt, pc 24/01/2012). Quite so. Orgasm raises oxytocin, which may well aid transport, but at no point is anyone claiming that it is a necessary condition for it. But the issue for the by-product advocate is made very difficult. It is worse than trying to explain how the integrated process came about accidentally. The task of the by-product theorist is daunting indeed. They now have to explain how a complex system that was

designed by natural selection to do one thing (in the male) now does a totally different set of things when a hormone that is not even available in men is administered to it, when that system finds itself (quite by accident) in women.

23 Levin, R. (2002). The physiology of sexual arousal in the human female: A recreational and procreational synthesis. *Archives of Sexual Behavior, 31,* 405–411.

24 It is possible that another hormone released during orgasm, prolactin, could capacitate sperm and increase fertility this way. Krüger, T. H., Haake, P., Hartmann, U., Schedlowski, M., & Exton, M. S. (2002). Orgasm-induced prolactin secretion: Feedback control of sexual drive? *Neuroscience & Biobehavioral Reviews, 26*(1), 31–44. See also Meston, C. M., Levin, R. J., Sipski, M. L., Hull, E. M., & Heiman, J. R. (2004). Women's orgasm. *Annual Review of Sex Research, 15*(1), 173–257.

25 See, e.g., Levin, R. J. (2009). Revisiting post-ejaculation refractory time—what we know and what we do not know in males and in females. *The Journal of Sexual Medicine, 6*(9), 2376–2389, or Turley, K. R., & Rowland, D. L. (2013). Evolving ideas about the male refractory period. *BJU International, 112*(4), 442–452, for discussions of these differences.

26 Huynh, H. K., Willemsen, A. T., & Holstege, G. (2013). Female orgasm but not male ejaculation activates the pituitary. A PET-neuro-imaging study. *Neuroimage, 76,* 178–182.

27 For more on reducing backflow in the farming industry, see Knox, R. V. (2001). *Artificial insemination of swine: Improving reproductive efficiency of the breeding herd.* Retrieved September 24, 2010, from www.gov.mb.ca/agriculture/livestock/pork/900pdf/bab13s04.pdf and Mezalira, A., Dallanora, D., Bernardi, M. L., Wentz, I., & Bortolozzo, F. P. (2005). Influence of sperm cell dose and post-insemination backflow on reproductive performance of intrauterine inseminated sows. *Reproduction in Domestic Animals, 40*(1), 1–5; and Steverink, D. W. B., Soede, N. M., Bouwman, E. G., & Kemp, B. (1998). Semen backflow after insemination and its effect on fertilisation in sows. *Animal Reproductive Science, 54,* 109–119.

28 E.g., in Baker, R. R., & Bellis, M. A. (2006). Human sperm competition: Ejaculate manipulation by females and a function for the female orgasm (1993). In *Sperm competition in humans* (pp. 177–210). Springer.

29 The method is detailed in King, R., Dempsey, M., & Valentine, K. A. (2016). Measuring sperm backflow following female orgasm: A new method. *Socioaffective Neuroscience & Psychology, 6*(1), 31927. We are somewhat proud of the fact that this method can be used with minimal specialist equipment and, we believe, can be carried out to measure elements of reproductive function in the comfort of one's own home. The initial idea to use the Mooncup™ in this way should be credited to Louise Spiers. The initial idea for measuring backflow (somewhat confusingly called "flowback" in some accounts) belongs to Baker and Bellis (e.g., Baker, R., & Bellis, M. A. (1993a). Human sperm competition: Ejaculate adjustment by males and the function of masturbation. *Animal Behavior, 46,* 861–885; and Baker, R., & Bellis, M. A. (1993b). Human sperm competition: Ejaculate manipulation by females and a function for the female orgasm. *Animal Behaviour, 46,* 887–909.)

30 Winston, R. (2010). *From Britain's leading fertility expert, an intriguing question . . . Is a woman more likely to conceive if she enjoys sex?* www.dailymail.co.uk/femail/article-1279841/From-Britains-leadingfertility-expert-intriguing-question_Is-woman-likely-conceiveenjoys-sex.html.

Similarly, a leading Czech fertility company, IVF Cube, presented data at the ESHRE conference in Helsinki in 2016, noting significant rises in fertility in couples who had high rates of orgasm in the woman receiving IVF. Thanks to Dr. Hana Vishnova, P.C. 09/11/2018.

8 The Myth of the Myth of the Vaginal Orgasm

The vast majority of women who pretend vaginal orgasm to their men are faking it to "get the job." In a new bestselling Danish book, *I Accuse*, Mette Ejlersen specifically deals with this common problem, which she calls the "sex comedy." This comedy has many causes. First of all, the man brings a great deal of pressure to bear on the woman, because he considers his ability as a lover at stake. So as not to offend his ego, the woman will comply with the prescribed role and go through simulated ecstasy.

—(Anne Koedt, 1968, p. 23)[1]

A vaginal orgasm is a dissolving in a vague, dark generalized sensation like being swirled in a warm whirlpool. There are several different sorts of clitoral orgasms, and they are more powerful (that is a male word) than the vaginal orgasm. There can be a thousand thrills, sensations, etc., but there is only one real female orgasm and that is when a man, from the whole of his need and desire takes a woman and wants all her response. Everything else is substitute and a fake, and the most inexperienced woman feels this instinctively, . . . "Do you know that there are eminent physiologists who say women have no physical basis for vaginal orgasm?" "Then they don't know much, do they?"

—(Doris Lessing, 1962, p. 200)[2]

Talk of female orgasm has frequently generated at least as much heat as light, and we can get a sense of some of why this is from the diametrically opposed attitudes expressed in the quotations opening this chapter. The two opening quotes capture two very different perspectives. On the one hand, that all (or nearly all) talk of vaginal (read: "coital") orgasm is fakery to pander to men's egos; while, on the other hand, that there is something especially noticeable about female orgasms generated in this way. This chapter is all about why female, but not male, orgasms are experienced and described with such astonishing variety, and what this tells us about their nature and function.

Before I turn to the data that lie behind the answers to these questions, let me deal with a pressing issue. Namely, why am I still using terms like "vaginal" and "clitoral" in reference to female orgasm at all? Did I not just spend an entire chapter (chapter six) demonstrating, in exhaustive detail, that clitorises are far larger than is often supposed, including sensitive and erectile areas that are inside the vagina, connected to a measurably different neural relay mechanism? And what about those people who claim that either no (or, at least, comparatively few) women can orgasm through purely penetrative sex at all? These are complex questions, partly because they involve something that is fiendishly difficult, namely: controlled physiological investigations of a behavior that is usually intimate and private. Can we do proper science on something where the mere act of observing it, changes it?

In sub-atomic physics, the Heisenberg uncertainty principle states that the accuracy of measurement of momentum of a particle is inversely proportional to the accuracy of measurement of its position. This truth is often (and somewhat misleadingly) understood to amount to the claim that the act of measurement always changes the thing being measured. I have even read interpretations of this truth of quantum mechanics that seem to imply that the universe knows that it is being watched. This does not appear to be usually true—at least, not of sub-atomic particles.[3] However, organisms *do* frequently know that they are being watched, and their behavior adjusts accordingly. Therefore,

DOI: 10.1201/9781003372356-8

as I said right at the start of the book, our only hope in behavioral science is to use a multiplicity of viewpoints to triangulate on our best explanations, and there is no clearer example of why this is necessary than here. So, I will have to proceed carefully, giving full voice to the dissenting, and often conflicting, voices that have been raised around these issues. There is no avoiding it; I will have to discuss some sexual politics.

8.1 SEXUAL POLITICS

In a 1972 review of some (then) recent books on sex research, the philosopher Irving Singer wrote that: "In these days of sexual politics it becomes necessary to separate the scientific study of female sexuality from its misuse by theorists who wish to establish strategic positions in the struggle between the sexes."[4] It would be foolhardy in the extreme to pretend that this observation could not be rolled out today (or tomorrow) with equal relevance. I will go further: It does not seem to me likely that we will *ever* completely rid ourselves of this aspect of our natures, any more than it would be reasonable to believe that total honesty and transparency will ever be the rule in business or politics.

Any academic discipline could be, with only some injustice, described as a set of individuals who have come to an agreement not to question the same assumptions.[5] Those of us involved in sex research sometimes talk (because we often have these ideals ourselves, and project them onto others) as if it is the norm for all humans to have the goal that each and every human is entitled to mutually rewarding sex lives. This is a worthy aim but a hopelessly naïve assumption, if one thinks that this goal is shared by all humans, or even *could* be shared by all the humans likely to ever exist. The evidence from the past is unequivocal; human history is replete with explicit and often gruesome examples of humans deliberately interfering with one another's sexual fulfillment via psychologically induced emotional mechanisms (such as guilt and shame), cognitive mechanisms (such as misinformation), physical mechanisms (such as the mutilation of the genitals of both sexes, foot-binding, and similar), and social mechanisms (such as purdah).

From an ethological perspective, it is obvious why these things happen: Humans, like all sexually reproducing species, compete for sexual access to mates of differently perceived value, and your most intense competitors are, therefore, members of your own sex. At the same time, the potential for exploiting members of the opposite sex is always available (though in different ways) to both sexes. We call these things *intra*sexual competition and *inter*sexual signaling in ethology. However, almost everyone I know has to run the prefixes "intra" and "inter" silently in their heads to remind themselves as to which is which. In recognition of that, I will try to use the terms "within sex" and "between sex" where possible confusion could occur between the various forms of competition, cooperation, and signaling that occur. It is important that this complex backdrop of competition is appreciated because, without it, it is itself something of a puzzle as to why female orgasm has remained a puzzle for so long. Other mysteries—not always obviously more technically challenging—have largely dissolved with the universal acid of Darwinian natural selection, so why not this one?[6] As a friend of mine is fond of saying, "We are tremendous creatures for getting in our own way." Nowhere is this truer than in the field of female orgasm.

8.1.1 DISPATCHES FROM THE FRONT LINE IN THE BATTLES OF THE SEXES

Let me begin with the first question: Is it not the case that all talk of "vaginal" and "clitoral" orgasms is hopelessly outdated, a relic of Freud's contentious (and long-debunked) views that sensitivity could (and should) transfer from the clitoris[7] to the vagina?[8] Yes, and no. Part of the answer is that many women spontaneously describe their orgasms using these terms—for example, when we asked them, without prompting, for descriptors of their experiences back in our study in 2012.[9] Others have replicated this finding, and it is a commonplace among sex therapists who, not unreasonably, tend to explain that anxiety about coital orgasm is likely to *itself* cause sexual dysfunction.[10] But another reason for the persistence of this terminology is, I fear I have to say, collateral damage in the battles of the sexes.

There is more than one battle of the sexes, because—as I said above—members of your own sex are not automatically your sexual allies—they are frequently competitors, and, even when they are not direct competitors, they can exploit you. Men compete with other men for access to women, and they also have a suite of strategies to attempt to control female sexuality, many of which I have mentioned, and will discuss further in later chapters. But women compete with other women for access to both men and (sometimes) women too. This latter strategy is possible because human female-female pairings can be adaptive under certain circumstances—as they are in other species such as gulls and bonobos.[11] However, these same-sex pairings can be vulnerable to destabilization because, at some point, they need male genetic input to be reproductively viable.[12] Thus, these same-sex relationships can be in need of policing to prevent defection to male partners who have supplied sperm and might supply other support. As we will see in chapter ten, the seeds for entirely sexually separatist cultures (or sub-cultures) are present in the human psyche, and are available as strategic options.[13]

It seems worth considering that our set of mixed human sexual strategies itself contributes to rival accounts of female sexual response. The spectrum of explanation here is between one extreme of claiming *either* that orgasms produced through penile penetration are impossible—and that therefore only lesbianism is a viable female sexual outlet—*or* that vaginal (read "penetrative") orgasms are psychologically healthier than ones produced in other ways—and that women who do not have them are therefore immature and frigid. This latter strategic trend, as we will see in later chapters, reaches its apotheosis in cultures that practice extreme forms of female genital mutilation/cutting (FGM/C), and deny any place for female sexual enjoyment at all. However, elements of this trend exist in western, supposedly sexually liberal, cultures too. Given that I am arguing that both these extreme views may be manifestations of sexual conflict being fought out in the scientific arena, I need to say something about both.

The title of this chapter comes from a famous paper written by feminist activist Anne Koedt. She made no secret about her goals of establishing separatist feminist communities, with men excluded. She laid out, in some detail, the reasons and methods supporting these goals. In her famous 1968 paper, "The Myth of the Vaginal Orgasm," she insisted that women's orgasming through penile penetration was, as near as makes no difference, impossible. Therefore, women claiming to experience it were either lying to please male egos or simply mistaken about anatomy. She drew on the work of Masters and Johnson—the "eminent physiologists" from Doris Lessing's quote—in support. That is one strategic position.[14]

Balancing the denial that vaginal orgasms were possible, others—recently, and notably, the neo-Freudian scholar, Stuart Brody—have asserted that, on the contrary, vaginal orgasms are associated with better mental health, and women who cannot experience them are psychologically less healthy.[15] In addition to this, male partners—because female orgasm is an important sexual signal to them—can put pressure on female partners to orgasm this way, which can result in their female partners faking orgasm to please them. The net result of all this is that these sexual encounters fail to give crucial feedback about what the female partner actually enjoys. Adding to the confusion, many (undoubtedly well-meaning) sexual advice experts may tell women that few (if any) can orgasm coitally, and to therefore not get stressed if they do not orgasm this way. However, this rarely makes the anxiety go away on its own.[16]

With all this talk of what women should or should not or can or cannot experience, no wonder so many of them are stressed about sex. So, I ask readers to put aside these insistent and somewhat extreme normative voices, at least for the rest of the chapter, while I explore what happens when others stop telling women what they *should* experience and, instead, try to make sense of what they *do* experience during sex.

8.2   REPORTING ON THE SELF?

Self-report can get a bad press in psychology.[17] It has even been referred to as "psychology's four-letter word."[18] It is true that humans are notoriously bad at accurately reporting their own motives—even (or perhaps *especially*) to themselves.[19] However, subjective conscious experience is a real part of the world, and a complete scientific account of that world therefore cannot leave it out.[20]

In a key summarizing paper helpfully titled, "Telling More Than We Can Know: Verbal Reports on Mental Processes," Nisbett and Wilson explored the use and abuse of self-report.[21] It turns out that, when we compare self-report with more objective validations, humans can (more or less) accurately report on their personal history, attentional focus, sensations, plans, and evaluations. However, we humans are notoriously bad at assessing our own motives, and have a noted tendency to behave as advocates, rather than as disinterested reporters.[22] But the truth of that—important though it is—should not blind us to judiciously using self-report at times. Indeed, the reason we know that humans can be unreliable narrators when it comes to motivation is that we can experimentally manipulate people to show that their reported motives can be in conflict and be sometimes shown to be rationalizations.[23]

It is important to realize what we have and have not achieved here: The very reason we know that self-report on motives *can* be misleading is that we have compared people's self-report with other methods, *including other self-reports*. If a range of people—especially reporting independently and from widely disparate cultures—spontaneously report closely related experiences and, furthermore, if we can correlate these reports with other measures, then we are entitled to add these measures to the scientific cannon. Experience has taught me that some examples will help in getting this point across, because an unwarranted degree of condescension towards self-report has become something of an epidemic in some scientific circles.

The great neurologist Ramachandran reported progress in understanding conditions like synesthesia, phantom limb, and Capgras' delusion, only because he (and others) took the self-report of sufferers of the conditions seriously. Other neurologists had dismissed the reality of these conditions out of hand.[24] Self-reports of cognitive risk (technically called "CSQ/DAS measures") are predictive of depression and schizophrenia in non-clinical populations in a way that no existing biological measure yet achieves.[25] Or, take the everyday experience of pain. Should comparing objective measures of pain such as blood-pressure or cortisol rising really take precedence over asking the patient that most subjective of questions, "does it hurt?"?[26]

8.2.1   OBJECTING TO THE SUBJECTIVE?

Our scientific quest for objectivity—sometimes a manifestation of misplaced physics envy—can go too far.[27] Subjective experience is a part of objective reality too, and any scientific account that does not have a place for it is incomplete. The philosopher Daniel Dennett calls this mixed approach, using both objective and subjective measures, "heterophenomenology," which is a bit of a mouthful. But what he means by the term is that things like self-reports and conscious actions are as much part of the real world as are brain scans and behavioral measures. Indeed, they *are* a kind of behavioral measure.[28]

Realizing this does not make subjective experience the final word on a phenomenon, but it does mean that we are entitled to refer to it in our science. One good reason for thinking this is that nature does not give free lunches. If conscious experience evolved to give organisms feedback about their world, the ability to reflect on these experiences, and plan accordingly (and at least that much has to be true), then we are doing science a disservice not to afford conscious experience the attention it deserves.

If, despite all I have said here, someone wants to insist that reporting on internal mental states is *always* unreliable, and must be rejected by a true scientist, then this becomes a classic case of sawing off the logical branch one is sitting on. After all, in relaying self-report's supposed inherent unreliability you are, of necessity, conveying your *own* self-report on your own mental states—itself a judgment about methodology—in doing so. You cannot, and remain in good logical standing, *tell* a person that talking is meaningless.

This fallacy has a name, the Ishmael Effect, named after the survivor of *Moby-Dick*, who, out of all the sailors said of himself, "and I only am escaped alone to tell thee," without explaining how this miracle was possible.[29] As Quine pointed out, we can hold any of our knowledge-gathering methods up to critical scrutiny, comparing it against other methods. However, we cannot be consistently and meaningfully skeptical about the entire enterprise of knowledge building without incoherence. In short, we can be skeptical about *anything*, but we cannot be skeptical about *everything*.

With any fears over the validity of self-report (one hopes) calmed somewhat, what do women report about their sexual experiences?

8.3   ASKING WOMEN ABOUT THEIR ORGASMS

According to ancient Greek legend, Tiresias, the sage who had lived as both a man and a woman, was blinded by the goddess Hera. This mutilation was in revenge for the crime of revealing—from personal comparative experience—that women had a greater capacity for sexual enjoyment than did men.[30] To the extent that we can learn from legends, this one seems to imply two things. First, that women's capacity for sexual enjoyment is greater than that of men. Second, that one might be wise not to reveal this fact too widely.

Sex researchers do not have that luxury.[31] For about a century now, we have been asking both sexes systematically about their experiences in bed together. The result has been that sexual scholars would be somewhat in agreement with Tiresias. More precisely, it has been something of a commonplace among sex therapists, and reporters of human sexual experience, that women, as compared to men, report a greater *range* of orgasmic experiences.[32] Not only that, but we have data from people who have transitioned from female to male, at least hormonally, and (sometimes) even back again, taking large doses of testosterone in the process. These people have reported on their orgasms, and they universally report that orgasms in the female state are longer lasting and fuller bodied.[33]

It is true that one study, in which both male and female students described their experience of orgasm, resulted in naïve observers not being able to clearly tell which verbal descriptions were female and which were male. But, and it is an important caveat, what this study failed to capture is the greater range of variability when taking women's descriptions both en masse and within-participant. Therefore, the methodology assumed (rather than demonstrated) a homogeneity of orgasmic experience within the subjects. If they had asked for the range of orgasmic experience within *each individual*, would they have obtained the same results? We will never know. However, more recently we have asked precisely those questions, and the answer is clearer. Women, on average, report a greater range of sexual response than do men. What do we mean by this, exactly?

It is true that, given a set of adjectives of a group of *single* orgasmic experiences (with each person sharing just one of these sets) it might be hard (in some cases) to tell which one came from a man and which one a woman.[34] But when you change the focus so as to be able to see the woods, rather than the trees then, taken as a group, a female set of orgasm descriptors show a much greater variability than does a set of male ones, both from one another and from other orgasm descriptions given by that one individual. This is a surprising finding, given how measurements of phenotypic variance in humans usually work. What do I mean by this? It will take a paragraph to explain.

In our species, where biological selection falls—as we saw in chapter three—more keenly on males than females, we typically find more variation in measured traits (psychological and physical) in males than in females. Phenotypic variance tracks reproductive variance.[35] Males are more reproductively variant than females in our species.[36] This means that (comparatively) some men reproduce a lot, quite a lot of men reproduce not at all, while most women fall somewhere in between these extremes. If we plot these results, then we get bell curves where the means are the same, but the shape of the curves is different—female bell curves are narrower. In our history the effect of this reality bit even more keenly. For instance, a comparison of mitochondrial DNA (from women) and Y-chromosome DNA (from men) in our genome revealed that, over evolutionary time, some 80% of women got to reproduce, compared to some 40% of men.[37]

In response to a Freudian explanation of female inorgasmia, the noted feminist scholar Germaine Greer angrily retorted, "I'm not frigid—you're boring."[38] I could not have put the point I am trying to make in this chapter, or throughout the book really, more clearly. One striking possibility, that the discrepancy between male and female reproductive variance prompts us to consider, is that the variation in female sexual responses is tracking the variability in male partners. This further prompts the thought that female orgasm is a (semi) cryptic selection mechanism. This insight, if it proved true, would provide part of the answer to the question of why female orgasm has been so elusive to study under controlled conditions. It would also, rather neatly, help explain a number of other puzzles, including male obsession over female orgasm and the feeling of pressure women might feel to express it (including faking it). However, the consequences of all these findings will have to wait for later chapters. To start things off, I need to say something about how scientists and others have tried to develop insights into the subjective experience of female orgasm.

At first sight, there might seem to be little consensus. One prominent physiologist wrote that, although "more has been written about . . . it than practically any other aspect of human sexuality [it would be true to say that] it is largely a will-o-the-wisp that defies accurate scientific description."[39] But that is just one viewpoint, and one rooted in studying orgasms produced by masturbation in laboratories. When we ask about the experience of orgasm beyond those conditions, females, in contrast to males, report that their orgasms differ not only in intensity but also in location, phenomenology, and emotional components. Indeed, this insight is widely appreciated by sex researchers and sex therapists. Ironically, it is sometimes the very same researchers, who have drawn attention to the variability of reported female experience outside of the laboratory, that have (surprisingly) neglected to apply the implications of their own insights to the study of orgasm under lab conditions.[40]

One major attempt to create a generally accepted industry standard for codifying the potentially wide range of female orgasmic experience was the Warner Peak of Sexual Response Questionnaire.[41] This was a commendable attempt to provide a value-free, self-report framework but, as noted sex researchers Kenneth Mah and Yitzchak Binik note, this measure suffers from a probable confounding of orgasm with other sexual arousal.[42] This confounding might be especially pronounced in those who may never have experienced orgasm. This leads us inevitably to talk of inorgasmia.

8.3.1   A Note about Female Inorgasmia

The absence of female orgasm during sex has often been pathologized, adding to this confounding just mentioned. This can result in inorgasmia being discussed in inappropriately narrow medical terms.[43] This book you are reading is not a clinical work, and those suffering inorgasmia may be doing so for a variety of reasons that a gynecologist or other medical specialist may be of assistance with. Obvious things that a specialist doctor might look at include hormone levels, drug interactions, or a history of trauma. I will have almost nothing to say about these things, since they are not my field, and are not the subject of this book. I focus on function, not dysfunction. However, just as we should be cautious about sweeping statements about what normal women are (or are not) capable of, we should be cautious about rushing to pathologize something—lack of orgasm in particular cases—which may well be responding (or not) to a suite of evolved mechanisms. As Germaine Greer put it above, we need to rule out partner characteristics before we invoke "frigidity" or any other pathologizing constructs. In short, we need to be sure that the system is not working precisely as it was designed to do.

Pathologizing frequently ignores salient evolutionary perspectives on issues. The evolutionarily minded psychiatrist Randy Nesse has been drawing attention to this oversight in recent years.[44] He notes that medical doctors all too infrequently take notice of the importance of evolutionary biology in diagnosis and treatment.[45] For example, evolutionary processes have begun to generate a strain of antibiotic-resistant super-bugs in ways with which we are all (alas) becoming familiar. And, of course, pandemics have made most of us aware of how fast mutant strains of viruses can adapt to our vaccines. But, failing to distinguish between adaptive responses and pathologies can itself cause real damage. Take the example of fever. For years well-meaning people thought the helpful thing to do was dramatically lower the temperature of the fever patient. Indeed, not far from where I am writing this, George Boole, one of the founding heroes of my university, died as a direct result of his well-meaning (and equally genius-level intellect) wife doing exactly this to him.[46] But, even if the cooling does not kill you, the pathogen your body was reacting to might. Fever is an adaptive response to certain infections, raising the temperature of the body until it becomes inhospitable to the invader. Cooling the patient down is the last thing that they need.

Female orgasmic dysfunction, which subsumes a lot of different things, was historically the most commonly presented sexual problem, at least in westernized countries.[47] However, there is reason to believe that the situation is improving with many self-help and de-stigmatizing books and videos available. It seems that more and more women and their partners are now perfectly aware of how to create orgasm individually, or paired up, and are now increasingly interested in the different types of orgasm that can be obtained.[48]

In this respect, it is well worth noting that a major theme of the clinical interventionist literature in respect of female inorgasmia is fantasy. It is noted among sex therapists that when women can relax, express their needs and desires without guilt (including, for example, not needing to be drunk in order to express such desires), and they can fantasize freely, these all dramatically improve the incidence of female orgasm.[49] In brief, mood matters.[50] Well, how could it not? We are creatures who typically have sex in private, intimate, moments sometimes over a course of hours or more, with carefully chosen partners—with whom we might (and often do) fall in love, sometimes on a permanent basis. Imagine how bizarre it would be if all the salient features of this (and more besides) were easily recreated in five minutes of masturbation in a laboratory. It would fly in the face of everything we know about our evolved natures if we regularly embarked upon all this effort for no particular reason. This is obviously a very important topic, and I will return to it later in more detail. For now, I need to complete the historical review.

8.3.2   Orgasm Research in the 1970s

As a result of Masters and Johnson's influence, much research into female orgasm, in the latter half of the twentieth century, tended to favor merely examining it dichotomously (occurrence/non-occurrence) only, or (at best) to include fringe assessments as to orgasmic frequency or female sexual satisfaction. Not all researchers in the 1960s and 1970s agreed with Masters and Johnson, however, preferring to ask women what they have experienced, rather than deciding in advance as to what was possible. For example, in the late 1970s, Bentler and Peeler found that women readily distinguished both type and intensity of their orgasms, readily referring to where they occurred: either deep inside or on the surface.[51]

At around the same time as Masters and Johnson were reporting on their research, other sex researchers also disagreed with their assertion that only one type of female orgasm existed. For example, Carol Butler reported that many women readily distinguished the types of sexual climaxes they experienced. Interestingly, the women she reported on also appeared to get more discriminating with age. Forty-two percent of her participants, aged 29 and under, reported that they could easily distinguish vaginally induced orgasmic spasms from clitorally induced ones, while 57% of participants over age 30 made the same distinction. Their descriptions of vaginally induced orgasms, compared to ones induced by the (surface) of the clitoris, included such phrases as "more internal," "deeper," "fuller, but not stronger," and "more subtle."[52] Similarly, Fisher found, first, that women could distinguish the patterns of sensations they experienced from direct vaginal vs. direct clitoral glans stimulation and, second, that they often reported strong preferences for one or the other type of orgasm. It follows from this that they must have been able to distinguish between types. However, perhaps surprisingly, it did not always mean that one type was always preferred over another. Fisher found that a majority of women (64%) preferred the masturbatory orgasm over the coital one and equated clitoral glans stimulation with higher excitement.[53] It should be added that Fisher did not distinguish orgasms that were brought about by simultaneous manual and penile stimulation. Thus, some of these women may be referring to what the team of Josephine and Irving Singer in the early 1970s had referred to as *blended* orgasms.

In a comprehensive review of the literature then to date, the Singers' sex therapist team concluded that there was good reason to believe that there were not two, but *three* types of female orgasm, which they labeled *vulval*, *uterine*, and *blended*. Mindful of the connotations of previous nomenclature, they said, "The terms 'clitoral orgasm' and 'vaginal orgasm' have taken on so many confusing and value-laden connotations that surely they ought to be avoided in scientific discourse whenever possible." Their vulval (or masturbatory) orgasms were reported as felt on the surface of the vagina, "like a trickle of sweet pleasure," sometimes with no satisfying release.[54] However, they did involve the rhythmic contractions of the orgasmic platform that Masters and Johnson noted. The same participants described their coital—or uterine—orgasms as being like the opening of a dam that floods the body with pleasure, leaving a feeling of deep release and satisfaction. These orgasms involved interaction between the penis and the head of the cervix and regularly included reports of a sudden involuntary tightening of breath (technically called *apnoea*), body folding (also involuntary), and felt subjectively deeper inside and with attendant emotional elements, which I will discuss in more detail later.[55] Finally, the Singers described *blended* orgasms as ones containing elements of the previous two.

In a similar vein, Robertiello reported that women routinely made subjective distinctions between masturbatory and coital orgasms. Masturbatory orgasms were described as (1) very intense, (2) relatively brief, and (3) building to a rapid crescendo and then falling off rapidly. By contrast, coital orgasms (1) were reached during intercourse, (2) rose more slowly, (3) didn't reach such a sharp peak, (4) lasted for longer, (5) subsided much more slowly, and (6) resulted in much deeper and fuller feelings of satisfaction. Women who had experienced both types made clear and regular distinctions.[56]

8.3.3   BACK UP TO DATE

That was a brief flavor of some the major themes of female orgasm research in the latter part of the twentieth century. It was characterized by some researchers insisting that women routinely (though not universally) made distinctions between their types of orgasms, while other researchers insisted that they must be mistaken. Naturally, scientists were not happy with this state of affairs and have been trying to untangle the issues ever since.

Notable among such attempts was a fairly large (503 participants) survey by Ken Mah and Irv Binik in 2002. They did not find any substantial differences in orgasmic report dependent on mode of orgasm creation, whether by masturbation or coition. However, they did find differences in degree of satisfaction, typically being greater with a partner, and they put this down to the emotional effects of intimacy.[57]

Their methodology required subjects to rate the appropriateness of 60 adjectives (such as "general spasming" or "ecstasy") in terms of describing their orgasm. This is a perfectly sound way to approach things. For example, in the early days of personality research, scientists did an analysis of dictionary definitions of adjectives describing people, to get a sense of the range of the measurable phenomena.[58] However, a high degree of agreement (technically called "inter-subjective reliability") is required in the scoring of requisite adjectives so that they reliably[59] describe underlying variables. In other words, to be useful as a measuring instrument, what one person means by a word needs to be reliably similar to what another person means by it. However, even if participants are at a loss to put those differences into words (as is often the case with diagnostic criteria), it does not follow that they would say that their experiences were the same.

For example, the descriptors of orgasmic location were not those typically used by women in their own self-report. Thus, Mah and Binik's survey had different codings for sensations experienced vulvally and for those on the surface of the vagina. This may have been confusing to some participants, and this effect might easily have swamped distinctions as to location of the orgasm that were there. Mah and Binik were not looking for the same things that we were interested in subsequently—e.g., the potential for female sexual response to vary according to partner characteristics and behaviors. When we came across their data, we approached them and asked them if we could re-analyze their data to test some hypotheses generated by looking at female orgasm through an evolutionary lens. They very kindly agreed. The result was my first paper in the field of sex research.[60]

8.4   A TYPOLOGY OF ORGASMS?

Although Mah and Binik did not collect their data with the aim of identifying different types of orgasms or to advance understanding of the possible adaptive function of female orgasms, we thought we might use their data in a new way. Specifically, inspired by the clinical and therapeutic works mentioned earlier, we planned to test the proposition that there might be different types of female orgasms, with some characteristics consistent with fitness-related features such as oxytocin release, insuck, and sperm selection. One of the key analytic tools we used was latent class analysis. This statistical technique is very versatile in that you can feed pretty much any type of data (or mix of data) in at one end, turn the statistical crank, and it disgorges ways in which the data can be meaningfully grouped, even if those groupings were not immediately apparent.[61] Hence the *latent* part. This gave us a pattern of four kinds of orgasm, divided up in terms of pleasure and sensations. The orgasms could also be grouped in terms of where they were centered—on the surface or deeper inside. We then took the resultant classes of orgasmic experience and tested them against some predictions that evolutionary theory might suggest.

First, we asked the data whether any types of the heterosexually induced coital orgasms differed phenomenologically from what was experienced during masturbation. Next, we tested whether different types of orgasms experienced during intercourse differed with respect to select aspects of heterosexual activity; we tested specific hypotheses involving orgasmic sensations, emotions, sexual behavior (such as oral sex), and location of orgasm. What we found, in brief, was that partnered orgasms showed much more variability than masturbatory ones, with some partnered orgasms scoring very high compared to masturbatory ones, while some other ones dipped below partnered ones in terms of both pleasure and sensation. Some (but not all) of the partnered orgasms had sensations that involved the whole body, including sensations deep inside, which might (but only *might*) be indicative of some sort of internal insuck action, as described in physiological detail in chapter seven. Apparently, at least in terms of the orgasm experience itself, sometimes sex with oneself is more physically pleasurable than sex with a male partner, even when the latter provides sufficient sexual arousal to generate an orgasm.[62]

Several things could have been going on here. First, there may be a confusion of *intensity* with sexual *satisfaction*. Masturbatory orgasms are frequently described as more intense and localized in comparison to coital ones. However, it also transpired that intensity and localization did not exhaust all the differences between participants' descriptions. Participants were not comparing their own orgasms with other ones that they had had; *we* were comparing their experiences *between* different people. This is a crucial difference. If, as we suspected, female orgasm acts partly as a test bed (pun intended) for partner characteristics, then it was important that the next step was to be comparisons being made between the sexual responses that each woman might have to variation in partners and behaviors. This is what we turned to next.

8.4.1 DEEP AND SURFACE ORGASMS

To get ourselves started, we asked a number of women, in semi-structured interviews, what sort of things scientists should be asking them about their orgasmic experiences. They were very forthcoming, but a frequent first response to this question was "which ones do you mean?" We wanted to know which ones *they* meant, and we also wanted to be very careful not to build in assumptions (for example, we did not introduce terms like "clitoral" or "vaginal") in advance of data collection.

All the sensations and experiences that we subsequently asked about (such as "internal sucking sensations") were prompted by these initial, largely free-form, pilot interviews. From there we constructed a survey that was engaged with by 265 participants, mostly from North America and Europe, but also including Asia and Africa.[63] Given both that female sexual orientation is known to be more fluid than that of males, and that lesbian sexual interactions leading to orgasm may well have evolutionary significance all their own, we decided to include those of all declared orientations in the study.[64]

We first asked women to think back to their most recent orgasm and to describe it by responding to the series of questions about that particular orgasm. The questions asked were: (1) presence/degree of internal sucking sensations; (2) clarity of thought after orgasm; (3) relaxedness following orgasm; (4) relaxedness prior to orgasm; (5) whether the orgasm was localized; and (6) amount of noise made (by self) during orgasm. There were also questions with either/or responses, pertaining to (7) where the orgasm was centered; (8) whether there were any post-orgasm floating sensations or (9) any apnea (involuntary breathlessness); (10) sense of loss of self; (11) ejaculation; or (12) sensation akin to urination. One final question concerned (13) the length of time it took to bring about the orgasm.[65]

We were careful to build the question sequence such that there was no way a participant would know at this point that she would be given the opportunity to describe, subsequently, orgasms that differed from that initial one. No participant described more than four types, and by a large margin the most commonly reported number of types of orgasm was two. Remember, this meant that women were spontaneously offering this typology, because it was not one we were uncovering at this stage by statistical analysis or by imposing (or implying) a typology from the way we asked the questions. This is important.

We then wondered if that versatile statistical tool of latent class analysis would independently confirm what the women spontaneously reported—that orgasms could be meaningfully grouped into two distinct types. It did, and they could. Based on the sorts of things women were saying about location and sensation, we tentatively labeled these two types as "deep" and "surface," to get away from the "clitoral" and "vaginal" confusions.

That said, it was starting to look clearer why people had used such terms in the past. In general, *deep orgasms* as compared to *surface orgasms* were experienced as occurring deeper inside, often with sensations of internal pulsing, involuntary apnea, and subsequent feelings of pleasure, trust, and mental floating, even to the extent of losing a sense of self. These effects are all associated with the actions of oxytocin and prolactin, discussed in chapter seven, although it must be emphasized that we were not directly measuring those here.[66] Deep orgasms were also more likely than the other type of orgasm to be accompanied by ejaculation, or the sense that ejaculation might occur. This phenomenon is worth talking about separately, and I will do so later on in the chapter.

Having established meaningful differences between two broad types of female orgasm, the next step was to do some rough-and-ready validation: Specifically, to see if these types of orgasm identified were predictive of partner presence, penetrative sexual activity, and time taken to occur. If our functionality inspired predictions were to be supported, we expected the first two events to predict deep orgasms but not the last (time taken to bring about).

This sort of thing is called *discriminative validity* in behavioral science, and it is a clue that the measures you have constructed are representing something real that underlies them. Each of the predictions was statistically significantly associated with the type of orgasm in the ways we predicted, e.g., penetrative, partnered sex was much more likely to bring about a deep orgasm, but time was not a factor here.

A further thing we found was that both types of orgasm could occur in a range of sexual positions. This flew in the face of the predictions of the Masters and Johnson model, whereby female orgasm can only be brought about by friction with the clitoral glans.[67] Positions such as rear entry (known colloquially as "doggy style") were perfectly able to bring about deep orgasms, without additional stimulation. I need to stop here and remind readers that you are not reading a sexual "how-to" manual. That said, it would seem that there is good scientific warrant for the common practice, endorsed by all good sex therapists, for partners trying a range of different sexual positions. Indeed, it could be argued that this very willingness to experiment and play, paying close attention to partner preferences, is a large part of what mutually satisfying sexual intercourse is for.

A further point worth emphasizing is that it would be a mistake to conclude from our findings that some orgasms (i.e., deep ones) were regarded as universally better than others (i.e., surface ones) on all measures. Surface ones were often reported as being more relaxing afterwards, as well as being more intense and localized during the orgasm itself. Some—notably some neo-Freudian scholars—have argued that one type of female orgasm is psychologically (or perhaps even physically) healthier, or otherwise better than another.[68] We found no warrant for that claim, but it does appear that there are a discernible set of differences.[69]

From this discussion it might be argued, reasonably enough, that dividing the female experience into two types of orgasm is needlessly reductive and simplistic. We would agree wholeheartedly! In respect of that, the obvious reply would be that what we were uncovering here is a range of interesting and inter-connected functions, many of which have no known male correlate, some of which are detectable by the experiencer, and—most crucially—some of which seem to have clear, or at least plausible, connections to reproductively salient functions. That we were describing all this in simplistic typological terms should not blind us to those facts.

8.5 SOME IMPLICATIONS AND PREDICTIONS OF THE COMPLEXITY OF FEMALE ORGASM

The balance of these findings is that women routinely, though not universally, report that coital and masturbatory orgasms are different, and in repeatable ways. In the last few years some have worried about the so-called replication crisis in psychology while others (and I include myself among this latter group) see these so-called failures as normal science proceeding exactly as it should.[70] Whatever the truth of this, it is gratifying to note that a lot of the findings here, in terms of reports of orgasmic experience, are replicating (and extending) findings from a diverse range of fields stretching back decades. This sort of triangulation, findings using diverse methods all pointing in similar directions, is a good clue that one is on the right track, especially in a field as noisy as behavioral science.

Furthermore, the phenomenology of some of these orgasms imply some reproductive effects (and I have not even discussed the role of them producing rewarding affiliative behaviors yet). That is interesting in and of itself, but it also fits nicely with some other findings that show that female sexual complexity is doing its own, rather different set, of jobs than the male sexual response. To start with, if there are genuinely distinct types of female orgasm, does that not imply that there are different neurological pathways in females than in males? In addition, are there events during sexual activity in women only that seem to have very different roles than seemingly homologous events in males? Both predictions turn out to be true. First, women (but not men) who have undergone complete spinal cord transection can still experience a type of orgasm. Second, the existence of the female homologue to the prostate gland—the Skene's or paraurethral gland—shows interesting activity that may have reproductive consequences. I shall address each of these distinctively female phenomena in turn.

8.5.1   SPINAL CORD PATIENTS

Through a variety of mishaps, some women have had their spinal cords severed. If these unfortunate individuals had been men, then the pathways leading from the penis to the brain would also have been severed, leading to a severely reduced sexual response. However, interestingly (and surprisingly), there exist a variety of neural pathways in the vaginal area, including deep inside as far as the cervix, which can also trigger female orgasm, quite independent of the clitoral pathway.[71] As we saw in chapter three, far from its being the case that women's brains are pale reflections of men's brains, women have their own complex, region-specific reward pathways linked to their own complex reproductive systems. Barry Komisaruk and Beverly Whipple's work, especially using MRIs to scan brain regions during orgasm, have extended and highlighted these facts.[72]

Once again, it was by initially listening to patients, and taking their self-report seriously, that these discoveries were made. The team wanted to "validate the experiences of certain groups of women who report that they feel orgasms . . . despite their health professionals denying the possibility due to the women's neurological condition, for example, complete spinal cord injury." Following on from earlier reports of female orgasm from spinal injury patients, Komisaruk, Whipple, and their team studied women who had suffered complete spinal cord injury at the tenth thoracic vertebra (T10) or higher. This means that the nerve pathways from the clitoris were completely disconnected from the brain. Despite this, vaginal-cervical mechanical self-stimulation (VCSS) resulted in reports of orgasm. These self-reports were backed up by fMRI scans, which showed that various areas of the brain were activated during the women's orgasm reports:

> Brain regions activated during orgasm included the hypothalamic paraventricular nucleus, amygdala, accumbens-bed nucleus of the striaterminalis-preoptic area, hippocampus, basal ganglia (especially putamen), cerebellum, and anterior cingulate, insular, parietal and frontal cortices, and lower brainstem (central gray, mesencephalic reticular formation, and NTS.[73]

Of particular significance was the fact that a region of the medulla oblongata, to which the vagus nerves project (the Nucleus of the Solitary Tract or NTS) is activated by VCSS: "We conclude that the vagus nerves provide a spinal cord-bypass pathway for vaginal-cervical sensibility and that activation of this pathway can produce analgesia and orgasm."[74]

In addition to their vindication of the experiences of their patients in the face of skepticism from the scholarly neurological community, Komisaruk and Whipple found evidence to distinguish orgasms that involve the so-called G-spot from either clitorally based or cervically stimulated ones. They even found evidence of females who could create orgasmic experience merely by appropriate visualization in the absence of physical stimulation. Visualization is something we will revisit in later chapters.[75]

8.5.2   EJACULATION, G-SPOTS, AND THEIR RELATIONSHIPS TO FEMALE ORGASM

In the 1950s Gräfenberg documented a particularly sensitive area in the anterior wall of the vagina, which has subsequently—and perhaps confusingly—become known as the *G-spot*.[76] One of the most significant popularizers of the discovery of this region (and this naming) was the 1982 book by Alice Ladas, Beverley Whipple, and John Perry.[77] What they described then was an area, not really a *spot* at all, but a region involving clitoral nerves and the paraurethral (or *Skene's*) gland. This gland, the homologue of the male prostate gland, is to be found in most women about an inch to two inches up inside the vagina, on the anterior wall.

Although one ultrasound report, over a decade ago, failed to find this gland,[78] there is really little debate about its existence among anatomists.[79] The existence of a distinct gland was confirmed by the *Federative International Committee on Anatomical Terminology* 2001, and detailed in their journal, *Histology Today*. The leading anatomist/pathologist Milan Zaviačič has shown that the human version of this gland is, on average, 5.2 g in weight and contains up to forty-eight separate ducts, compared to the male prostate gland measurements of 23.7 g (average) and two ducts.[80] The male prostate is back towards the bladder, and both the male and the female prostate secrete prostate-specific antigens, as well as zinc and sugar.[81]

The existence of a specifically female mammal prostate gland was noted in Europe in the seventeenth century, but the female ejaculate it produces was prized by some cultures well before that—as detailed following.[82] This structure is not unique to humans, but found in other mammals too. For example, female rodent paraurethral glands have been extensively studied and compared in structure to those of our species.[83] They are similarly complex and (apparently) functional. Some researchers have drawn attention to the likely fitness benefits in the form of prostatic fluid that the Skene's gland secretes during sex that could aid sperm survival. This prostatic fluid—female human ejaculate—has been studied in the lab for some time and, more recently, the various functions and immune-system roles of the prostatic fluid have recently started to be investigated directly in humans and in other animals.[84]

Why are these fascinating and well-established facts not more widely known? One reason seems to be that the G-spot has sometimes been confused with the clitoris, which (as we saw in chapter six) is itself a structure that has generated frequent confusion. Although these above-mentioned researchers have been at pains to emphasize that a large number of areas of female genitalia (and beyond) can be erotically charged, it seems that the mystique of some kind of magic button for producing female orgasm is something that always attracts interest. To be blunt, men—especially those lacking in sexual experience—are likely to be motivated to seek a quick route to generate orgasm in their partners. Later I will argue that this very enthusiasm is evidence that female orgasm appears to be fitness related and part of a complex, between-the-sexes signaling system. Alas, such overeager male partners are likely to be disappointed. For just the same reason (as noted in chapter two) that there are no known male behaviors that universally work to grab female sexual attention, women are very likely to present a moving target here as well. There are not likely to be shortcuts, for, if there were, they could be subverted.

That said, knowledge does count for something. Stimulation of the paraurethral region is capable of contributing to female orgasmic experience in ways quite distinct in feeling from other types of female orgasm. In addition, there is the potential to produce a measurable and easily detectable physical effect, namely female ejaculate.[85] This is liquid expelled from the female urogenital system, often accompanied by pleasant feelings of release, and often attendant on other orgasmic feelings. It was characteristic to (although not identical with) feelings of deep orgasm in our (2012) study.

Some have suggested—with scant evidence—that female ejaculate is urine.[86] Indeed, in 2014, the UK government banned its portrayal in pornographic films on precisely these grounds.[87] A popular science site also recently trumpeted this incorrect notion. They claimed that out of seven participants brought to (ejaculatory) orgasm in the lab, two had chemicals present similar to those in urine, and that this, according to some over-enthusiastic pop science blogs, constituted "Proof that Female Ejaculation is Just Pee." Such conclusions are premature.[88] In all these seven cases, the liquid had come from the bladder and refilled before orgasm, despite all the women having previously emptied their bladders prior to the study. Therefore, it could not be the case that, even as reported, ejaculation was simple urinary incontinence. In addition, five of the seven women were reported as producing prostatic-specific antigen, a substance not normally present in urine, along with the other liquid. They are separate systems.[89]

People who are enthusiastic about female ejaculation (and plenty such exist[90]) are not, by that token, necessarily aroused by female urination. Evidence for such enthusiasts goes back some considerable years, too. For example, those previously mentioned 1,000-year-old statues at Karnataka,[91] sixteenth-century Japanese woodcuts,[92] and accounts from Margaret Mead among the Polynesians[93] all record the production, valuing, storing, and sometimes ingestion of female ejaculate. Classical Daoism refers to it as the "third water."[94] There even exist bowls from the Japanese EDO period made specifically to collect female ejaculate, dating back to the sixteenth century. Modern enthusiasts report a sweet-tasting substance, without any particular connection to urine, and investigation confirms that female ejaculate does indeed contain fructose.[95]

One final aspect should make us realize that a simplistic equating of ejaculate with urine is misguided. There is a second component to some manifestations of it—a milky-white substance that emerges from the Skene's gland, that has absolutely no connection with the urinary tract whatsoever. What does this substance do? At present we do not know, although some scholars think that the combination of chemicals just referred to likely can aid sperm survival, and thereby promote fertility.[96] Clearly, as the saying goes, further study is warranted. But even the ejaculate that does contain liquid from the bladder—as does urine—has a different chemical composition to urine. A big clue to this is that female ejaculate (unlike urine) is odorless, although some people report a sugary taste.

For someone to move from "the liquid in ejaculate is produced from the bladder" to "because the liquid in urine comes from the bladder, they must be the same liquid" is to make a very elementary error. Other chemicals (including urea, of course) are added to water to make it urine. If someone were to conclude, on the basis that orange juice is 90% water, and that beer is 90% water that, thereby, orange juice is the same as beer, then the error would be obvious. Not only is this school-child error misleading, it has consequences. One thing humans, especially female humans, do not need is any more reason to feel self-conscious during sex, and telling women that their ejaculatory orgasms are really urinary incontinence is not a helpful contribution to the sum of human happiness.

This is how Beverly Whipple and her team (and they are not alone) have described this distinct, milky white fluid that can emerge directly from the paraurethral gland during orgasm:

> Specifically, the ejaculate contains high levels of prostatic acid phosphatase, prostatic specific antigen, glucose, and fructose, but low levels of urea and creatinine. The chemical composition of healthy urine is the opposite of female ejaculate, so they are easily differentiated from each other. That is, healthy urine contains high levels of urea and creatinine and no prostatic acid phosphatase or glucose.[97]

In our 2012 study, many women reported ejaculatory experiences along with their (deep) orgasms. Some described the feeling as being, initially, somewhat like urination, but easily distinguishable from it once both experiences were compared. Might the chemical constitution of female ejaculate contribute to making the reproductive tract less hostile to sperm?[98] At the moment we do not know. However, it is well worth investigating. It is an experience that persistently occurs in a sizable minority of women and, furthermore, is regarded as important by them. It would be a shame if squeamishness or misinformation hindered both personal enjoyment and scientific advance.[99]

8.6 CONCLUSIONS

Without prompting, women frequently, though not universally, describe their orgasms as being different in type, and those differences are backed up by carefully consulting the subjective measures and lining these up with objective ones. One type of female orgasm seems to be specifically generated by deep penetrative sexual activity, with effects that are consistent with the known effects of several hormones, but particularly with the effects of oxytocin. Even when orgasm is generated by other action, penetration is often desired in addition, prompting the thought that a sperm selection mechanism is at play. Given that we know that oxytocin generates a peristaltic pump action for some considerable time following orgasm, these actions are fully consistent with one another and what we know of other mammals. Although the range of sexual help from therapists, books, and videos is considerable, there is still considerable coyness as regards female sexual pleasure. For a recent example, consider the Osé™ personal massager. This was chosen to be the prestigious CES 2019 Innovations Awardee. Interestingly, having won this award, the inventors were not allowed to showcase it at the annual Las Vegas–based trade fair. Those in charge said it was "immoral, obscene, indecent, profane, or not in keeping with CTAs image."[100] This is particularly noteworthy given that the year before, a male-focused sex doll had been featured prominently in the show. In a pleasing confirmation of our research into female orgasm typology, the chief engineer of the Osé™ project, whose robotics allows for penetration, described the resultant (female) orgasms as "blended ones."

The intriguing possibility has been raised that women's orgasmic variability is not a bug—it is a feature. Furthermore, it may be a feature that non-accidentally tracks characteristics and behaviors of sexual partners and the sexual circumstances themselves. The details of this are the subject of the next chapter. To answer this question, we are going to have to consider in detail what sex is for in the first place, and then how sex provides tests for with whom we are going to share our genes.

NOTES

1 Koedt, A. (1970). The myth of the vaginal orgasm. *Radical Feminism: A Documentary Reader*, 371–377. (Re-issue of 1968 paper).

2 Lessing, D. (1962). *The golden notebook*. Simon & Schuster.

3 For as accessible a discussion of quantum mechanics as is possible without delving too much into the mathematics, see Gell-Mann, M. (1995). *The Quark and the Jaguar: Adventures in the simple and the complex*. Macmillan.

4 Singer, I. (1972). Anti-climax: Reviews of *The Nature and Evolution of Female Sexuality*, by Mary Jane Sherfey, and *The Female Orgasm: Psychology, Physiology, Fantasy*, by Seymour Fisher. New York Review of Books (November 30th 1972 Issue). www.nybooks.com/articles/1972/11/30/anti-climax/ last accessed 15/04/2019.

5 I do not think that this thought is original to me, but a diligent search has returned no other origin. If someone recalls who said it first, then please share this with me. If it was me, then I am happy to stand by it.

6 For the first person to use the metaphor of evolution by natural selection as a Universal Acid that dissolves problems, see Dennett, D. C. (1995). Darwin's dangerous idea. *The Sciences*, *35*(3), 34–40.

7 It should be clear by now that what he was referring to is what we now know to be the clitoral *glans*.

8 Freud, S. (1953). Three essays on the theory of sexuality (1905). In *The standard edition of the complete psychological works of Sigmund Freud, volume VII (1901–1905): A case of hysteria, three essays on sexuality and other works* (pp. 123–246).

9 King, R., & Belsky, J. (2012). A typological approach to testing the evolutionary functions of human female orgasm. *Archives of Sexual Behavior*, *41*(5), 1145–1160.

10 The advice to not allow oneself to get stressed during sex is assuredly good sex advice, by the way, but it's unhelpful unless accompanied by practical input. The research of some fertility clinics also supports the idea that mood and stress are important factors in fertility. Some are starting to openly explore the impact of stress and artificiality on insemination rates. This is an important topic, and I will deal with it further towards the end of the book.

11 For a discussion of stable female-female pairings in gulls, see Pierotti, R., Annett, C. A., & Hand, J. L. (1997). Male and female perceptions of pair-bond dynamics: Monogamy in western gulls, larus occidentalis. In *Feminism and evolutionary biology* (pp. 261–275). Springer.

   For a discussion of female-female pair bonds in bonobos, see Parish, A. R. (1996). Female relationships in bonobos (Pan paniscus). *Human Nature*, *7*(1), 61–96.

   For a general discussion of how an appreciation of female human sexual fluidity has fed back into ethology, see Radtke, S. (2013). Sexual fluidity in women: How feminist research influenced evolutionary studies of same-sex behavior. *Journal of Social, Evolutionary, and Cultural Psychology*, *7*(4), 336.

12 At least they have had to up until very recently. This has already changed and may change even further in the future.

13 Whiting, J. W., & Whiting, B. B. (1975). Aloofness and intimacy of husbands and wives. *Ethos*, *3*(2), 183–207. See also Luoto, S., Krams, I., & Rantala, M. J. (2019). A life history approach to the female sexual orientation spectrum: Evolution, development, causal mechanisms, and health. *Archives of Sexual Behavior*, *48*(5), 1273–1308.

14 In her highly influential book, Elizabeth Lloyd documented the various rates of female orgasm generated through vaginal intercourse by way of various surveys. She did not go as far as Koedt in saying that such orgasms were impossible, only that they were highly unlikely. The earliest of the surveys she documented was done in 1921, the most recent 1994. Over thirty-five studies were discussed, with respondents mostly from the United Kingdom and the United States. Sample sizes ranged from a low of thirty-twoi to a high of 100,000, with some surveys (from popular magazines) not stating a sample

size at all. How many women orgasm during penetrative intercourse? The evidence seems to be that the answer to this question depends very much on how it is asked. That is what the rest of this chapter is about. Lloyd, E. (2005). *The case of the female orgasm: Bias in evolutionary science.* Harvard University Press.

15 Brody, S., & Costa, R. M. (2008). Vaginal orgasm is associated with less use of immature psychological defense mechanisms. *The Journal of Sexual Medicine*, *5*(5), 1167–1176.

16 For a history of framing female inorgasmia as frigidity, see Angel, K. (2010). The history of 'female sexual dysfunction' as a mental disorder in the twentieth century. *Current Opinion in Psychiatry*, *23*(6), 536. She argues that the concept of frigidity as regards vaginal orgasm goes beyond what Freud said in his writings. A detailed exegesis is not part of this book, but the point is well made.

17 This may not seem to be the place for an extended discussion of scientific method, but given that being patronizing to self-report has become something of an article of faith among some self-consciously scientific commentators, something needs to be said in its defense. For an extended discussion of how any procedure or assumption in science can be subject to criticism and skepticism, essentially becoming the independent variable in another study while the dependent variable becomes some measure of validity of the data, see Quine, W. V. O. (1953). *From a logical point of view.* Harvard University Press.

18 Haeffel, G. J., & Howard, G. S. (2010). Self-report: Psychology's four-letter word. *American Journal of Psychology*, *123*(2), 181–188.

19 See, for example Trivers, R. (2011). *Deceit and self-deception: Fooling yourself the better to fool others.* Penguin. For an earlier discussion of the same defences from a clinical, rather than an evolutionary perspective, see Freud, A. (1992). *The ego and the mechanisms of defence.* Karnac Books.

20 A similar point is made in Nagel, T. (1989). *The view from nowhere.* Oxford University Press.

21 Nisbett, R. E., & Wilson, T. D. (1977). Telling more than we can know: Verbal reports on mental processes. *Psychological Review*, *84*(3), 231.

22 Haidt, J. (2001). The emotional dog and its rational tail: A social intuitionist approach to moral judgment. *Psychological Review*, *108*(4), 814.

23 A classic example of motives being used in the service of self-image rather than more objective features was shown by Festinger, L. (1962). Cognitive dissonance. *Scientific American*, *207*(4), 93–106. Festinger showed that people given a financial motive to lie were actually *more* truthful with themselves than those who were not given such an extrinsic motive. This surprising finding is not, despite its frequent misrepresentation in popular culture, an example of believing inconsistent things, but rather it was a demonstration of how there is an internal drive to achieve narrative consistency concerning motives.

24 Synesthesia is cross-modal experiencing of sensations such as hearing colors or seeing sounds. Phantom limb is the experience of sensation in an amputated limb. Capgras' delusion is a condition where the sufferer believes their loved ones to have been replaced by look-alikes such as aliens or robots. See Ramachandran, V. S., Blakeslee, S., & Shah, N. (1998). *Phantoms in the brain: Probing the mysteries of the human mind* (p. 224). William Morrow, for an accessible discussion of these and other neurological mysteries.

25 Haeffel, G. J., Gibb, B. E., Metalsky, G. I., Alloy, L. B., Abramson, L. Y., Hankin, B. L., . . . Swendsen, J. D. (2008). Measuring cognitive vulnerability to depression: Development and validation of the cognitive style questionnaire. *Clinical Psychology Review*, *28*(5), 824–836.

26 Howell, C. J. (1999). Epidural versus non-epidural analgesia for pain relief in labour. *The Cochrane Database of Systematic Reviews*, *3*.

27 It is very easy to confuse accuracy with objectivity, but they are not the same thing at all. Take this example that the philosopher of science James Ladyman shared with me (pc 11/4/2018).

    The sociologist Douglas Porpora (Porpora, D. V. (2015). *Reconstructing sociology: The critical realist approach.* Cambridge University Press, p. 14.) gave the example of the following three statements:

    1) In WWII 6 million Jews *died*.

    2) In WWII 6 million Jews *were killed*.

    3) In WW II 6 million Jews *were murdered*.

    Ladyman contends (and I agree with him) that the first has the semblance of objectivity, but the latter descriptions actually contain more genuine information. For example, the first statement could be consistent with these Jewish victims dying by accident. George Orwell makes similar points about how so-called objectivity can obscure salient facts via political speech in Orwell, G. (2002). Politics and the English language. *British Army Review-London-Ministry of Defence*, 52–56.

28 Dennett, D. (2003). Who's on first? Heterophenomenology explained. *Journal of Consciousness Studies*, *10*(9–10), 19–30.

29 With no awareness of irony, I have been occasionally lectured by sober scientists (who presumably can rely on their own internal mental states about this issue) to the effect that reporting on internal mental states is inherently unreliable. As that curmudgeonly diagnostician House would have it, "Everybody lies." Really? Including him? Did he lie when he was telling us that? Everybody can lie, but do they really and reliably lie all the time? That cannot be true else lying would no longer exist.

    The Australian philosopher D. C. Stove coined the term "The Ishmael Effect" to refer to philosophical positions that help themselves to an unwarranted escape route from horrors that they condemn all other thinking to. Stove, D. C. (1991). *The Plato cult and other philosophical follies.* Blackwell Publishers.

30 Brisson, L. (1976). *Le mythe de Tirésias: essai d'analyse structurale.* Brill.

31 Although something like one-third of people, when asked, say that they would like to be reincarnated as the opposite sex if they could retain memories. Byers, E. S., Goldsmith, K. M., & Miller, A. (2016). If given the choice, would you choose to be a man or a woman? *The Canadian Journal of Human Sexuality, 25*(2), 148–157.

32 Some representative sex therapists who talk about female orgasmic variability would include Fisher, S. (1973). *The female orgasm: Psychology, physiology, fantasy.* Basic Books; and Sundahl, D. (2003). *Female ejaculation and the G-spot.* Hunter House.

33 See Hooven, C. (2021). *T: The story of testosterone, the hormone that dominates and divides us.* Henry Holt and Company, for numerous case studies reporting this.

34 Vance, E. B., & Wagner, N. N. (1976). Written descriptions of orgasm: A study of sex differences. *Archives of Sexual Behavior, 5*(1), 87–98. Is the study using descriptions of orgasm. One factor that did stand out was the possibility of multiple orgasms in the female (but not the male) respondents.

35 Cronin, H. (1993). *The ant and the peacock: Altruism and sexual selection from Darwin to today.* Cambridge University Press.

36 There have been some recent attempts to downplay this variance across species, for example, Harrison, L. M., Noble, D. W., & Jennions, M. D. (2022). A meta-analysis of sex differences in animal personality: No evidence for the greater male variability hypothesis. *Biological Reviews, 97*(2), 679–707. This resulted in some, frankly, sensationalist headlines, which (perhaps unconsciously) revealed an underlying motive for such alleged "debunkings." For example, one press release was titled "Sexist sexplanation for men's brilliance debunked" and concentrated on male IQ variance, compared to female. Alas for such sensationalism, the original article had looked only at *non-human animal personality variance* and (one hopes this is needless to point out) did not involve administering IQ tests to halibuts. As pointed out in this chapter, sex research and sex-differences research are bedeviled with this sort of recursive problem: Sex researchers being themselves (sexual) organisms involved in the various sex wars and, somehow, unable to stop their own desires intruding into their findings. See Del Giudice, M., & Gangestad, S. W. (2023). No evidence against the greater male variability hypothesis: A commentary on Harrison et al.'s (2022) meta-analysis of animal personality. *Evolutionary Psychological Science,* 1–8, for a detailed technical commentary on why the original meta-analysis does not show what the authors wish that it did.

37 Wilder, J. A., Mobasher, Z., & Hammer, M. F. (2004). Genetic evidence for unequal effective population sizes of human females and males. *Molecular Biology and Evolution, 21*(11), 2047–2057.

38 Greer, G. (2014). Interview: *The Rebels of Oz: Germaine, Clive, Barry and Bob* [TV Programme]. *BBC Four.* Aired 1st July 2014, Serendipity Productions.

39 The quote comes from Levin, R. (1981). The female orgasm-a current appraisal. *Journal of Psychosomatic Research,* 121.

40 The noted physiologist Roy Levin has repeatedly drawn attention to the range of reported female orgasmic experience, e.g., Levin, R. (2001). Sexual desire and the deconstruction and reconstruction of the human female sexual response model of Masters and Johnson. In W. Everard, E. Laan, & S. Both (Eds.), *Sexual appetite, desire and motivation: Energetics of the sexual system* (pp. 63–93). Royal Netherlands Academy of Arts and Sciences; Levin, R. J. (2004). An orgasm is . . . who defines what an orgasm is? *Sexual and Relationship Therapy, 19*(1), 101–107; Surprisingly he has not drawn the obvious conclusion from this: that orgasms generated by laboratory-based masturbation (e.g., Levin, R. J., & Wagner, G. (1985). Orgasm in women in the laboratory—quantitative studies on duration, intensity, latency, and vaginal blood flow. *Archives of Sexual Behavior, 14*(5), 439–449.) might not capture the full range of the phenomenon.

41 Warner, J. (1998). Peak of sexual response questionnaire. In *Handbook of sexuality-related measures* (p. 256). SAGE.

42 Mah, K., & Binik, Y. M. (2002). Do all orgasms feel alike? Evaluating a two-dimensional model of the orgasm experience across gender and sexual context. *Journal of Sex Research, 39*(2), 104–113.

See also, Bensman, L., Hatfield, E., & Doumas, L. A. (2011). *Two people just make it better: The psychological differences between partnered orgasms and solitary ogasms* [Doctoral dissertation, University of Hawai'i at Mānoa].

43 Lavie, M., & Willig, C. (2005). "I don't feel like melting butter": An interpretative phenomenological analysis of the experience of Inorgasmia. *Psychology & Health*, *20*(1), 115–128.

44 For the book that founded the field of evolutionary medicine, see Nesse, R. M., & Williams, G. C. (2012). *Why we get sick: The new science of Darwinian medicine*. Vintage. For some of the founding ideas, see Nesse, R. M., Williams, G. C., & Mysterud, I. (1995). Why we get sick. *Trends in Ecology and Evolution*, *10*(7), 300–301.

45 Nesse, R. M. (2019). *Good reasons for bad feelings: Insights from the frontier of evolutionary psychiatry*. Dutton.

46 MacHale, D., & Cohen, Y. (2018). *New light on George Boole*. Cork University Press.

47 Kaplan, H. (1974). *The new sex therapy: Active treatment of sexual dysfunctions*. Brunner-Routledge.

48 Some self-help books include Chalker, R. (2000). *The clitoral truth: The secret world at your fingertips*. Seven Stories Press; and Sundahl, D. (2003). *Female ejaculation and the G-spot*. Hunter House.
Barbach, L. G. (1975). *For yourself: The fulfilment of female sexuality*. Doubleday.
There is also the excellent website, OMGYES.com-The Science of Women's Pleasure https://start.omgyes.com/join last accessed 14/08/2023, which contains masses of excellent technical and anecdotal information.

49 For the effects of pharmacology on orgasm, see Jani, N. N., & Wise, T. N. (1988). Antidepressants and inhibited female orgasm: A literature review. *Journal of Sex & Marital Therapy*, *14*(4), 279–284; and Buffum, J. (1986). Pharmacosexology update: Prescription drugs and sexual function. *Journal of Psychoactive Drugs*, *18*(2), 97–106.

50 Nims, J. P. (1975). Imagery, shaping, and orgasm. *Journal of Sex & Marital Therapy*, *1*(3), 198–203.

51 Sex researchers who have emphasized a typology of female orgasm would include Hite, S. (1979). The Hite report. *Journal of School Health*, *49*(5), 251–254; Bentler, P. M., & Peeler, W. H. (1979). Models of female orgasm. *Archives of Sexual Behavior*, *8*, 405–423; Singer, J., & Singer, I. (1972). Types of female orgasm. *Journal of Sex Research*, *8*, 255–267.

52 Butler, C. A. (1976). New data about female sexual response. *Journal of Sex & Marital Therapy*, *2*(1), 40–46.

53 Fisher, S. (1973). *Understanding the female orgasm*. Basic Books.

54 The quote is from Singer and Singer (1972, p. 259).

55 This has been hypothesized to stimulate the Ferguson reflex, a release of oxytocin. Ferguson, J. K. (1941). *A study of the motility of the intact uterus at term. Surgical Gynecology and Obstetrics*, *73*, 359–366.

56 Robertiello, R. C. (1970). The "clitoral versus vaginal orgasm" controversy and some of its ramifications. *Journal of Sex Research*, *6*(4), 307–311.

57 Mah, K., & Binik, Y. M. (2002). Do all orgasms feel alike? Evaluating a two-dimensional model of the orgasm experience across gender and sexual context. *Journal of Sex Research*, *39*(2), 104–113. More recent research has confirmed that partnered orgasms tend to feel different. See, for example,
Bensman, L., Hatfield, E., & Doumas, L. A. (2011). *Two people just make it better: The psychological differences between partnered orgasms and solitary orgasms* [Doctoral dissertation, University of Hawai'i at Mānoa].

58 See, for example, Cattell, R. B. (1946). *Description and measurement of personality*.

59 "Reliable" means "produces the same measured outputs for the same inputs." To be contrasted with "valid," which means "measuring what we hope to measure." Measures can be reliable, valid, both, or neither.

60 King, R., Belsky, J., Mah, K., & Binik, Y. (2011). Are there different types of female orgasm? *Archives of Sexual Behavior*, *40*(5), 865–875.

61 There are four types of data: nominal (yes/no), ordinal (first, second, third . . .), interval (1, 2, 3), and ratio (like interval data but with a meaningful zero). It is arguable whether behavioral science ever produces the latter type. However, all these data can be subjected to statistical analysis. Whatever you may have been told, numbers do not know that they are ordinal.

62 We are not the only people to note that even partnered orgasms may not be as good as those achieved through masturbation. See, for example, Chadwick, S. B., Francisco, M., & van Anders, S. M. (2019). When orgasms do not equal pleasure: Accounts of "bad" orgasm experiences during consensual sexual encounters. *Archives of Sexual Behavior*, 1–25.

63 For more on how the internet has greatly expanded our capacity to do sex research (with important safeguards and caveats), see Ochs, E. P., Mah, K., & Binik, Y. M. (2002). Obtaining data about human sexual functioning from the Internet. In A. Cooper (Ed.), *Sex and the Internet: A guidebook for clinicians* (pp. 245–262). Brunner-Routledge.

64 King, R., & Belsky, J. (2012). A typological approach to testing the evolutionary functions of human female orgasm. *Archives of Sexual Behavior*, *41*(5), 1145–1160.

65 We also asked about sexual practices. These were reported as: (1) clitoral stimulation (self), (2) manual clitoral stimulation (partner), (3) vaginal stimulation (self), (4) vaginal stimulation (partner), (5) clitoral stimulation (external vibrator), (6) vaginal stimulation via dildo/vibrator, (7) oral stimulation, (8) anal penetration, (9) breast stimulation, and (10) talking dirty. Also, various (mutually exclusive) sexual positions were endorsed for the sex with a partner condition at time of orgasm: (11) missionary, (12) missionary with legs raised, (13) missionary with legs bent back over head, (14) doggy style, (15) cowboy (woman on top), and (15) reverse cowboy (woman on top facing towards partner's feet). There was also a write-in option for other positions, which returned a couple of interesting replies. By far the most common accompanying sexual behavior was stimulation of the breasts. Vaginal self-stimulation was the least common activity, but this increased considerably when partner action and vibrator/dildo use were factored in.

66 For more on the effects of oxytocin, see Insel, T. R. (2010). The challenge of translation in social neuroscience: A review of oxytocin, vasopressin, and affiliative behavior. *Neuron*, *65*(6), 768–779. For more on the effects of prolactin following partnered orgasm see Krüger, T. H., Haake, P., Hartmann, U., Schedlowski, M., & Exton, M. S. (2002). Orgasm-induced prolactin secretion: Feedback control of sexual drive? *Neuroscience & Biobehavioral Reviews*, *26*(1), 31–44.

67 Masters, W. H., & Johnson, V. E. (1966). *Human sexual response* (p. 185). Little, Brown & Co.
       What remain of Masters and Johnsons data: www.kinseyinstitute.org/newsletter/fall2011/masters-johnson.html.

68 Those who have argued that some types of orgasm are inherently better than others include Stuart Brody. For example, see Brody, S., & Costa, R. M. (2008). Vaginal orgasm is associated with less use of immature psychological defense mechanisms. *The Journal of Sexual Medicine*, *5*(5), 1167–1176. Brody is a neo-Freudian, trying to resurrect Freud's distinction of clitoral orgasms being immature compared to vaginal ones. See Jannini, E. A., Rubio-Casillas, A., Whipple, B., Buisson, O., Komisaruk, B. R., & Brody, S. (2012). Female orgasm (s): One, two, several. *The Journal of Sexual Medicine*, *9*(4), 956–965, for a review.

69 Some scholars, notably Cynthia Jayne, have argued that Freud has been mistranslated here. She argues that Freud was not asserting a typology but rather the supremacy of orgasms generated one way rather than another. Jayne, C. (1984). Freud, Grafenberg, and the neglected vagina: Thoughts concerning an historical omission in sexology. *Journal of Sex Research*, 212–215.
       Although we certainly found evidence for an orgasm typology, we do not think we have any reason to argue that one type is more physically or psychologically healthier than other. As far as we can tell, they are all good for us, however brought about. As far as we can tell, masturbation is healthy, both as a solitary and a joint activity. Partners of both sexes would be well-advised to pay close attention to how their partner masturbates and learn accordingly. The mutual sharing and knowledge of masturbation is not only good for the individual but also for couples. That said, if masturbation re-created *all* the same sensations and responses as sexual intercourse, then we would likely never do the latter. It would be far too much trouble. So, there is something else going on there. Not necessarily (always) better, but different.

70 See, for example, Maxwell, S. E., Lau, M. Y., & Howard, G. S. (2015). Is psychology suffering from a replication crisis? What does "failure to replicate" really mean? *American Psychologist*, *70*(6), 487.

71 Researchers have also pointed out that the Ferguson reflex, the release of oxytocin through *vagino-cervical stimulation* (Ferguson, J. K. (1941). A study of the motility of the intact uterus at term. *Surgical Gynecology and Obstetrics*, *73*, 359–366), is a key feature of sexual intercourse in mammals (Komisaruk et al., 2004).

72 Key papers that highlight reports of orgasm from spinal cord patients include:
       Komisaruk, B. R., Whipple, B., Crawford, A., Grimes, S., Liu, W. C., Kalnin, A., & Mosier, K. (2004). Brain activation during vaginocervical self-stimulation and orgasm in women with complete spinal cord injury: fMRI evidence of mediation by the Vagus nerves. *Brain Research*, *1024*(1–2), 77–88.
       Komisaruk, B. R., Whipple, B., Gerdes, C. A., Harkness, B., & Keyes, J. W., Jr. (1997). Brainstem responses to cervical self-stimulation: Preliminary PET-scan analysis. *International Behavioral Neuroscience Society Annual Conference Abstract Book*, *6*, 38.
       Komisaruk, B. R., Sansone, G., (2003). Neural pathways mediating vaginal function: The vagus nerves and spinal cord oxytocin. *Scandinavian Journal of Psychology*, *44*, 241–250.
       Komisaruk, B. R., Gerdes, C. A., & Whipple, B. (1997). Complete spinal cord injury does not block perceptual responses to genital self-stimulation in women. *Archives of Neurology*, *54*(12), 1513–1520.

Cueva-Rolón, R., Sansone, G., Bianca, R., Gómez, L. E., Beyer, C., Whipple, B., & Komisaruk, B. R. (1996). Vagotomy blocks responses to vaginocervical stimulation after genitospinal neurectomy in rats. *Physiology & Behavior*, *60*(1), 19–24.

Komisaruk, B. R., Bianca, R., Sansone, G., Go, L. E., Cueva-Rolo, R., Beyer, C., & Whipple, B. (1996). Brain-mediated responses to vaginocervical stimulation in spinal cord-transected rats: Role of the vagus nerves. *Brain Research*, *708*(no. 1–2), 128–134.

73 As noted below, the "NTS" is the Nucleus of the Solitary Tract.

74 The quotes come from Komisaruk, B. R., & Whipple, B. (2005). Functional MRI of the brain during orgasm in women. *Annual Review of Sex Research*, *16*(1), 62–86, p. 62.

75 For more on orgasms generated by visualization alone, see Whipple, B., Ogden, G., & Komisaruk, B. R. (1992). Physiological correlates of imagery-induced orgasm in women. *Archives of Sexual Behavior*, *21*(2), 121–133.

76 Grafenberg, E. (1950). The role of the urethra in female orgasm. *International Journal of Sexology*, *3*, 145–148.

77 Ladas, A. K., Whipple, B., & Perry, J. D. (1982). *The G spot and other recent discoveries about human sexuality.* McDougal.

78 Despite an ultra-sound study (Gravina, G. L., Brandetti, F., Martini, P., Carosa, E., Di Stasi, S. M., Morano, S., . . . Jannini, E. A. (2008). Measurement of the thickness of the urethrovaginal space in women with or without vaginal orgasm. *The Journal of Sexual Medicine*, *5*(3), 610–618) that claimed that some females do not possess G-spots, but a more likely explanation is that the gland was insufficiently stimulated in these females and thus was inaccessible. It would indeed be strange if a gland whose existence is common knowledge among anatomists was missing in a significant number of women, and the lack of appropriate stimulation was not an issue addressed in the Gravina study. The researchers refer to the area as the "[H]uman clitoris-urethrovaginal complex also known as the G-spot" (Gravina et al., 2008, p. 610), but this is an unusual usage, and not in keeping with current anatomical practice. Certainly, the coiners of the term "G-spot" (Whipple et al., 2008) disagree with this usage. Whipple, B. (2008). *Interview with new scientist discussing the Gravina et al. (2008) study of the G-spot.* www.newscientist. com/article/mg19726444.100-ultrasound-nails-location-of-the-elusive-g-spot.html last accessed 20/2/2008.

Having said this, it is also possible that modern use of hormone contraceptive has had an effect on the female prostate, whose development is known to be linked to hormone homeostasis (Santos, F. C. A., & Taboga, S. R. (2006). Female prostate: A review about the biological repercussions of this gland in humans and rodents. *Animal Reproduction*, *3*(1), 3–18.)

79 Zaviačič, M., Jakubovská, V., Belošovič, M., & Breza, J. (2000). Ultrastructure of the normal adult human female prostate gland (Skene's gland). *Anatomy and Embryology*, *201*(1), 51–61.

Zaviačič, M., Zajíčková, M., Blažeková, J., Donárová, L., Stvrtina, S., Mikulecký, M., . . . Breza, J. (2000). Weight, size, macroanatomy, and histology of the normal prostate in the adult human female: A minireview. *Journal of Histotechnology*, *23*(1), 61–69.

80 Zaviačič, M., & Ablin, R. J. (1999). *The human female prostate: From vestigial Skene's paraurethral glands and ducts to woman's functional prostate.* Slovak Academic Press. This was reconfirmed by another team over a decade later. Ostrzenski, A., Krajewski, P., Ganjei-Azar, P., Wasiutynski, A. J., Scheinberg, M. N., Tarka, S., & Fudalej, M. (2014). Verification of the anatomy and newly discovered histology of the G-spot complex. *BJOG: An International Journal of Obstetrics & Gynaecology*, *121*(11), 1333–1340. A student of medical history could have a fine time explaining why (uniquely) a female piece of anatomy has to keep being "reconfirmed."

81 Addiego, F., Belzer, E. G., Jr., Comolli, J., Moger, W., Perry, J. D., & Whipple, B. (1981). Female ejaculation: A case study. *Journal of Sex Research*, *17*(1), 13–21.

82 De Graaf, R. (1672). *De mulierum organis generationi inservientibus tractatus novus: demonstrans tam homines & animalia caetera omnia, quae vivipara dicuntur, haud minus quàm ovipara ab ovo originem ducere.*

83 Shehata, R. (1974). Urethral glands in the wall of the female urethra of rats, mice and closely related rodents. *Cells Tissues Organs*, *90*(3), 381–387.

Shehata, R. (1975). Female prostate in Arvicanthis niloticus and Meriones libycus. *Cells Tissues Organs*, *92*(4), 513–523.

Gross, S. A., & Didio, L. J. (1987). Comparative morphology of the prostate in adult male and female Praomys (Mastomys) Natalensis studied with electron microscopy. *Journal of Submicroscopic Cytology*, *19*(1), 77–84.

Dos Santos, F. C. A., Carvalho, H. F., Góes, R. M., & Taboga, S. R. (2003). Structure, histochemistry, and ultrastructure of the epithelium and stroma in the gerbil (Meriones unguiculatus) female prostate. *Tissue and Cell*, *35*(6), 447–457.

Flamini, M. A., Barbeito, C. G., Gimeno, E. J., & Portiansky, E. L. (2002). Morphological character-ization of the female prostate (Skene's gland or paraurethral gland) of Lagostomus maximus maximus. *Annals of Anatomy-Anatomischer Anzeiger*, *184*(4), 341–345.

Dos Santos, F. C. A., & Taboga, S. R. (2006). Female prostate: A review about the biological repercus-sions of this gland in humans and rodents. *Animal Reproduction*, *3*(1), 3–18.

84 Dos Santos, F. C. A., Carvalho, H. F., Góes, R. M., & Taboga, S. R. (2003). Structure, histochemistry, and ultrastructure of the epithelium and stroma in the gerbil (Meriones unguiculatus) female prostate. *Tissue and Cell*, *35*(6), 447–457.

Custodio, A. M., Santos, F. C., Campos, S. G., Vilamaior, P. S. L., Oliveira, S. M., Goes, R. M., & Taboga, S. R. (2010). Disorders related with ageing in the gerbil female prostate (Skene's paraurethral glands). *International Journal of Experimental Pathology*, *91*(2), 132–143.

85 Perry, J. D., & Whipple, B. (1981). Pelvic muscle strength of female ejaculators: Evidence in support of a new theory of orgasm. *Journal of Sex Research*, *17*(1), 22–39.

Addiego, F., Belzer, E. G. Jr, Comolli, J., Moger, W., Perry, J. D., & Whipple, B. (1981). Female ejacu-lation: A case study. *Journal of Sex Research*, *17*(1), 13–21.

86 *Discover Magazine*. (non attributed). On popular reporting of the idea that female ejaculate is urine. http://blogs.discovermagazine.com/seriouslyscience/2018/08/22/7025/ last accessed 08/03/2019.

87 Electronic Regulations. (2014). [UK Statute Law]. 2014 legislation banning (among other things) por-trayals of urination, bondage and abusive language. www.legislation.gov.uk/uksi/2014/2916/pdfs/uksi_20142916_en.pdf last accessed 15/03/2019.

88 The original article is Salama, S., Boitrelle, F., Gauquelin, A., Malagrida, L., Thiounn, N., & Desvaux, P. (2015). Nature and origin of "squirting" in female sexuality. *The Journal of Sexual Medicine*, *12*(3), 661–666.

89 In addition, the opposite side of the brain region associated with urination is activated during ejacula-tion in women. Huynh, H. K., Willemsen, A. T., Lovick, T. A., & Holstege, G. (2013). Pontine control of ejaculation and female orgasm. *The Journal of Sexual Medicine*, *10*(12), 3038–3048.

90 For some discussions of this, see Cusack, C. M. (2012). Obscene squirting: If the government thinks it's urine, then they've got another thing coming. *Texas Journal of Women & Law*, *22*, 45. Interestingly, female ejaculation seems more common with asphyxiophilia but, as yet, the full story of what is occur-ring here is unknown. Zaviačič, M. (1994). Sexual asphyxiophilia (Koczwarism) in women and the biological phenomenon of female ejaculation. *Medical Hypotheses*, *42*(5), 318–322.

91 Anand, M. R. (1958). *Kama Kala: Some notes on the philosophical basis of Hindu erotic sculpture*. Skilton.

92 Uhlenbeck, C., & Winkel, M. (2005). *Japanese erotic fantasies: Sexual imagery of the Edo period*. Hotei Publishing.

93 Mead, M. (1963). *Sex and temperament in three primitive societies*. Morrow.

94 For a discussion, see Sundahl, D. (2003). *Female ejaculation and the G-spot*. Hunter House.

95 Whipple, B. (2015). Ejaculation, female. *The International Encyclopedia of Human Sexuality*, 1–4.

96 It has been claimed that stimulation of the G-spot is a technique well known to midwives in order to produce a lubricating fluid that aids birthing (Odent, 1999). The baby's birth might, under normal cir-cumstances, milk the female prostate, and thus we would all be born in a flood of ejaculate waters. This may well be linked to the Ferguson reflex (Ferguson, J. K. (1941). A study of the motility of the intact uterus at term. *Surgical Gynecology and Obstetrics*, *73*, 359–366), in which the pressure of the fetus on the cervix causes a release of oxytocin, which aids uterine peristalsis and thus aids parturition. See also Odent, M. (1999). *The scientification of love*. Free Association Books, on the link between orgasm and childbirth via oxytocin.

97 Whipple, B. (2015). Ejaculation, female. *The International Encyclopedia of Human Sexuality*, 1–4.

98 Zaviačič, M., Doležalová, S., Holomán, I. K., Zaviačičová, A., Mikulecký, M., & Brázdil, V. (1988). Concentrations of fructose in female ejaculate and urine: A comparative biochemical study. *Journal of Sex Research*, *24*(1), 319–325.

99 Achille, L., & Wilkinson, C. (2001). Submission to the BBFC: Female ejaculation: Research contrary to BBFC ruling and replies. *An open letter detailing objections to banning portrayals of female ejacula-tion on film. On behalf of Feminists against Censorship*. Retrieved August 16, 2010.

100 www.theguardian.com/technology/2019/jan/08/ces-dildo-gender-sex-toy-ose-personal-massager Article on the withdrawal of the Osé massager award last accessed 11/01/2019.

9 Picky, Picky, Picky

"The pleasure is momentary, the position ridiculous, and the expense damnable."

(Lord Chesterfield, in a letter of advice to his son, concerning the matter of sex)

9.1 WHAT IS SEX FOR?

Why do we have sex at all? I know what you are thinking: What could be more obvious? But, it is not obvious. Even the obvious (proximate) physical pleasures of partnered sex are not as reliable as are those of masturbation, and there are always risks. This riskiness is not unique to our species; sex is a biological problem to explain. When we take a gene's-eye perspective (and that is the only rational game in town), then there is a puzzle as to why organisms go to all the trouble and expense of passing on their genes that way. Maynard Smith calls this issue the "two-fold cost of males."[1] Not only are female genes halving their chances of being represented in the next generation, but they might not find a mate at all.

If you thought that online dating was a rough world, then imagine trying to find a compatible sexual partner at 6,000 meters deep in the inky blackness of the Pacific Ocean. Down there it is so dark that fully half the time a male giant squid ends up mating with other males, because he cannot discern the females at all.[2] However, despite these kinds of drawbacks, every complex creature (technically *eukaryote*), barring one rather interesting exception, reproduces sexually. They did not have to have done this. They could have cloned themselves instead.

Cloning themselves is how that one exception, the *bdelloid rotifers*, reproduce, and their story is not simply one of uniqueness. The evidence is that *bdelloid rotifers*—beautiful microscopic organisms that look a bit like elongated glass orchid vases—took to cloning themselves *after* having previously reproduced sexually. The evolutionary biologist Olivia Judson calls this situation so unusual as to constitute a "scandal."[3] That may be true, but it does not help us understand why all the other eukaryotes, at least some of the time, go to all the trouble and expense of finding a mate and splicing their genes with it. There are two viable contenders for explaining the puzzle of sexual reproduction: the Red Queen hypothesis and the mutation/selection hypothesis. To explore which of these is right, we need to briefly recap some of the implications of a species dividing itself in two in the first place.

All the way back in chapter three, I pointed out that sex determination can be incredibly complex and very varied between species. Some critters are XY heterogametic[4] (like us) while birds have ZO, ZW, or complex ZW determinants. There are literally dozens of ways to determine sex (and, thereby, sexual reproduction), and this should command our attention, for much the same reason as does the fact that eyes have emerged from evolution fifty independent times or more.[5] Being able to gain information from electromagnetic scatter, over a particular range common on this planet, is so incredibly useful that even to be able to do it a little bit has been selected for—multiple times—and in creatures that never shared a common ancestor that also had eyes. Eyes are, thereby, much less of a puzzle than they used to be. The advantage of sexual reproduction is, however, trickier to pin down.

Our best account is that sex all began a billion or so years ago, when one independent bacterium engulfed another. So far, so much the norm, in the primordial soup. This engulfing had happened countless billions of times before, but this time was different. This time, both entities somehow benefited from the engulfment and, crucially, could pass on that benefit. Our reason for thinking that this happened is that (modern) eukaryotic cells have two separate positive genomes in them: nuclear DNA and mitochondrial DNA.[6] So, sex in terms of splicing together of two entities happened, but

DOI: 10.1201/9781003372356-9

why did it persist? To put this another way, what adaptive benefit did sexual reproduction convey? As I said before, two answers have been proposed: mutation/selection and the Red Queen hypothesis.

In the mutation/selection picture, mutations in the DNA accumulate in every generation, mostly reducing fitness. As anyone with committee experience will confirm, most ways of changing something make it worse. However, occasionally, the change can be an improvement. Sexual reproduction increases the variance between parents and offspring, multiplying with each generation, and so selection weeds out these mutations. "Mutation proposes, selection disposes" is a slogan of evolutionary biology.

The Red Queen had to run to stay still in Lewis Carroll's *Through the Looking-Glass*.[7] From this character's peculiar mode of being, we get the name for the Red Queen hypothesis. This proposes that the ultimate purpose of sex is that of there being a direct benefit to offspring being different from their parents.[8] That direct benefit is to keep immune systems one step ahead of rapidly reproducing pathogens that are every organism's bitter enemy (and, usually, eventual nemesis).

Which of these is the right way to think of it? The mutation/selection hypothesis predicts a higher *variance in reproductive success* in sexual rather than asexual offspring.[9] The Red Queen hypothesis predicts that *mean reproductive success* will be higher for sexual rather than asexual offspring. Both theories agree that there is an advantage to be gained by an offspring's being somewhat different from its parents. We have evidence for both.

9.1.1 KEEPING ONE JUMP AHEAD

In an influential paper, a team of scholars led by Stuart West proposed that we do not have to choose between these options.[10] Both happen. Asexual reproduction is ubiquitous in single-celled organisms, while sexual reproduction is (nearly) ubiquitous in large, complex animals. Some organisms switch strategies depending on local conditions and in ways that predictably maximize fitness. What is the upshot of all of this in relation to the question that opened the chapter? It implies the following: All sexually reproducing species—and we are no exceptions—are, in William Hamilton's memorable formulation, "Guilds of genotypes committed to free, fair exchange of biochemical technology for parasite exclusion."[11]

Sounds romantic! However, we have to remember that genes cannot see these important qualities directly. Instead, genes build bodies to advertise these qualities, and then send them out into the world to compete for attention with other genetically designed vehicles.[12] We call the resultant feelings that these vehicles have been equipped with (among other things) "Love," "Desire," and "Arousal." A major theme of the book so far has been that women are not responding to these things capriciously, randomly, or passively. Now it is time to really explore how female orgasm makes sense as a response to patterns of appropriate displays of these key features in sexual partners. I will focus on male sexual partners for the moment, but keep in mind that female-female pairings are also adaptive under many conditions and also deserve attention.[13] Rather than being broken, or a poor imitation of male sexuality, the female sexual response system—much of which predates human existence—has been honed by tens of millions of years of evolution, to be a masterpiece of sensitive calibration and measurement, in a noisy and confusing world.[14]

9.2 FISHING FOR COMPLIMENTS?

The great biologist Amotz Zahavi once told me that if anything in nature seemed too ornate, confusing, or baroque, as needed to do its job, the answer was likely sexual selection.[15] And, as a quick glance at the bright colors, loud calls, and exotic dances, produced by male primates, birds, and insects across the planet will confirm, that usually means *female* sexual selection. But what are those females choosing, exactly? Another way of putting this is to ask: What do those male ornaments signify? Once again, there are two main schools of thought here.

The first suggestion, proposed by Ronald Fisher, is that females profit from copying the preferences of other females.[16] Assuming the (male) quality preferred is a heritable one, and that the preference for it can also be passed on, then any sons you have will be likewise sexy (to other females). This can lead to a runaway quality in a trait. This idea was recently championed by Richard Prum who, somewhat surprisingly, called it "Darwin's Forgotten Theory."[17] I can assure readers that, as you will see from this chapter, female choice mechanisms are far from being forgotten in mainstream biology, or ethology. The question is as to what function such choices perform. Fisher's *Sexy Sons hypothesis* suggests that traits being selected for may be purely aesthetic. A long, colorful feather just looks pretty. However, there is a second option. In the 1970s, Amotz Zahavi proposed the *Honest Signals hypothesis*. This suggests that male ornaments—and, in humans, this includes our brains and the products thereof—evolved as honest, hard-to-fake signals of underlying quality. You cannot grow a big peacock's train (or a big human brain) if all your efforts have been spent fighting off pathogens. That large peacock's train is attractive precisely *because* under any other circumstances—such as escaping a hungry fox—it would be a handicap.

Which of these alternatives is correct? Are females selecting for arbitrary aesthetic qualities, feeding off one another's preferences, like so many groupies at a rock concert? Or are they careful shoppers, selecting for otherwise hard-to-spot features of deep worth? I think the answer is (as above) "both." In particular, the biologist Helena Cronin has argued that, in the limit, it is differential gene selection that matters to natural selection (of which sexual selection is an interesting subset), and our imposing of our judgments of beauty (or its absence) is beside the point.[18] I think she is right. In the cases I discuss as follows, a convincing case can be made for the Zahavian interpretation of female sexual selection— that an honest signal is being picked out of the noise. However, it is also likely true that such qualities will gain momentum in the preferences of all the other females. Hard-to-fake signals beget sexy sons. One thing is certain, when it comes to sexual selection, for women, it is anything but a casual affair.

9.3 CASUAL SEX?

The root of all noir is the gulf of empathy between a man and a woman.

— (Cathi Unsworth)[19]

The costs and benefits of sexual reproduction do not fall evenly in humans. Nowhere is this clearer than in attitudes toward how casually men and women can treat sex. For women the concept of casual sex means something very different from what it means to men. Therefore, the occasional claims to the effect that women are as open to casual sex as are men reveals that the word "casual" is doing double duty.[20]

In a famous study by Clark and Hatfield, a clear majority of college-age men, when approached with an offer of casual sex, said "yes," compared to absolutely no women.[21] That this was not simply a function of different attractiveness of the sexual proposers can be shown by the fact that about half the women approached agreed to dinner dates, and similar numbers agreed to go for a coffee. Video footage of the encounters shows men and women reacting completely differently. The men cannot believe their luck, while the women cannot believe that they are being asked this in the first place.[22]

Terri Conley attempted to debunk this male/female discrepancy in willingness to engage in casual sex by pointing out that these figures changed when women were asked if they would sleep with, say, Johnny Depp.[23] Almost nothing could more clearly illustrate the gulf between male and female understanding of the words "casual sex" than this attempted refutation. Does it really seem accurate to compare the ready (male) acceptance of an offer of actual sex with a stranger, distinguished by nothing more than being of reproductive age, to the (female) fantasy of sex with one of the richest and most famous actors on the planet, known for his exceptional good looks?

This is a perfect example of how (and why) men and women frequently talk (and act) past one another in matters of sex. Some have replied that women would be as willing to engage in casual sex as men if there were not risks of harm involved, an objection that conveniently ignores the fact that men are more likely to brush aside considerations of any potential harms.[24] But, a serious objection has been raised to the "men and women are different when it comes to casual sex" hypothesis, and it is possible for us to control for the fear element too.

Try the following thought experiment, if you dare: Assuming that you are heterosexual, conjure up (in your mind's eye) your ten most attractive friends of the opposite sex, that you have not had sex with. This is a pure thought experiment, so you do not have to be burdened with worries about any sorts of harm. We can put aside any risk of STDs, jealousy, embarrassed encounters after the office party, abandoned children, and any other socially conditioned roles, because this is a pure fantasy.[25] Let your imagination go. Now, still in the privacy of your own mind, separate off all the ones you would consider having sex with, with all those negative conditions just mentioned (and any others that occur to you) taken away by the power of imagination.

Now, if you are really daring, compare the size of the remaining crowd in your mind with that of a trusted (heterosexual) friend of the opposite sex, who has performed the same thought experiment. If the numbers are even close, then something highly unusual would have occurred. If someone replies that *even in their imagination* they cannot remove their so-called socially conditioned sense that fewer people could be sexual possibilities, then we are really stuck. We would be forced to admit that, even in the full privacy of their minds, men and women are measurably different. The onus is on any remaining skeptics to find any population on the planet where these patterns are reversed.[26]

However, the fact that men regularly say *yes* to the possibility of sex with strangers, while women regularly do not, implies that sometimes (albeit rarely) women must have been open to sex with strangers too. After all, the "this man is a stranger to this woman" relationship works both ways, and sometimes "strangers" (especially if also tall and dark) do strike a spark. Spam emails—another low-cost form of solicitation—might fail most of the time, but when they win, they win big. We know that both sexes are far more similar in terms of what psychologists call long-term mating (LTM) strategies than they are in short-term mating (STM) ones.[27] That admitted, both sexes *do* have STM strategies.

However, it would be foolhardy to think that the relative numbers match up, with an STM female partner for every STM male who wants one. That admitted, for men to have evolved an error management system that makes them much less likely to miss an opportunity for low-cost sex (and therefore be more likely to misperceive interest where it is absent) means that there have to have been at least *some* circumstances where this has occurred.[28] At least, they must have occurred in the past.

Careful replications of Clark and Hatfield's classic study have found that women sometimes are open to STM—including sex with strangers—but that context and attractiveness are major mediators of this, and the thresholds differ hugely from those of men.[29] Additionally, even when women do hook-up for STM, they are more likely to wake up to troubling feelings such as "was this just about sex?"[30] We do not do anyone's mental health any favors by denying sexual dimorphism when it comes to causal sex. Perhaps we should stop referring to "casual sex" at all, when these things mean very different things to each sex? For women, STM means the chance of gaining very high-grade genes, or perhaps, beguiling a partner into LTM (mate switching rather than short-term mating *per se*). The crucial thing to appreciate is that neither of these female goals is low-cost. For men, the option of a low-cost (to them) mating opportunity always exists.[31]

9.4   PICKINESS IS NOT A BUG, BUT A FEATURE

Our highest assurance of the goodness of Providence seems to me to rest in the flowers. All other things, our powers, our desires, our food, are all really necessary for our existence in the first instance. But this rose is an extra. Its smell and its colour are an embellishment of life, not a condition of it. It is only goodness which gives extras, and so I say again that we have much to hope from the flowers.

— (Arthur Conan Doyle, speaking through Sherlock Holmes, in *The Naval Treaty*)[32]

Sherlock Holmes was wrong. Beauty is not some embellishment on life; it is one of its primary conditions of generation. As stated earlier, a good principle to adhere to is that whenever something appears to be more ornate, bizarre, or complex, than is strictly required to do its job, the clever money is to bet on our not having understood that job properly.[33] Sexual selection is usually the key to that understanding. Human sex is a perfect storm of all three qualities of ornate, bizarre, and complex. To put this another way, efficiency is not the hallmark of good sex. At least, not of good *human* sex. Any man who wants to test this by putting "Come on over to my place for brisk and clinical sexual intercourse" on their dating profile is, of course, welcome to try.

Differences between men and women do not stop at differing meanings of what constitutes casual sex. Because the costs of sex do not fall evenly, women have made it tough for men to get *access* to sex, and then made themselves tough to satisfy *during* sex, and furthermore, tough to *keep satisfied* with the sex that they are having in ongoing relationships. As the IT specialists are fond of saying, "This is not a bug, it's a feature." Indeed, it is a sexual smorgasbord of evolved features.

The previous chapter showed the scientific evidence that female orgasms are not all the same. Ones brought about by external masturbation are experienced as different in location, intensity, and associated physical and phenomenological aspects. None of this is surprising when we consider the complex internal neural machinery, both inside women's sexual organs, in terms of genitalia, and that linked to that largest of all sexual organs, the brain. These latter details were the subjects of chapters three and six respectively. I reminded you that none of this is to downplay the importance of so-called foreplay, the clitoris, or masturbation and, indeed, every sexual partner would be well advised to learn about their partner's masturbatory preferences (of whatever sex), because one of the functions of shared orgasms (that I have not even begun to discuss yet) is of shared closeness and intimacy.

Not only that, but we also know that lots of prior sexual activity (including other orgasms) *before* intercourse are much more likely to bring female orgasm about *during* intercourse.[34] All that said, female orgasms brought about by penetration are regularly reported as feeling different and, furthermore, can be shown to have some differing effects. The fact that they are harder to bring about this way just adds to their biological charm. So, if we leave aside those purely masturbatory orgasms for the time being, what are the predictors of partnered orgasms in women, resulting from penile-vaginal intercourse?[35] Let me put this another way, a more ecological way. Given that female orgasm is somewhat harder to bring about through this method, what is the feedback about the partner that can be gleaned by our surveying of when and where this occurs? The answer is, once again, that women are picky. Let me count the ways . . .

. . . But first, a word of caution: One key difference in studies is whether we measure differences *between* groups or *within* groups. Let us say we have a group of people who regularly report that they have great sex with their physically fit partners, compared with another group, that regularly report poor sex with their couch-potato partners. We might naturally conclude that partner fitness was driving the sexual satisfaction, and it might be doing that. But there is another possibility— namely, that fitter people have fitter partners (maybe they met one another at the gym). Maybe, physically fitter people tend to have better sex because their bodies respond more effectively? So, we have to proceed cautiously, building up a picture from lots of separate puzzle pieces. I will flag up experimental design differences of studies as I go along, noting limitations or where corroborating evidence is available. Here is the crucial point: When the design is within the group, then by implication, women are making comparisons *between different* male sexual partners. And, if everything I have said so far has been right, this is a key feature of female orgasm.

9.5   PICKINESS ABOUT PENISES

An old man came courting me, hey ding-doorum down
An old man came courting me, me being young
An old man came courting me, fain he would marry me
Maids when you're young never wed an old man

Because he's got no faloorum, faliddle aye oorum
He's got no faloorum, faliddle aye ay
He's got no faloorum, he's lost his ding-doorum
So maids when you're young never wed an old man
                                        — (Trad. often associated with the *Dubliners* folk rock band)

Women's sexual pickiness includes being responsive to the health and vitality of their partners, and this is most certainly reflected in their sexual preferences for penises.

What did I mean earlier by saying that female orgasms were harder to bring about through penile penetration? For one thing, penile erections are costly signals.[36] I do not wish to come across as bragging here, but humans have the largest penises of any primates. Not the largest in proportion to their bodies mind you, the largest in every respect—length *and* girth. Adult male gorillas, who violently compete for females with their 400-pound bodies, have penises averaging two inches in length. Bonobos, our sexiest cousins, who are much less aggressive than chimpanzees, have erect penises more nearly approaching ours in length, but not nearly so thick.

It is technically possible, although not very likely, that all this human penis size is *not* the result of sexual selection by the females of our species. The argument for supposed lack of direct selection on penis size is that the size of the penis is related to the size of the vagina, and the size of *this* is large, because the heads of baby humans are large.[37] This is a possible explanation, but it does not fit the pattern of other primates. It is true that humans experience birth difficulty in relation to other primates of our size, and that this is related to the relative size of the head of our babies. However, lots of smaller monkeys and gibbons also have disproportionately large heads (and consequent birth difficulties), and none of them have disproportionately large penises.

The more direct evidence for female sexual selection on penises is that, not only do women show active preferences for larger penises (in length somewhat, but more particularly in girth) when asked,[38] but also that penetrative sex is more likely to result in their orgasm with larger ones.[39] Not only that, but penetrative vigor is predictive of female orgasm during intercourse.[40] The well-known male anxiety linked to this organ in terms of both penis size and what they do with it is not mysterious. Incidentally, there is no research I am aware of supporting the idea that intercourse going on for great lengths of time—what one noted expert referred to as "interminable pistoning"—is, ceteris paribus, something prized by women.[41] However, any who are experiencing size-based anxiety should take heart. For one thing, it is possible for penises to be too large, and additionally, there is evidence that most men underestimate their own size in relation to others.[42]

More to the point, humans can cheat. Just as a smaller fighter can learn judo, and thereby take on someone twice their size, there is no requirement for any of us to fight by the Marquis of Queensbury rules laid down by our genes in the sexual arena. We live in an age where sexual toys, skills, and sensuality can all step in and prevent us from having to generate orgasms purely through penile intromission, if we choose not to. In one study, those with micro-penises were shown to have perfectly acceptable rates of sexual satisfaction in their partners.[43] Skill, sensitivity, and attention to detail count. Or, as Carol Wade put it, "Sex is not a soccer game. The use of hands is permitted."[44] Humans are allowed to break the rules that genes lay down, but to break the rules, it is best if you first know them.[45]

Examples of this sort of "rule breaking" are not restricted to this day and age either. Humans have been adding to, subtracting from, and generally interfering with penises in order to impress or please others across both time and space. Notable examples include the Dayaks of Borneo, who are known to pierce the glans of the penis and then insert objects to increase partner stimulation, or the Topinama people of Brazil, who introduce snake venom to penises to make them enlarge.[46]

9.5.1 Bending the Rules

As well as sheer size, we humans have another interesting penile feature: We are one of the minority of mammals *not* to have a penile bone—the technical term is baculum—in our penises, and thus our erections give (reasonably) honest feedback to the partner about the man's health and level of arousal, in relation to themselves.[47] This curious feature of humans is one we have shared with other primates such as tarsiers and some platyrrhines. Another thing we have in common with these primate cousins is that, after sex, the females of our species do not typically get straight up and (possibly) start looking around for another potential partner.[48] So, it could be that the baculum evolved as a mechanism to prolong intercourse, and thereby delay the (now aroused, but possibly unsatisfied) female in looking for other possible playmates. This might mean that the ancestral females of our own species have therefore been selected to be especially picky about those who could keep them entertained for long enough (and intensely enough) that other sexual partnerships did not rapidly present themselves as an option.

This firm-but-flexible nature of the human penis fits well with what I said in chapter six. Recall the details about the sensitive areas of the internal parts of the clitoris and other structures being available for interaction with the uniquely long (five inches plus), thick (one and a quarter inches plus in diameter), and flexible (I directed you to some fMRI photos demonstrating this) male human organ.[49] By contrast, our primate cousins have much smaller (two to three inches), thinner (technically *filiform*) penises that are stiffened by a baculum and thus do not act as honest signals of health in the same way.[50]

Some have suggested that sperm competition—i.e., in terms of battles fought out between sperm simultaneously present in the female reproductive tract—may have driven this extra size. While possible, this is unlikely for three reasons. First, as I said earlier, we have failed to replicate the hypothesized "killer sperm" that these theories predicted. If it exists, we have not found it yet. Second, it is rare (but not impossible) for human female sexual behavior to reach the levels of promiscuity necessary for this form of sperm competition to occur. Finally, human testicle size, while much bigger than that of monandrous gorillas, is smaller than that of species where such direct sperm competition routinely occurs (such as chimpanzees).

As we saw in chapter six, human penises bend to interact with sensitive internal areas of the female genitalia, while retaining enough rigidity to generate internal pressure. Today, products such as Viagra™ can somewhat circumvent this honest signaling. However, this should not be taken to imply that a person who uses such performance enhancers is not genuinely attracted to their partner. It is worth reflecting that the person is typically aroused by their female partner (or planning to be so) before they take the drug. The drug does not create desire from nothing, but it allows the man to exhibit said desire. All that said, the fact that erections tend to decline with age and poor health means that their presence is (and has likely been in the past) an honest signal of these features over time.[51] As an aside, this is why the so-called female Viagra has been so disappointing, being found to be no more effective than a glass of wine in generating arousal in women, and with significant (down)side effects. The female equivalent to *erection* is *lubrication*, and if this is absent for a variety of reasons, then it can typically be supplemented with artificial lubricants. It is absence of desire that female Viagra™ does not treat, but no more so does male Viagra™. It treats blood flow issues, not desire issues. When it comes to desire, both men and women are more complex than a pill.[52]

9.6   PICKINESS ABOUT PULCHRITUDE[53]

He looked good, like sin in a suit.

— (Melissa Marr, *Wicked Lovely*)

In the nineteenth century, Darwin's cousin, Francis Galton, carried out one of the most infamous statistical studies of all time. Armed with a concealed counting device (which he dubbed "the pricker"), he walked the streets of British towns, recording whenever he encountered a woman whom he found to be either attractive, middling, or repellent.[54]

Women do not need such a measuring device. Furthermore, they do not need to scour the streets for mates. Dating is now often done over the internet, and this allows for the easy mechanization of metrics of attractiveness. OK Cupid™ recently revealed that the predictions of pickiness in women are mirrored on their dating site, which has about one million visits per month.[55] Most men found most women at least somewhat attractive, and the estimations formed a bell curve. The women's curve was, however, heavily positively skewed, with more than 80% of males falling into the bottom three categories (of seven) in attractiveness. That does not mean that all those males are consigned to reproductive oblivion. However, it does mean that, if they use this sort of method of meeting women, they had better be prepared to be swiped left more often than not.

Why does attractiveness matter anyway? In essence, it is because outward symmetry is a costly signal. It means that you have won a number of fights against the incursions of pathogens.

Attractiveness has been found to be a consistent feature of male partners with whom women orgasm during coition. But, as psychologists, we want to drill down into what lies behind this subjective judgment. As stated in chapter three, there is a way to objectively and mathematically model some elements of attractiveness, and perhaps surprisingly, this method is found to be valid across species. This is because being outwardly symmetrical is one of those all-important costly signals. In this case, it is a sign that your developmental trajectory was not horribly hijacked by some parasite.[56] Thus, your good looks display the fact that your immune system (or your luck) is impressive.

Whether you are a spider monkey,[57] a salmon,[58] a scorpion fly,[59] or a sage grouse,[60] your partners prefer you if you are more symmetrical. Even symmetrical *poppies* do better than less symmetrical ones.[61] That last one might be puzzling. How can it be that poppies find each other attractive or not? They do not. However, their pollinators, such as bees, appear to—being preferentially attracted to those flowers that have also demonstrated developmental stability. This property is technically known as low fluctuating asymmetry (FA). It can take a second or two to get one's head around this term, because what we are describing by it is a double-negative.

Asymmetry is lack of symmetry, and we expect to find a *low* level of it in those deemed beautiful. The upshot is that a low FA is a reproductively good thing to have.[62] And it, or its reasonable proxies, is a thing that we have found to happen far better than chance, in the partners of orgasmic women.[63] In a growing body of studies, symmetry, or other reasonable proxies of attractiveness such as muscularity, physical attraction to their partner, how much of a "catch" they rate their partner as being, and their partners' shoulder width (width-to-waist ratio being an indicator of high testosterone) all predicted significantly increased likelihood of orgasm though intercourse.[64]

9.7   PICKINESS ABOUT PERFUME

Follow the Duck, not the theory of the Duck.

— (William Charlesworth, ethologist)

Unlike ducks, humans get shy if you spy on them during their mating. However, one advantage that students of human ethology have over duck ethologists is that their observed species can tell you about things they have actually done, and about what really matters to them.[65] When we first started studying female orgasm, one of the first things that we did was ask women what scientists should be asking *them* about its nature. I detailed a lot of this in chapter eight, but we got a lot more interesting information. One thing that surprised us in terms of its ubiquity was, "Ask about partner smell." Whenever I give talks about our work and mention this fact, the women in the audience nod, and say things like "Of course!" and "Well, who doesn't know that?" while the men tend to look astonished. The smarter ones take notes. Of course, partner smell is important, but we really did not know how important until recently. As Jennifer Aniston said, "The best smell in the world is that man that you love." Why is attractive smell such a big part of the sexual mix? It turns out that women are especially picky about smell compared to men, because of the information that it conveys.

When it comes to olfactory bulbs—the part of the brain that is sensitive to smell—there are, on average, over 40% more cells in women's brains than there are in men's brains. While the average *size* of this brain structure is not significantly different between men and women (0.132 vs,

0.137 grams respectively), the *density* of neurons is hugely different. Imagine finding that women's eyes had 40% more light-receptive cells than men's did, or that women's tongues had 40% more tastebuds. Once again, Mother Nature does not hand out free lunches.[66] What is all this extra hardware for? If there is a complicated message decoder, that implies a complicated messenger. What message might men be selling via smell?

There is a lot of media talk about human pheromones, but much of this talk is misplaced.[67] Lots of animals use pheromone chemicals to signal alarm, sexual receptivity, or territory, but these signals do not distinguish among individuals. If you take the top off a beehive, you can see the bees put their abdomens in the air and start waving their wings over a white patch on their abdomens. This is spreading *Nasonov* pheromone, and its purpose is to call other bees to come back to the hive and defend it.[68] It is a general, chemical-based alarm shout-out, broadcast to all of their sisters. Pheromones are chemical ways to transmit persuasion, and sex pheromones advertise general sexual receptivity. It is part of live research whether humans do anything quite like this. Some studies, such as the ones mentioned earlier, have found evidence for a male pheromone (androstenone) that improves mood in women, especially around ovulation, or increases female sexual behavior generally. However, that sort of general effect does not seem to be detectable at the level of conscious individual sexual response. As any woman will happily tell you, not *all* men smell that great. As Ms. Aniston said, it is only the ones that you love.

One implication of this point is that excitable (and gullible) young men who rush for "hormone sprays" advertised in the back of men's magazines, to try to make themselves irresistible to women, should beware. I have it on reliable authority that any such sprays—if they are genuine pheromones at all—contain pig pheromones. Boarmate™ is certainly a thing, and its effect on female pigs is not in doubt. But unless someone's tastes are very unusual indeed, a human male is going to be very disappointed in the results of using it on himself.

9.7.1 TRUTH IN ADVERTISING

Pheromones are indiscriminate, but humans are choosy about with whom they mate. This is true of both sexes, but it is especially true for women because, as I hope is clear by now, the costs of any mistakes fall disproportionately on them. What does smell advertise? We are only beginning to understand this, but one theory with growing empirical support from diverse sources is that our smell advertises our major histocompatibility complex (MHC).[69] This is the part of our genome that advertises our particular and individual immune systems. Folk sometimes talk grandly of "the immune system," as if it was an organ with a simply understood function—like the heart or lungs. But the immune system is an incredibly complex set of interlocking defenses that draw on lots of disparate bodily resources. Scientists devote whole careers to studying sections of it, so I am not going to be able to do much more than scratch the surface of its importance here.

One crucial balance that reproducing organisms need to achieve in their offspring is between their being immune to those pathogens that are local—which means inheriting immunity from parents—but at the same time not producing inbreeding depression. Inbreeding multiplies any deleterious genetic mutations and rapidly becomes a reproductive dead-end. This is why some so-called pure dog breeds are so unhealthy. It is why we discourage cousins from marrying. On the other hand, who embodies the local resistance to pathogens better than your own parents? The fact that you exist at all is proof that they managed to survive the local diseases long enough to have you. So, anything that they advertise about their immune systems needs to be part of the immune picture. Part, but not whole, because of the aforementioned inbreeding problem.

What this seems to amount to in practice (for women) is wanting a smell that is a bit like daddy's, but not too much like daddy. That "not too much" means the amount by which we differ on six key markers of the MHC. It turns out that specific differences here predict how sexy a woman will find a man's smell. Scientists have tested this in various ways, but one of the most entertaining is the smelly T-shirt paradigm. In this test, the males supply some appropriate DNA (via saliva) and also wear a T-shirt at night for a few days (no deodorants or perfumes allowed). Said T-shirts are placed in sealed containers, and then women come along, break the seals, and rate the attractiveness of the smells. It works the other way around too—with female DNA and male smelling.[70]

Interestingly, women on the contraceptive pill did not experience this effect. Some have even suggested that women who are getting serious about a partner should vary their contraceptive use, because pill use can change the perception of partner attractiveness considerably.[71] This may be because the oral contraceptive works (roughly) by fooling the body into thinking that it is already pregnant, and this may alter one's perceptions of who is the most attractive partner at the time. The corollary is also true. At the most fertile phase of their ovulatory cycle, women report higher orgasm frequency with mates who are genetically compatible in this way.[72]

Work is still ongoing, and the evidence so far is mixed but highly promising. For example, is it immune system *compatibility* or general immune system *strength* that is being sought after by women? Claus Wedekind's research team did not find an effect compatible with simply random mixing (technically called "heterozygosity") that humans find generally attractive.[73] However, Thornhill and Gangestad found that more symmetrical men had T-shirts rated as more attractive by women.[74] Symmetry—what we experience as beauty—implies a generally robust immune system. It could even be that what is being sought varies somewhat from place to place. After all, we know that women prefer more muscular men in areas of high pathogen load, and muscularity indicates a generally robust immune system, because testosterone is an honest signal of immunocompetence. In areas where disease is less of a threat, maybe other systems come into play more prominently? I discuss some of this in greater detail later.[75]

Recently, it has been found that odor preference does not covary with a generally stronger immune system or one recently buffered through vaccines.[76] But this is unsurprising, unless we ever thought that humans sent out generalized pheromonal signals of untargeted receptivity. It is precisely because we do *not* smell equally pleasant to one another that it seems worth exploring the case for compatibility of immune systems revealed through odor, rather than mere strength.[77]

Although Wedekind's team did not find women to be especially sensitive, when compared to men, in their study, others have done so. For instance, another team found not only that women were more sensitive in general to odor than men, but also that this sensitivity was more susceptible to disruption by perfume.[78] The corollary is also true. As Diana Fleischmann has shown in a number of studies, women are more likely to be disgusted by all sorts of stimuli—including smell—than are men.[79] Disgust, like fear, is a protection against potentially dangerous interactions, and we ignore it at our peril. There is thus a real chance that we may be getting in our own way by not allowing ourselves to be led by the nose. The reverse is also true: Sadly, those who lose their sense of smell are very likely to experience a decrease in sexual desire and enjoyment.[80] The anosmia brought about by SARS-Cov2 may bring about a mini epidemic of this, and there have been reports of partnership break-ups as a result.[81]

In our own (2012) study, we found that an attractive partner smell was the single most reliable predictor of female orgasmic response from intercourse.[82] We believe that we were the first to find such an association, and it is very exciting, because it suggests that a major feature of female sexual responsiveness is tied to a key reproductive component, namely, that of potential offspring immune systems. James Sherlock's 2016 study confirmed our phenomenological findings relating to smell.[83] Both of these studies explicitly looked at women's differing sexual responses to different men and their behaviors, making the story of female orgasm being a choice mechanism that much more likely. In between these two studies (in 2014), Gordon Gallup's research team measured the effect of a compatible MHC directly, rather than indirectly, by asking about smell. By measuring the degree of immune system similarity in three key genes of the MHC—recall that, like fluctuating asymmetry, less is more here—the amount of compatibility was found to decrease sexual responsiveness in women (but not in men).[84] Having a too-similar set of markers also decreased women's attraction to their own partner and increased the chances of affairs.[85] These effects were amplified during the most fertile parts of the woman's ovulatory cycle, and this fits with what we suspect about women's propensity to look for new genes at such times. In summary, attractive partner smell—probably indicating immune system compatibility—has turned out to be one of the key predictors of female orgasmic response from intercourse. This has profound implications.

9.8 PICKINESS ABOUT PERSONALITY

Chick: *Are you coming right home after work? I need sex so badly.*
Guy: *Yeah, I can tell . . . Why don't you use your Valentine's Day gift?*
Chick: *The Rabbit? It's not the same.*
Guy: *Why isn't it the same?*
Chick: *Well, it doesn't talk.*
Guy: *Wait—so if it talked, you wouldn't need me at all? Is that what you're saying?*
Chick: *Um . . . No?*

(Overheard in New York)[86]

Everything said so far in this chapter might be taken to imply that men are little more than walking adverts for their genes, using the advertising billboards of scent, muscular bodily symmetry, and healthy working genitalia. But that is not true. Men have personalities too. Personalities are also highly heritable, of course, so they advertise a particular subset of genetically related desiderata.[87] But, *hang on a second*, you might be thinking. Is it not true that sexual selection implies that there is just one way of being attractive and that the (alpha) male who has the longest, shiniest, most colorful version of . . . whatever that is, will be the most successful? And is it not the case (you might go on to think) that personality is about how we are all different?

No, and yes. In peacocks it is indeed the peacock with the longest train who will attract the females. But humans are not peacocks.[88] Not only are we a mutually sexually selecting species, but even in those species that are not mutually selecting, there is not only a single male way of reproductively succeeding. Let me illustrate this key issue with the tale of the side-blotched lizards.

9.8.1 Rock, Paper, Scissors in the Lizard World

Side-blotched lizard males come in three types, or colors, if you prefer. They can be orange, blue, or yellow. Each of these varieties is definitely the same species—they can all (potentially) mate with all the females—but their strategies, both physical and behavioral, for achieving this goal are markedly different.[89] The orange males are large and aggressive. They have high testosterone levels and defend a territory that attracts females. They do not form lasting pair bonds, but they will aggressively defend this territory and their resultant harem from any interlopers. But, one of those interlopers has come up with a rather cunning strategy. They look female.

The yellow male is small and passes for female, so when he sneaks into an orange's territory, the bigger male welcomes him in. He is not here to mate with the male, however, but rather the females. The females seem to like this just fine, finding his sneaky cunning just as attractive (in its own way, perhaps) as the dramatic demonstrations of his more bombastic orange brother.

Adjacent to this orange/yellow warfare is the blue side-blotched lizard. He is bigger than the yellow version and will chase yellow off if he spots him. But blues are smaller than oranges, and they will not pick a fight with them. On the other hand, the blue lizard faithfully commits to his one single female partner and will not seek out others like the orange guys do. He also sees off yellow rivals if they stray into *his* domain. All of these three strategies work, in their own terms, and they are held in a dynamic equilibrium, like a genetic game of rock, paper, scissors. Orange males push blue ones from their territories, while blue males get together with females and pair bond, blocking out the yellows. But the yellows get their own back by sneak-mating on the orange males. Round and round the genetic game goes, with no one strategy winning out overall.

In case it is not yet obvious, let me spell out that humans are at least as complicated, personality-wise, as side-blotched lizards, so we should not expect there to be one single set of female desiderata that always work. Freud once complained that, "The great question that has never been answered, and which I have not yet been able to answer, despite my thirty years of research into the feminine soul, is 'What does a woman want?' "[90] Freud expected women to present a static target to male suitors. If he had appreciated the implications of differential parental investment—and hence female choosiness—then he would have been less surprised that females demand hard-to-fake signals. A static target is that much easier to game with fakery, so we should not expect women to not keep men guessing anytime soon. In a species like ours, which has a big brain rather than big feathers, it is easier to fake personality aspects if you know in advance what target is being sought. We know better than Freud could have known. We expect to find a complex set of interlocking needs, sensitive to local conditions and exquisitely tuned. Or, as humans prefer to think of it, love and courtship, and the traditions surrounding these.

9.8.2 Oceans Away: The Importance of Variety

Personality is one way in which the rock-paper-scissors routine of frequency-dependent selection for genes works its way out in human beings. This is a huge topic, and I can do little more than introduce some of the major themes here, as they relate to female orgasm. There is broad consensus in behavioral science that five reasonably stable axes of personality exist, and this is true across species, not just of humans.[91] The "Big Five" are easily remembered by the acronym OCEAN. "O" stands for openness to experience, "C" stands for "conscientiousness," "E" stands for extraversion, "A" stands for agreeableness, and "N" stands for neuroticism. Some years ago, a student pointed out to me that the main characters in Winnie the Pooh embody these personality traits, making them easier to remember. Piglet is openness—he will try any adventure. Kanga, the archetypal mom, is conscientiousness. Extraversion is Tigger. Agreeableness is Pooh himself, leaving poor Eeyore as the neurotic. So, is it the case that selection tends to maximize each of these traits in the species? No, because each is an axis along which fitness is increased in some circumstances but not in others. As the saying goes, the early bird may get the worm, but the second mouse gets the cheese. What counts as the best strategy in each circumstance varies, and thus personality is held in dynamic equilibrium in the human, just like lizard, populations.[92]

For example, those with high trait openness are more likely to try new sexual partners, but find it hard to ignore new stimuli, and therefore can get distracted. Those with a low trait conscientiousness tend to have riskier sex, and the fitness aspects of this can go either way, depending on things like disease prevalence.[93] Extraverted men are more likely to approach women, but are also more likely to die in car crashes or other mishaps.[94] Women seem to benefit less (sexually at least) from possessing extraversion, in reproductive terms. It is, for example, a cliché among female rock stars that the male ones get all the groupies. There are (of course) plenty of male groupies who would be only too happy to have sex with those talented female extraverts; however, the talented female extraverts are pickier than their male counterparts.[95]

On the other hand, agreeable men (the opposite trait being aggression) may be attractive in some circumstances but seen as pushovers in others. Finally, those high in neuroticism (some prefer to put this the other way around and call this being low in emotional stability) have more sexual anxiety, which can manifest in moodiness. This is not just restricted to those cultures with which we are more familiar. In non-WEIRD populations,[96] the findings about extraversion in both sexes are replicated.[97] In addition, high neuroticism in women from this traditional population increased the number of children that they had but deceased the amount they invested in them. This decreased fitness if they had poorer partners. The upshot is that it pays to be a demanding female partner, but only if you can use this demandingness to get investment from a high-status, and resources, male.

9.8.3 INTELLIGENCE

I require three things in a man: he must be handsome, ruthless, and stupid.

— Dorothy Parker

Do women sexually select for intelligence? What is *intelligence* anyway? Let us take the last question first, because the repeated efforts to debunk the concept over the last few decades have, in some cases, become absurdly overwrought and (frankly) scaremongering.[98] So much so that Scooby Doo and his friends might suspect that someone has a vested ideological interest in keeping those meddling kids from something very interesting down at that old disused fairground.

The bogeyman here is racism, of course. It is not that racism is not a real threat, but the idea that banning all talk of intelligence, for fear that it will be used by racists, and that doing this will somehow dispel racism, is incorrigibly naïve. First, it is charmingly arrogant on the part of academics to think that racists are typically getting their steer from formal academic papers, and that this racism would evaporate if only the concept of intelligence would go away. Second, in those rare cases where (white) racists *do* talk openly about IQ research, they cheerfully admit that their particular racial group does not score the highest on such tests and that this is *further* reason for racial separatism, given that each group would (obviously) be happier with their so-called own.[99] Perhaps we need to look elsewhere than academic papers for a cure for human bigotry?

Group differences in IQ—whether or not they exist—are not the issue for us here, in any case. We are concerned with individual selection of intelligence by sexual partners, some of which is captured by the most validly tested psychometric trait ever examined, intelligence quotient (IQ). It is a cliché that measured IQ, meaning formal psychometric intelligence test data, does not capture everything we might mean by "intelligence," but even saying that amounts to admitting that intelligence is, at least, something we know when we see it—else how could we know when something was missing from it? The response that there are so-called different intelligences should not blind us to the fact that, to the extent that different sub-classes within intelligence can be identified, they get dwarfed by an over-riding (or G) factor, or into thinking that there is no underlying trait. The fact that we can readily see that the strength of a ballet dancer is different from the strength of a power lifter does not leave anyone confusedly thinking that Arnold Schwarzenegger is indistinguishable from Pee-wee Herman.

The evidence here for female sexual selection (whether or not mediated by orgasm) is somewhat mixed. Geoffrey Miller made it the centerpiece of his book, *The Mating Mind*, arguing that mutual sexual selection for intelligence was the turbo-charger for human brain evolution.[100] High intelligence is an oft-repeated desiderata in partners among women when they are asked,[101] and there may even be sapiosexuals, who seem to be aroused directly by intelligence.[102] However, even when this is studied, extremely high IQs were rarely found attractive, and there are even more alarming findings for proponents of intelligence-based sexual selection. For example, one study, by *Carolyn Halpern*'s team, revealed a curvilinear relationship between intelligence and likelihood of having sex. It showed the, frankly shocking, revelation that men with an IQ of 60 were more likely to have had sex than men with an IQ of 130 in the previous year.[103] This could be a reporting issue, but this is not usually the case with other traits.

A moment's reflection should make this finding striking. An IQ of 60 is right at the threshold of someone who can function in modern society, without specialized supports in place. Any job that they would hold would be menial in the extreme and not require initiative. An IQ of 130 is fully two standard deviations above the mean, and you would expect to find such people with responsible jobs, and the capacity to progress effectively in consumerist society.

So, what can we conclude from this? Is it perhaps the case that women say the socially desirable thing to interviewers—that they find intelligence attractive—while privately agreeing with Dorothy Parker, or Francoise Sagan, when she said that, "I like men to behave like men. I like them strong and childish."

How to reconcile this disparity? Is this one of those glaring discrepancies between what women want and what they say they want? Not necessarily. It is worth noting that in the case of both Parker and Sagan, none of the three husbands each had (separately) could remotely be described as "stupid," whatever may be said of the numerous lovers they both had. As I noted earlier, humans have more than one mating strategy, which, for simplicity's sake, can be divided across an axis of short and long term. Once we remember this, the apparent paradox goes away. A diminishingly small number of women desire low intelligence in a long-term partner. The Carolyn Halpern study, while certainly striking, was on adolescents, who are in an exploratory phase of sexuality, not settling down with a lifetime partner. However, things like low IQ and, frankly, downright anti-social traits like Machiavellianism, narcissism, sadism, or an actual history of murder *can* all be found attractive to at least some women, sometimes. This will be explored in more detail later.

9.8.4   ARE DIAMONDS REALLY A GIRL'S BEST FRIEND?

All of this may go some way to explaining one puzzle we have found in orgasm research: The sexual response of women to partners with wealth, which often correlates positively with IQ—at least, up to a point. There is a well-known set of reasons why we might expect women to respond sexually to partner wealth. The evolutionary story seems to be that status, resources, and the willingness to share them would correlate pretty strongly with offspring survival in ancestral conditions. Some have discovered an association between (male) status and reproductive success even in modern populations—where pressures are not nearly so acute.[104] The result of all of this would be that women over time became more attuned to these qualities and the traits associated with wealth acquisition and targeted generosity. This may even result in the so-called "Lady Chatterley" strategy, whereby a woman gains (long-term) resources from one male but genes from another, short-term, contributor.

What happens when we ask women about the pattern of orgasmic response where partner wealth is the key variable? The evidence for the association between female orgasm and richer partners is mixed. Daniel Nettle and Tom Pollett found just such a (positive) relationship in 2009, but then retracted their findings the next year, unsatisfied with the data crunching. James Sherlock's team did not find an association between what they termed "high orgasm" experiencers and high-income (or high *potential* income) partners in 2016.[105] On the other hand, Gordon Gallup's 2014 surveys *did* find an association between high income of the man and likelihood of orgasm by the woman.[106]

How to make sense of all this? The truthful answer is that none of us are sure at the moment, but it seems likely that a number of factors are at play. High income may be seen as a marker of other desirable male qualities, such as confidence, ruthlessness, or drive. These are, after all, regularly found as desiderata in interviews, surveys, and female-driven erotica.[107] Maybe, as the protagonist of that paean of praise to greed, *Liar's Poker*, put it: "money was just a way of keeping score"?[108] If that is the case, then it is likely that the recognition of other male traits is doing the real work, and actual partner income is producing noise in the system.

After all, wealth per se is quite a modern invention—a modern way of "keeping score." But, keeping score of what, exactly? Recall that in our own 2012 study we found that women reported that those men who were both more dominant (which was distinct from being aggressive) *and* more considerate, were more likely to generate orgasm in partners.[109] Those characteristics may be taken as proxies for the erotically charged partner features already described—both the features needed to obtain resources and the willingness to share them with their female partner. This may happen irrespective of whether the person had translated this into finances (yet), and ours was a within-groups finding, so the point is that these women were making explicit comparisons between male partner characteristics, using sex as a test bed.

A number of personality characteristics seem more directly, and unambiguously, related to female orgasmic response, and they are unlikely to surprise anyone too much. What is the personality profile of the preferred male lover? Along with the physical characteristics described earlier, they would typically have a good sense of humor, be sexually dominant, while also considerate, creative, and confident.

A possible objection to these findings is that there might exist a "halo effect" whereby those who are enjoying sex with a partner tend to exaggerate their other positive qualities. That is always possible, but there are reasons to think not. One particularly good reason is that a large bunch of socially desirable qualities are *not* associated with increased female orgasm. These include (surprisingly, perhaps) height, aggressiveness, muscularity, and intelligence. It is always possible that future research may unpack some of these interesting apparent paradoxes.[110] What can we conclude from all this?

9.9 CONCLUSIONS: TRY BEFORE YOU BUY?

At the beginning of this book, I said that the key division in adaptationist accounts of female orgasm was between sire choice, and mate choice. Women want good genes *and* good fathers. Cad and dad qualities. The only non-starter is the idea that female orgasm happens randomly and capriciously. We know this for a number of reasons, but one pressing one is that we have discovered clear patterns in female sexual response. Women orgasm in (somewhat) predictable ways, but is this in response to indicators of good genes, or indicators of good investors? There is evidence that could be taken to be for both.

This is certainly far from settled, but I want to propose an alternative to prematurely trying to settle this part of the issue one way or the other. Perhaps female orgasm functions in both ways? We could call this a "try before you buy" strategy. If the evidence, from the perspective of the female experience, has been that female orgasm responds in patterned ways to particular mate characteristics and, furthermore, that the best established of these include indices of general health, immune system compatibility, and a personality profile associated with the potential to invest, if not actual investment itself, then maybe we are strait-jacketing ourselves by forcing our option to be bimodal—harvesting genes versus persuading male commitment? In brief, women's orgasms are multifactorial and complex because the same is true of their mating strategies.

Perhaps a better way to think of sexual encounters is as a test bed where a complex array of physical and psychological factors are being displayed and assessed, with a view to potentially taking the reproductive plunge. In the immortal words of Charles Darwin, "We shall further see, and this could never have been anticipated, that the power to charm the female has been in some few instances more important than the power to conquer other males in battle."[111]

A reasonable response to what has been said so far is that there has been a lot of talk about how, in ethology, behavioral responses are keyed to local patterns of threat and opportunity, and there has been comparatively little talk of such cultural variations. It is to that topic I now need to turn.

NOTES

1 Maynard-Smith, J. (1978). *Models in ecology*. CUP Archive.
2 Hoving, H. J., Bush, S. L., & Robison, B. H. (2012). A shot in the dark: Same-sex sexual behaviour in a deep-sea squid. *Biology Letters*, 8(2), 287–290.
3 Judson, O. P., & Normark, B. B. (1996). Ancient asexual scandals. *Trends in Ecology & Evolution*, 11(2), 41–46, a description repeated in Judson, O. (2002). *Dr. Tatiana's sex advice to all creation: The definitive guide to the evolutionary biology of sex*. Macmillan.
4 *Hetero* meaning "different" and *gametic* referring to sex cells, eggs and sperm.
5 Fernald, R. D. (1997). The evolution of eyes. *Brain, Behavior and Evolution*, 50(4), 253–259.
6 Margulis, L. (1981). *Symbiosis in cell evolution: Life and its environment on the early earth*.
7 Carroll, L. (1917). *Through the looking glass: And what Alice found there*. Rand, McNally.
8 E.g., Lively, C. M., & Dybdahl, M. F. (2000). Parasite adaptation to locally common host genotypes. *Nature*, 405(6787), 679–681.
9 E.g., Goddard, M. R., Godfray, H. C. J., & Burt, A. (2005). Sex increases the efficacy of natural selection in experimental yeast populations. *Nature*, 434(7033), 636–640.

10 West, S. A., Lively, C. M., & Read, A. F. (1999). A pluralist approach to sex and recombination. *Journal of Evolutionary Biology*, 12(6), 1003–1012.

11 William Hamilton's memorable phrase is quoted in Trivers, R. (1985). *Social evolution*. Benjamin/ Cummings. The quote is from p. 324.

12 Dawkins, R. (2006). *The selfish gene: With a new introduction by the author*. Oxford University Press. (Originally published in 1976)

13 And they will get attention—but one thing at a time! See Kuhle, B. X., & Radtke, S. (2013). Born both ways: The alloparenting hypothesis for sexual fluidity in women. *Evolutionary Psychology*, 11(2).

14 Tens of millions of years? Yes, a lot of these systems are older than humans. Evolutionary psychologists tend to focus on the last couple of millions of years—and that is fine for some mechanisms. Other mechanisms are far older than that though, and in the case of the magnocellular oxytocin system, we are talking about something that predates humans by a long way. Another way of putting this is that we share elements of the system with creatures for which our most recent common ancestor is a long time ago. See, for example. Knobloch, H. S., & Grinevich, V. (2014). Evolution of oxytocin pathways in the brain of vertebrates. *Frontiers in Behavioral Neuroscience*, 8, 31.

15 Amotz Zahavi, personal communication. (2010). PhD competition presentation to Amotz Zahavi, organized by *London Evolutionary Research Network*.

16 Fisher, R. (1958). A. 1930. *The genetical theory of natural selection*, 247–262.

17 Prum, R. O. (2017). *The evolution of beauty: How Darwin's forgotten theory of mate choice shapes the animal world-and us*. Anchor.

18 Cronin, H. (1993). *The ant and the peacock: Altruism and sexual selection from Darwin to today*. Cambridge University Press.

19 Cathi Unsworth (2012) in an interview in the *Guardian*, www.theguardian.com/books/2012/jun/29/cathi-unsworth-on-women-and-noir last accessed 25/03/2023.

20 For example, the marvellous and erudite Dr. Zhana in her Casual Sex Project: www.drzhana.com/about/ last accessed 25/03/2023. If you are a guy, try my "Clark and Hatfield" thought experiment, detailed on her page, and see where you get.

21 Clark, R. D., & Hatfield, E. (1989). Gender differences in receptivity to sexual offers. *Journal of Psychology & Human Sexuality*, 2(1), 39–55.

22 A recent video replication was done by Lord Winston for the *Human Instinct* series. It can be viewed here: www.youtube.com/watch?v=oTMCM4SIH4k last accessed 11/12/2010.

23 Conley, T. D. (2011). Perceived proposer personality characteristics and gender differences in acceptance of casual sex offers. *Journal of Personality and Social Psychology*, 100(2), 309.

24 Fine, C. (2017). *Testosterone rex: Unmaking the myths of our gendered minds*. Icon Books, raises this objection.

25 If someone objects that social roles are so deeply ingrained that even in fantasy we cannot unburden ourselves from them, then they are by implication admitting total defeat at the hands of an all-powerful social programmer, and we may as well stop bothering to try to change things. They also have to explain how they are the only ones to escape this effect, like Ishmael did in *Moby Dick*. See Eagly, A. H., & Wood, W. (2013). The nature–nurture debates: 25 years of challenges in understanding the psychology of gender. *Perspectives on Psychological Science*, 8(3), 340–357, for more on social roles theory.

26 Clark and Hatfield's results have been replicated, controlling for risk perception, by (among others) Baranowski, A. M., & Hecht, H. (2015). Gender differences in and similarities in receptivity to casual sex invitations: Effects of location and risk perception. *Archives of Sexual Behavior*, 44, 2257–2265. https://doi.org/10.1007/s10508-015-0520-6.

27 Buss, D. M., & Schmitt, D. P. (1993). Sexual strategies theory: An evolutionary perspective on human mating. *Psychological Review*, 100(2), 204.

28 For more on error management theory, see Haselton, M. G., & Buss, D. M. (2000). Error management theory: A new perspective on biases in cross-sex mind reading. *Journal of Personality and Social Psychology*, 78(1), 81.

29 Hald, G. M., & Høgh-Olesen, H. (2010). Receptivity to sexual invitations from strangers of the opposite gender. *Evolution and Human Behavior*, 31(6), 453–458.

30 E.g., Townsend, J. M. (1995). Sex without emotional involvement: An evolutionary interpretation of sex differences. *Archives of Sexual Behavior*, 24(2), 173–206, and Townsend, J. M., & Levy, G. D. (1990). Effects of potential partners' physical attractiveness and socioeconomic status on sexuality and partner selection. *Archives of Sexual Behavior*, 19(2), 149–164.

31 The research into female STM as an adaptive strategy is considerable, with evolutionary psychologists (as opposed to traditional social psychologists) being more inclined to see it as adaptive and selective rather than pathological. This is a book about orgasm, rather than mating per se. That said, those who wish to pursue this line might wish to check out the following reading list:

Buss, D. M., & Schmitt, D. P. (1993). Sexual strategies theory: An evolutionary perspective on human mating. *Psychological Review, 100*(2), 204–232.

DeBruine, L. M. (2014). Women's preferences for male facial features. In *Evolutionary perspectives on human sexual psychology and behavior* (pp. 261–275). Springer.

Gangestad, S. W., Garver-Apgar, C. E., Cousins, A. J., & Thornhill, R. (2014). Intersexual conflict across women's ovulatory cycle. *Evolution and Human Behavior, 35*(4), 302–308.

Gangestad, S. W., Thornhill, R., & Garver-Apgar, C. E. (2010). Fertility in the cycle predicts women's interest in sexual opportunism. *Evolution and Human Behavior, 31*, 400–411.

Gildersleeve, K., Haselton, M. G., & Fales, M. R. (2014a). Do women's mate preferences change across the ovulatory cycle? A meta-analytic review. *Psychological Bulletin, 140*, 1205–1259.

Gildersleeve, K., Haselton, M. G., & Fales, M. R. (2014b). Meta-analyses and p-curves support robust cycle shifts in women's mate preferences: Reply to Wood and Carden (2014) and Harris, Pashler, and Mickes (2014). *Psychological Bulletin, 150*(5), 1272–1280.

Grammer, K., Renninger, L., & Fischer, B. (2004). Disco clothing, female sexual motivation, and relationship status: Is she dressed to impress? *Journal of Sex Research, 41*, 66–74.

Greiling, H., & Buss, D. M. (2000). Women's sexual strategies: The hidden dimension of short-term mating. *Personality and Individual Differences, 28*, 929–963.

Guéguen, N. (2009). Menstrual cycle phases and female receptivity to a courtship solicitation: An evaluation in a nightclub. *Evolution and Human Behavior, 30*, 351–355.

Haselton, M. G., & Miller, G. F. (2006). Women's fertility across the cycle increases the short-term attractiveness of creative intelligence. *Human Nature, 17*, 50–73.

Hughes, S. M., Farley, S. D., & Rhodes, B. C. (2010). Vocal and physiological changes in response to the physical attractiveness of conversational partners. *Journal of Nonverbal Behavior, 34*, 155–167.

Kenrick, D. T., Groth, G. E., Trost, M. R., & Sadalla, E. K. (1993). Integrating evolutionary and social exchange perspectives on relationships: Effects of gender, self-appraisal, and involvement level on mate selection criteria. *Journal of Personality and Social Psychology, 64*, 951–969.

Kenrick, D. T., Sadalla, E. K., Groth, G., & Trost, M. R. (1990). Evolution, traits, and the stages of human courtship: Qualifying the parental investment model. *Journal of Personality, 58*, 97–116.

Larson, C. M., Pillsworth, E. G., & Haselton, M. G. (2012). Ovulatory shifts in women's attractions to primary partners and other men: Further evidence of the importance of primary partner sexual attractiveness. *PLoS One, 7*, e44456. https://doi.org/10.1371/journal.pone.0044456.

Li, N. (2007). Mate preference necessities in long- and short-term mating: People prioritize in themselves what their mates prioritize in them. *Acta Psychologica Sinica, 39*, 528–535.

Li, N. P., & Kenrick, D. T. (2006). Sex similarities and differences in preferences for short-term mates: What, whether, and why. *Journal of Personality and Social Psychology, 90*, 468–489.

Little, A. C., Jones, B. C., Penton-Voak, I. S., Burt, D. M., & Perrett, D. I. (2002). Partnership status and the temporal context of relationships influence human female preferences for sexual dimorphism in male face shape. *Proceedings of the Royal Society of London B, 269*, 1095–1103.

Pawlowski, B., & Jasienska, G. (2005). Women's preferences for sexual dimorphism in height depend on menstrual cycle phase and expected duration of relationship. *Biological Psychology, 70*, 38–43.

Penton-Voak, I. S., Little, A. C., Jones, B. C., Burt, D. M., Tiddeman, B. P., Perrett, D. I. (2003). Female condition influences preferences for sexual dimorphism in faces of male humans (Homo sapiens). *Journal of Comparative Psychology, 117*, 264–271.

Pillsworth, E. G., & Haselton, M. G. (2006). Male sexual attractiveness predicts differential ovulatory shifts in female extra-pair attraction and male mate retention. *Evolution & Human Behavior, 27*, 247–258.

Pipitone, R. N., & Gallup Jr, G. G. (2008). Women's voice attractiveness varies across the menstrual cycle. *Evolution and Human Behavior, 29*, 268–274.

Provost, M. P., Kormos, C., Kosakoski, G., & Quinsey, V. L. (2006). Sociosexuality in women and preference for facial masculinization and somatotype in men. *Archives of Sexual Behavior, 35*, 305–312.

Provost, M. P., Troje, N. F., & Quinsey, V. L. (2008). Short-term mating strategies and attraction to masculinity in point-light walkers. *Evolution and Human Behavior, 29*, 65–69.

Puts, D. A. (2006). Cyclic variation in women's preferences for masculine traits: Potential hormonal causes. *Human Nature, 17,* 114–127.

Quist, M. C., Watkins, C. D., Smith, F. G., Little, A. C., DeBruine, L. M., & Jones, B. C. (2012). Sociosexuality predicts women's preferences for symmetry in men's faces. *Archives of Sexual Behavior, 41,* 1415–1421.

Regan, P. C. (1998). Minimum mate selection standards as a function of perceived mate value, relationship context, and gender. *Journal of Psychology and Human Sexuality, 10,* 53–73.

Regan, P. C. (1998). What if you can't get what you want? Willingness to compromise ideal mate selection standards as a function of sex, mate value, and relationship context. *Personality and Social Psychology Bulletin, 24,* 1294–1303.

Regan, P. C., & Berscheid, E. (1997). Gender differences in characteristics desired in a potential sexual and marriage partner. *Journal of Psychology and Human Sexuality, 9,* 25–37.

Regan, P. C., Levin, L., Sprecher, S., Christopher, F. S., & Cate, R. (2000). Partner preferences: What characteristics do men and women desire in their short-term and long-term romantic partners? *Journal of Psychology and Human Sexuality, 12,* 1–21.

Regan, P. C., Medina, R., & Joshi, A. (2001). Partner preferences among homosexual men and women: What is desirable in a sex partner is not necessarily desirable in a romantic partner. *Social Behavior and Personality, 29,* 625–633.

Sacco, D. F., Jones, B. C., DeBruine, L. M., & Hugenberg, K. (2012). The roles of sociosexual orientation and relationship status in women's face preferences. *Personality and Individual Differences, 53,* 1044–1047.

Schmitt, D. P., Couden, A., & Baker, M. (2001). Sex, temporal context, and romantic desire: An experimental evaluation of sexual strategies theory. *Personality and Social Psychology Bulletin, 27,* 833–847.

Schmitt, D. P., Jonason, P. K., Byerley, G. J., Flores, S. D., Illbeck, B. E., O'Leary, K. N., & Qudrat, A. (2012). A reexamination of sex differences in sexuality: New studies reveal old truths? *Current Directions in Psychological Science, 21,* 135–139.

Scheib, J. E. (2001). Context-specific mate choice criteria: Women's trade-offs in the contexts of long-term and extra-pair mateships. *Personal Relationships, 8,* 371–389.

Shackelford, T. K., Weekes, V. A., LeBlanc, G. J., Bleske, A. L., Euler, H. A., & Hoier, S. (2000). Female coital orgasm and male attractiveness. *Human Nature, 11,* 299–306.

Stewart, S., Stinnett, H., & Rosenfeld, L. B. (2000). Sex differences in desired characteristics of short-term and long-term relationship partners. *Journal of Social and Personal Relationships, 17,* 843–853.

Simpson, J. A., Gangestad, S. W., Christensen, P., & Niels, K. (1999). Fluctuating asymmetry, sociosexuality, and intrasexual competitive tactics. *Journal of Personality and Social Psychology, 76,* 159–172.

Thornhill, R., & Gangestad, S. W. (2003). Do women have evolved adaptation for extra-pair copulation? In *Evolutionary aesthetics* (pp. 341–368). Springer.

Part of the point of this extended list is to show that evolutionarily inflected research, in this case lasting over three decades, is the only program dedicated to exploring women's STM in an adaptive way. The overall pattern is that women's STM is guided by reproductive decisions (related to fertility cycles) and responsive to signifiers of high mate value. Recent research has reframed a lot of women's STM as mate *switching*, and I will have more to say about this point in the last chapter.

Wiederman, M. W., & Dubois, S. L. (1998). Evolution and sex differences in preferences for short-term mates: Results from a policy capturing study. *Evolution and Human Behavior, 19,* 153–170.

Scheib, J. E. (1994). Sperm donor selection and the psychology of female mate choice. *Ethology & Sociobiology, 15,* 113–129.

Seal, D. W., Agostinelli, G., & Hannett, C. A. (1994). Extradyadic romantic involvement: Moderating effects of sociosexuality and gender. *Sex Roles, 31,* 1–22.

Simpson, J. A., & Gangestad, S. W. (1992). Sociosexuality and romantic partner choice. *Journal of Personality, 60,* 31–51.

32 Conan Doyle, A. (1893). The naval treaty. In *The memoirs of Sherlock Holmes* (pp. 238–282).

33 Eberhard, W. G. (1985). *Sexual selection and animal genitalia.* Harvard University Press, argues that female choice mechanisms are pretty much the only explanation for the variation among male genitalia.

34 For details of how prior sexual activity is predictive of female orgasm, see, for example, Richters, J., de Visser, R., Rissel, C., & Smith, A. (2006). Sexual practices at last heterosexual encounter and occurrence of orgasm in a national survey. *Journal of Sex Search, 43*(3), 217–226.

35 *Not*, I hope it is clear, merely heterosexual intercourse. As to be discussed later, plenty of female-female sexual pairings can involve penetration too, often (but not always) using artificial aids.

36 Dawkins, R. (2006). The Selfish Gene: With a new Introduction by the Author. He explores this idea on pp. 307–308. See also Eberhard, W. G. (1985). *Sexual selection and animal genitalia*. Harvard University Press.

37 Dixson, A. F. (1999). *Primate sexuality: Comparative studies of the prosimians, monkeys, apes, and human beings*. Oxford University Press. Dixon not only provides comparative measures of primate genitalia but also advances the possibility that penises are the length they are in humans because of the necessary size of the birth canal relative to the baby's head. The argument goes that human penises have to be long because the evolution of a large cranium in relation to the birth canal locates the cervix a long way from the vaginal entrance. Thus, this is an argument that physical constraints trump selection. However, it may not be necessary to choose one over the other. There is a well-known trade-off here in that human babies have to be born somewhat prematurely, or otherwise the mother would be injured during childbirth, or worse. However, if this was the main (or the only) selectional driver, then we would expect other primates to follow it. It is true that our similarly sized primate cousins, gorillas, chimpanzees, and orang-outans, do not appear to experience a great difficulty in childbirth. However, other smaller primates, such as many monkeys and gibbons, have just as large a head in relation to the birth canal as humans do, and they do not need disproportionate penises to inseminate the females. See Rosenberg, K., & Trevathan, W. (2002). Birth, obstetrics and human evolution. *BJOG: An International Journal of Obstetrics & Gynaecology*, *109*(11), 1199–1206, for a review of the data.

38 For more on how penis size is probably under selection by female choice mechanisms, see Mautz, B. S., Wong, B. B., Peters, R. A., & Jennions, M. D. (2013). Penis size interacts with body shape and height to influence male attractiveness. *Proceedings of the National Academy of Sciences*, *110*(17), 6925–6930, and Stulhofer, A. (2006). How (un)important is penis size for women with heterosexual experience? Letter to the editor. *Archives of Sexual Behavior*, *35*, 5–6.

39 For (yet) more on the fact that larger penises are associated with orgasms produced by penetration (and hence may be under selection), see Costa, R. M., Miller, G. F., & Brody, S. (2012). Women who prefer longer penises are more likely to have vaginal orgasms (but not clitoral orgasms): Implications for an evolutionary theory of vaginal orgasm. *The Journal of Sexual Medicine*, *9*(12), 3079–3088. The authors follow Freud's somewhat confusing distinction between clitoral and vaginal orgasms, the nomenclature and normativity of which I explored in chapter eight. Stuart Brody, in particular, advances the claim that vaginal orgasms are healthier in some way. I have to confess that I find there to be too many confounding variables to find this conclusion compelling. That said, it is certainly true that humans actively run interference on one another's sexual pleasure.

40 Penetrative vigor was predictive of what we termed "deep" (rather than "vaginal") orgasms in our 2012 study. See King, R., & Belsky, J. (2012). A typological approach to testing the evolutionary functions of human female orgasm. *Archives of Sexual Behavior*, *41*(5), 1145–1160, for more details.

41 The term "interminable pistoning" is owed to Maggie McNeil, whose site "The Honest Courtesan" should be required reading for all sex researchers and therapists. https://maggiemcneill.wordpress.com/2019/01/31/time-management/#comments last accessed 01/02/2019.

42 For the intriguing story of a penis that was too large, see Kompanje, E. J. O. (2006). Painful sexual intercourse caused by a disproportionately long penis: An historical note on a remarkable treatment devised by Guilhelmius Fabricius Hildanus (1560–1634). *Archives of Sexual Behavior*, *35*, 603–605. The "remarkable treatment," for those agog to know the answer, was a sort of washer arrangement.

43 van Seters, A. P., & Slob, A. K. (1988). Mutually gratifying heterosexual relationship with micropenis of husband. *Journal of Sex & Marital Therapy*, *14*(2), 98–107.

44 Carol Wade is quoted in Tavris, C. (1992). *The mismeasure of woman*. Simon & Schuster.

45 For more on the general healthiness of orgasms, however generated, see Prause, N., Kuang, L., Lee, P. M., & Miller, G. F. (2016). Clitorally stimulated orgasms are associated with better control of sexual desire, and not associated with depression or anxiety, compared with vaginally stimulated orgasms. 'Clitoral' versus 'vaginal' orgasms: False dichotomies and differential effects. *Journal of Sexual Medicine*, *13*(11), 1676–1685. https://doi.org/10.1016/j.jsxm.2016.08.014.

46 Talalaj, J. (1994). *The strangest human sex, ceremonies and customs*. Hill of Content Publishing Company Pty Limited, for a review of the ways penises have been interfered with across time and space.

47 The details and implications of the hydraulics are explored here: Sheets-Johnstone, M. (1990). Hominid bipedality and sexual selection theory. *Evolutionary Theory*, *9*(1), 57–70, Short, R. V. (1980). The origins of human sexuality. In C. R. Austin & R. B. Short (Eds.), *Reproduction in mammals* (Vol. 8, pp. 1–33). Cambridge University Press.

48 "Get straight up and start looking for another potential partner." The technical term for this is a "high degree of post-copulatory sexual selection." See Brindle, M., & Opie, C. (2016). Postcopulatory sexual selection influences baculum evolution in primates and carnivores. *Proceedings of the Royal Society B: Biological Sciences, 283*(1844), and Dixson, A. F. (2012). *Primate sexuality: Comparative studies of the prosimians, monkeys, apes, and human beings* (2nd ed.). Oxford University Press.

49 For fMRIs of the interactions between penises and vaginas, see Schultz, W. W., van Andel, P., Sabelis, I., & Mooyaart, E. (1999). Magnetic resonance imaging of male and female genitals during coitus and female sexual arousal. *British Medical Journal, 319*, 1596–1600. What you will see is that penises bend, much like modern sex toys, as described in chapter six.

50 For a more thorough review, see Miller, G. F. (1998). How mate choice shaped human nature: A review of sexual selection and human evolution. In *Handbook of evolutionary psychology: Ideas, issues, and applications* (pp. 87–129). Psychology Press.

51 Braun, M., Wassmer, G., Klotz, T., Reifenrath, B., Mathers, M., & Engelmann, U. (2000). Epidemiology of erectile dysfunction: Results of the 'cologne male survey'. *International Journal of Impotence Research, 12*(6), 305. It is worth noting that the rates of male erectile impotence—ranging from 20% to as high as 53% in some populations—is never put forward as evidence that the male uro-genital system was "not designed for sexual intercourse." Contrast this with the ready acceptance of the notion that (allegedly) low rates of female orgasm from intercourse indicate "dysfunction" or a by-product.

52 The medicalization of so-called female sexual dysfunction and the call for a pink Viagra is explored here: Tiefer, L. (2006). Female sexual dysfunction: A case study of disease mongering and activist resistance. *PLoS Medicine, 3*(4), e178. This is a large topic, and this is not a book about medicine per se. However, it is worth pointing out that lack of desire is not something we might feel entirely comfortable about treating with a pill, even if that were possible.

53 For a fuller list of how low partner FA interacts with female reproductive cycles (and other factors), see note 19.

54 Galton, F. (1908). *Local association for promoting eugenics*. Eugenics Education Society. For the curious, Galton rated London as having the most attractive women, Aberdeen the least.

55 https://theblog.okcupid.com/tagged/data OK Cupid regularly publish and discuss metadata. This was retrieved on 09/12/2019.

56 Møller, A. P., & Pomiankowski, A. A. (1993). Fluctuating asymmetry and sexual selection. *Genetica, 89*(1–3), 267. See also Thornhill, R., Gangestad, S. W., & Comer, R. (1995). Human female orgasm and mate fluctuating asymmetry. *Animal Behaviour, 50*(6), 1601–1615 for evidence that orgasm frequency is predicted by (independently rated) measures of partner attractiveness.

57 Dixson, A. F. (1999). *Primate sexuality: Comparative studies of the prosimians, monkeys, apes, and human beings*. Oxford University Press.

58 Leary, R. F., Allendorf, F. W., & Knudsen, K. L. (1984). Superior developmental stability of heterozygotes at enzyme loci in salmonid fishes. *The American Naturalist, 124*(4), 540–551.

59 Thornhill, R., & Sauer, P. (1992). Genetic sire effects on the fighting ability of sons and daughters and mating success of sons in a scorpionfly. *Animal Behaviour, 43*(2), 255–264.

60 Vehrencamp, S. L., Bradbury, J. W., & Gibson, R. M. (1989). The energetic cost of display in male sage grouse. *Animal Behaviour, 38*(5), 885–896.

61 Møller, A. P., & Eriksson, M. (1994). Patterns of fluctuating asymmetry in flowers: Implications for sexual selection in plants. *Journal of Evolutionary Biology, 7*(1), 97–113.

62 Møller, A. P., & Pomiankowski, A. A. (1993). Fluctuating asymmetry and sexual selection. *Genetica, 89*, 267–279.

63 Lee, A. J., & Zietsch, B. P. (2011). Experimental evidence that women's mate preferences are directly influenced by cues of pathogen prevalence and resource scarcity. *Biology Letters*, rsbl20110454.

64 Gallup, G. G., Jr., & Frederick, D. A. (2010). The science of sex appeal: An evolutionary perspective. *Review of General Psychology, 14*(3), 240.
        Thornhill, R., Gangestad, S. W., & Comer, R. (1995). Human female orgasm and mate fluctuating asymmetry. *Animal Behaviour, 50*(6), 1601–1615.

65 Sometimes web-based surveys have been specifically criticized as lacking diversity and seriousness. However, a review and comparison between studies has largely addressed these worries. See Gosling, S. D., Vazire, S., Srivastava, S., & John, O. P. (2004). Should we trust web-based studies? A comparative analysis of six preconceptions about internet questionnaires. *American Psychologist, 59*(2), 93, for a review. In our case, our web-based studies enhanced and were consistent with interviews. It is possible that everyone is lying about everything, all the time. But if so this is not a problem for sex research per

se, but for all research with humans. As Tinbergen 1963 put it, "Contempt for simple observation is a lethal trait in any science, and certainly in a science as young as ours (p. 412). We are not too old for simple observation yet.

66 Oliveira-Pinto, A. V., Santos, R. M., Coutinho, R. A., Oliveira, L. M., Santos, G. B., Alho, A. T., . . . Pasqualucci, C. A. (2014). Sexual dimorphism in the human olfactory bulb: Females have more neurons and glial cells than males. *PLoS One, 9*(11), e111733.

67 Mostafa, T., El Khouly, G., & Hassan, A. (2012). Pheromones in sex and reproduction: Do they have a role in humans? *Journal of Advanced Research, 3*(1), 1–9. https://doi.org/10.1016/j.jare.2011.03.003, and McCoy, N. L., & Pitino, L. (2002). Pheromonal influences on sociosexual behavior in young women. *Physiology & Behavior, 75*(3), 367–375. https://doi.org/10.1016/S0031-9384(01)00675-8 have investigated the effects of androstenone.

68 Schmidt, J. O., Slessor, K. N., & Winston, M. L. (1993). Roles of Nasonov and queen pheromones in attraction of honeybee swarms. *Naturwissenschaften, 80*(12), 573–575.

69 Garver-Apgar, C. E., Gangestad, S. W., Thornhill, R., Miller, R. D., & Olp, J. J. (2006). Major histocompatibility complex alleles, sexual responsivity, and unfaithfulness in romantic couples. *Psychological Science, 17*(10), 830–835.

70 Allen, C., Cobey, K. D., Havlíček, J., & Roberts, S. C. (2016). The impact of artificial fragrances on the assessment of mate quality cues in body odor. *Evolution and Human Behavior, 37*(6), 481–489.

71 Alvergne, A., & Lummaa, V. (2010). Does the contraceptive pill alter mate choice in humans? *Trends in Ecology & Evolution, 25*(3), 171–179.

72 Garver-Apgar, C. E., Gangestad, S. W., Thornhill, R., Miller, R. D., & Olp, J. J. (2006). Major histocompatibility complex alleles, sexual responsivity, and unfaithfulness in romantic couples. *Psychological Science, 17*(10), 830–835.

73 Wedekind, C., & Füri, S. (1997). Body odour preferences in men and women: Do they aim for specific MHC combinations or simply heterozygosity? *Proceedings of the Royal Society of London B: Biological Sciences, 264*(1387), 1471–1479.

74 Thornhill, R., & Gangestad, S. W. (1999). The scent of symmetry: A human sex pheromone that signals fitness? *Evolution and Human Behavior, 20*(3), 175–201.

75 See, for example, Garver-Apgar, C. E., Gangestad, S. W., Thornhill, R., Miller, R. D., & Olp, J. J. (2006). Major histocompatibility complex alleles, sexual responsivity, and unfaithfulness in romantic couples. *Psychological Science, 17*(10), 830–835, and Jones, B. C., Feinberg, D. R., Watkins, C. D., Fincher, C. L., Little, A. C., & DeBruine, L. M. (2013). Pathogen disgust predicts women's preferences for masculinity in men's voices, faces, and bodies. *Behavioral Ecology, 24*(2), 373–379.

76 Schwambergová, D., Sorokowska, A., Slámová, Ž., Fialová, J. T., Sabiniewicz, A., Nowak-Kornicka, J., . . . Havlíček, J. (2021). No evidence for association between human body odor quality and immune system functioning. *Psychoneuroendocrinology, 132*, 105363.

77 Schwambergová, D., Sorokowska, A., Slámová, Ž., Fialová, J. T., Sabiniewicz, A., Nowak-Kornicka, J., . . . Havlíček, J. (2021). No evidence for association between human body odor quality and immune system functioning. *Psychoneuroendocrinology, 132*, 105363.

78 Allen, C., Cobey, K. D., Havlíček, J., & Roberts, S. C. (2016). The impact of artificial fragrances on the assessment of mate quality cues in body odor. *Evolution and Human Behavior, 37*(6), 481–489.

79 Fleischmann, D. (2016a). Sex differences in disease avoidance. In T. K. Shackelford & V. Weekes-Shackelford (Eds.), *Encyclopedia of evolutionary psychological science.*

80 Schäfer, L., Mehler, L., Hähner, A., Walliczek, U., Hummel, T., & Croy, I. (2018). Sexual desire after olfactory loss: Quantitative and qualitative reports of patients with smell disorders. *Physiology & Behavior.*

81 King, R. (2022). "Nose Job": Possible side effects of Sars-Cov-2. *Archives of Sexual Behavior, 51*(2), 705–705.

82 King, R., & Belsky, J. (2012). A typological approach to testing the evolutionary functions of human female orgasm. *Archives of Sexual Behavior, 41*(5), 1145–1160.

83 Sherlock, J. M., Sidari, M. J., Harris, E. A., Barlow, F. K., & Zietsch, B. P. (2016). Testing the mate-choice hypothesis of the female orgasm: Disentangling traits and behaviours. *Socioaffective Neuroscience & Psychology, 6*(1), 31562.

84 Gallup, G. G., Jr., Ampel, B. C., Wedberg, N., & Pogosjan, A. (2014). Do orgasms give women feedback about mate choice? *Evolutionary Psychology, 12*(5).

85 Garver-Apgar, C. E., Gangestad, S. W., Thornhill, R., Miller, R. D., & Olp, J. J. (2006). Major histocompatibility complex alleles, sexual responsivity, and unfaithfulness in romantic couples. *Psychological Science, 17*(10), 830–835.

86 Overheard in New York: Retrieved from overheardinnewyork.com August 2008.

87 Plomin, R., Caspi, A., Pervin, L. A., & John, O. P. (1990). Behavioral genetics and personality. In *Handbook of Personality: Theory and Research* (Vol. 2, pp. 251–276). Guildford Press.

88 For a review of why humans are mutually sexually selecting see Stewart-Williams, S., & Thomas, A. G. (2013). The ape that thought it was a peacock: Does evolutionary psychology exaggerate human sex differences? *Psychological Inquiry*, *24*(3), 137–168, and commentaries on the paper.

89 Alonzo, S. H., & Sinervo, B. (2001). Mate choice games, context-dependent good genes, and genetic cycles in the side-blotched lizard, Uta stansburiana. *Behavioral Ecology and Sociobiology*, *49*(2–3), 176–186.
    Corl, A., Davis, A. R., Kuchta, S. R., Comendant, T., & Sinervo, B. (2010). Alternative mating strategies and the evolution of sexual size dimorphism in the side-blotched lizard, Uta stansburiana: A population-level comparative analysis. *Evolution: International Journal of Organic Evolution*, *64*(1), 79–96.

90 Jones, E. (1953). *Sigmund Freud: Life and work*. Hogarth Press. The quote is from Vol. 2, Pt. 3, Ch. 16, p. 421. In a footnote Jones gives the original German, "*Was will das Weib?*

91 John, O. P., & Srivastava, S. (1999). The Big Five trait taxonomy: History, measurement, and theoretical perspectives. *Handbook of Personality: Theory and Research*, *2*, 102–138.

92 Gillespie, D. O., Russell, A. F., & Lummaa, V. (2008). When fecundity does not equal fitness: Evidence of an offspring quantity versus quality trade-off in pre-industrial humans. *Proceedings Biological Science*, *22*(275), 713–722.

93 Figueredo, A. J., Sefcek, J. A., Vasquez, G., Brumbrach, B., King, J. E., & Jacobs, W. J. (2005). Evolutionary personality psychology. In D. M. Buss (Ed.), *Handbook of evolutionary psychology* (pp. 851–877). Wiley.

94 Nettle, D. (2005). An evolutionary approach to the extraversion continuum. *Evolution and Human Behavior*, *26*(4), 363–373.

95 There is a discussion of it at www.salon.com/2011/02/22/neko_case_male_groupies/ last accessed 28/02/2019. The explanation offered, that female rock stars "do not get hit on" seems unlikely in the age of #metoo. What is more likely is that, like the characters in *Sex and the City* in their famous scene surrounded by (male) waiters and talking about how there are "literally no men in New York," they simply see such men as beneath their notice.

96 Henrich, J., Heine, S. J., & Norenzayan, A. (2010). The weirdest people in the world. *Behavioral and Brain Sciences*, *33*(2–3), 61–83.

97 Alvergne, A., Jokela, M., & Lummaa, V. (2010). Personality and reproductive success in a high-fertility human population. *Proceedings of the National Academy of Sciences*, *107*(26), 11745–11750.

98 See, for example, Gould, S. J. (1997). *The mismeasure of man* (Rev. and expanded ed.). W.W. Norton. Despite numerous calls to revise the repeated misrepresentations of research in this book, it stands as a testament to an ideological commitment to not applying scientific methods to human nature. For more details on this misrepresentation, see Lewis, J. E., DeGusta, D., Meyer, M. R., Monge, J. M., Mann, A. E., & Holloway, R. L. (2011). The mismeasure of science: Stephen Jay Gould versus Samuel George Morton on skulls and bias. *PLoS Biology*, *9*(6), e1001071.

99 For example, avowed (white separatist) racist Jared Taylor has this to say on the matter: "I think Asians are objectively superior to Whites by just about any measure that you can come up with in terms of what are the ingredients for a successful society." He repeats this idea in an interview with Cenk Uygur, available Taylor, J. (2008). *TV interview for young Turks about his racism with Cenk Uygur*. www.youtube.com/watch?v=oxhsSLos8hk last accessed 03/04/2023.

100 Miller, G. (2011). *The mating mind: How sexual choice shaped the evolution of human nature*. Anchor.

101 For example, Shackelford, T. K., Schmitt, D. P., & Buss, D. M. (2005). Universal dimensions of human mate preferences. *Personality and Individual Differences*, *39*(2), 447–458. These findings seem robustly cross-cultural, Buss, D. M. (2007). The evolution of human mating. *Acta Psychologica Sinica*, *39*(3), 502–512.

102 Gignac, G. E., Darbyshire, J., & Ooi, M. (2018). Some people are attracted sexually to intelligence: A psychometric evaluation of sapiosexuality. *Intelligence*, *66*(C), 98–111. The findings here were that sapiosexuality was rare indeed and that extremes of intelligence were also rarely valued. Thus, selection in this trait may be somewhat stabilizing in nature rather than producing runaway selection.

103 Halpern, C. T., Joyner, K., Udry, J. R., & Suchindran, C. (2000). Smart teens don't have sex (or kiss much either). *Journal of Adolescent Health*, *26*(3), 213–225.

104 Fieder, M., Huber, S., Bookstein, F. L., Iber, K., Schäfer, K., Winckler, G., & Wallner, B. (2005). Status and reproduction in humans: New evidence for the validity of evolutionary explanations on basis of a university sample. *Ethology*, *111*(10), 940–950.

105 Sherlock, J. M., Sidari, M. J., Harris, E. A., Barlow, F. K., & Zietsch, B. P. (2016). Testing the mate-choice hypothesis of the female orgasm: Disentangling traits and behaviours. *Socioaffective Neuroscience & Psychology, 6*(1), 31562.

106 Gallup, G. G., Jr, Ampel, B. C., Wedberg, N., & Pogosjan, A. (2014). Do orgasms give women feedback about mate choice? *Evolutionary Psychology, 12*(5).

107 For example, Ogas, O., & Gaddam, S. (2011). *A billion wicked thoughts: What the world's largest experiment reveals about human desire.* Dutton, Penguin Books.

108 Lewis, M. (2010). *Liar's poker.* W.W. Norton.

109 King, R., & Belsky, J. (2012). A typological approach to testing the evolutionary functions of human female orgasm. *Archives of Sexual Behavior, 41*(5), 1145–1160.

110 These findings are explored in King and Belsky (2012), Sherlock et al. (2016), and Gallup et al. (2014).

111 Darwin, C. (1888). *The descent of man and selection in relation to sex* (Vol. 1). Murray. The quote comes at the very end of chapter eight.

10 Getting Cross about Culture

Every instinct is an impulse. . . . Man has a far greater variety of impulses than any lower animal; and any one of these impulses, taken in itself, is as 'blind' as the lowest instinct can be; but, owing to man's memory, power of reflection, and power of inference, they come each one to be felt by him, after he has once yielded to them and experienced their results, in connection with a foresight of those results.

— (William James, *Principles of Psychology*, chapter XXIV)

No culture, that we know of, is neutral about female orgasm. Some celebrate it, to the extent of (literally) enshrining techniques to generate it in the living stone of sacred places. Long after the pornography of the West has crumbled to dust, the detailed instructions on how everyone could enjoy sex to the utmost will remain on the walls of the Rajput Kingdom temples.[1] At the other extreme, some cultures fear female orgasm so much that they alter women's bodies in the most eye-watering ways, in order to try to prevent its happening at all. In between these extremes we have every flavor of denial, misinformation, and manipulation by guilt, alternating with celebration, mythologizing, and honest instruction. Why is this? The truthful response is that we do not have all the answers, and in any case, a full analysis of human cultural variation is beyond the scope of a book about female orgasm.

However, it is simply illegitimate, as a behavioral scientist, to look at this variance and throw up one's hands in an admission of defeat. There *are* patterns here, if one is patient and takes data from a multiplicity of sources. There are lessons to be learned, and hypotheses to be tested. That is what this chapter is about. The sheer weight of evidence and previous depth of explanatory power, both between and within species, suggests that life history theory is going to be the most fruitful place to look for the source of many of these variances. In addition, cultural variance allows us to test our ideas about how humans maximize fitness in different settings, using female orgasm as a selection mechanism.[2]

10.1 WEIRD SCIENCE

Humans are weird. More to the point, humans are WEIRD.[3] Even more to the point, the bulk of those humans, whom we have studied in our behavioral sciences, are White, Educated, Industrialized, Rich, and Democratic.[4] And that sort of WEIRD-ness is weird, because none of those five traits are, or have ever been, common to humans over most of our history and geography.[5] However strongly committed one is to the doctrine of the psychic unity of humanity, and one should be so committed, it is hard to escape the thought that the bulk of human behavioral science has been conducted on atypical humans.[6]

For these, and many other reasons, it is important to study target behaviors in a range of cultures. This sort of cross-cultural study has often been the purview of anthropology, but increasingly, the methods and theories of other branches of behavioral science have been applied cross-culturally.[7] In this chapter I am going to discuss what happens when we apply this cross-cultural requirement to the study of female orgasm. But, before I get to that, what do we mean by cultural differences anyway?

10.1.1 NURTURE IS SOMETHING THAT NATURE DOES

For a variety of reasons, some of which go very deep into our attempts to understand ourselves, and some of which seem to be mere historical accident, the study of human behavior has been split

DOI: 10.1201/9781003372356-10

down two paths. One of these paths, the biology of human beings (pace the wilder fantasies of post-modernism), has been assumed to be fixed across the species—analogously to computer hardware. At the same time, the things that make us different from one another—cultural variation—has been taken to be something akin to computer software.

One (relatively dull) reason for this split has been that it is simply mathematically easier to model human biological data—genetics, for the most part—in terms of what population geneticists call the neutral theory.[8] At the level of magnification of population genetics, a species is a collection of genes, not really a collection of individuals at all, and if that collection can be modeled as mating randomly with one another, any genetic changes can be put down to neutral drift, rather than selection. From this research bias we sometimes get the curious notion that evolution has somehow stopped for humans. However, humans show little evidence of mating randomly in real life.[9]

On the other hand, treating human brains as being simple (but highly absorbent) cultural sponges—the blank slate model—neatly explains, for a certain, limited definition of "explains," human cultural differences.[10] However, if you were to go on to ask where all this culture ultimately comes from, then blank slatists can only respond with blank stares.

How can we make sense of the limitations of each of these models? A useful first step is to realize that they *are* models, and that the myth of Pygmalion warns of the danger of falling in love with the models we make, mistaking them for reality itself.[11] "All models are false, some models are useful" is a handy thing to remember, if one is in danger of reifying one's methods and mistaking them for reality itself.[12] There is nothing inherently wrong with simplifications and placeholders in science. Physics, for example, has plenty of placeholders such as "dark matter" or "dark energy," which are conceptual containers for things we know are exerting some influence but we, as yet, do not understand.[13]

Recent advances in genetics have made some of the simplifying assumptions of both the biological and cultural black boxes more and more untenable. In due course, a lot of cultural anthropology and evolutionary psychology will need revising, but our discussion here needs consideration as to what, given the argument that female orgasm is a (semi) cryptic selection mechanism, might be said to be under selection.[14] Here we are talking about female sexual selection, of course, but these revisions I mention are going to be more far reaching than that, eventually. The number of genes that directly code for proteins—the building blocks of our bodies—is less than 30,000 and, according to some accounts, considerably less than that.[15] This means that only a few hundred coding genes separate humans from chimpanzees.[16] If these are doing all the work of variance, then that is a lot of heavy lifting being off-loaded onto "culture," and anyone who sincerely believes that it is mainly upbringing that means that Washoe the clever chimpanzee is not writing (or, indeed, reading) this book must accept that the burden of proof is on them.[17]

One of the implications of all this is that, to anyone who has given it more than a few minutes' thought, we have miscounted the units of selection. As the great AI researcher Marvin Minsky is said to have once quipped, "Don't believe any genetics research published in your lifetime." The source of this miscounting is the roughly 98% of the human genome which, until recently, was described as "junk DNA." This was always a placeholder description, and in recent years it has become more and more obvious that this so-called junk has all sorts of regulatory functions, turning other genes on and off in response to, for example, threats and opportunities, in a lifelong dance. These genes are still under selection, however, complicating the simple "hardware versus software" model mentioned earlier.[18]

There is no clear consensus yet on how to fully resolve these issues in the behavioral sciences. We know that humans look, and sound very different across the planet, but we are not yet in full agreement about how deep these differences really go. Languages and cultures *seem* very different, but all are equally complex, and (for example) express admiration of very similar human virtues, across time and space.[19] For example, no culture denigrates cooperation, or values cowardice, although, of course, they can vary enormously in how they express these things or to whom they apply them.[20]

Linguistic differences provide interesting twists on local emphases, not fundamental gulfs that defy translation (and thus mutual understanding) between peoples.[21] Culture provides a kind of information—like DNA does—but altered at a much more rapid pace. While some scholars liked to emphasize the fact that behavior, unlike bones, does not fossilize, DNA provides much more than that; it is an unbroken chain of successful information stretching back millions of years.[22]

Where does the information space that culture provides belong in our ethological picture? Traditional ethology does not make hard-and-fast distinctions between physical and behavioral traits: They are all subject to the same four types of questions that Tinbergen outlined sixty years ago.[23] On the other hand, some would argue for a dual-inheritance model, which roughly means that culture is believed to consist of an information stream that is independent from the information stream of our DNA.

Some call this cultural information "memes," but others do not.[24] Proponents of the varieties of dual-inheritance models would draw attention to good tricks such as fire, the wheel, and knowledge of local foods and poisons that definitely contribute to survival and reproductive fitness but are not plausibly directly encoded in human DNA.[25] On the other hand, those scholars who oppose the idea of dual inheritance would emphasize that many supposed cultural differences amount to differently evoked instincts in differing ecologies, combined with specialized (proximate) learning mechanisms.[26] These scholars would remind us that human skills are not acquired evenly, with some being acquired especially rapidly, effortlessly, and with little (or even no) explicit instruction. As the pioneering psychologist William James said in the above quote, humans have more instincts than other animals, not fewer. However, we need to remember that instincts are canalized learning mechanisms that evolved to (on average) generate fitness-producing outputs from a variety of inputs. They are not, with a few interesting exceptions, unbendable reflexes.

10.2   NATURE AND NURTURE

The nature/nurture opposition, redolent as it is with connotations of good versus bad, changeable versus inevitable, and learned versus innate, has so much baggage attached to it that it has ceased to be a useful conveyor of information,

For example, a question that seems to excite people hugely is whether humans are naturally monogamous or polygamous, and I think the question deserves an answer. It is: Yes. We are naturally monogamous, and naturally polygamous.[27] I am not trying to be facetious. The evidence—some of which I will review in this chapter—is strongly that humans have a suite of reproductive strategies available to them, dependent on individual differences, ecological exigencies, and personal opportunities. Some of this variance is captured by the concept of *sociosexuality*—broadly, the willingness to engage in commitment-free sex, but not all—because patterns of multiple long-term matings, polygamy, are also part of the human condition.[28] Furthermore, the evidence is that this polygamy includes cultures with stable polyandry (multiple males per female[29]), polygyny (multiple females per male),[30] and considerable degrees of sexual separatism, with attendant homosexuality. If we have not always been aware of this—or if we have been in denial about it—this is partly because we are victims of our own propaganda. This propaganda includes our own attempts to convince ourselves, as well as others' attempts to convince us.

If human mating were simple, then there would be no poetry, no love songs, no stories of star-crossed lovers. All of these things—whatever some sociologists may have bizarrely claimed—are cross-cultural universals.[31] Everywhere there are humans, they declare undying love for each other, and everywhere they cheat. Everywhere there are humans, there are poems praising their lovers to the stars. Does this imply that said lovers still need convincing? Everywhere there are humans, there are rituals that bind people together in the eyes of their friends, their families, and their wider community.[32] Why bother, if such things just flowed naturally and effortlessly? Love is hard work. The good news is that scientific evidence is that love is a very real thing, but it a lot more complicated than we might have initially thought.

10.2.1 A Note about Instincts

Early ethologists, rather like Freud, had a hydraulic model of instincts—they were seen as pressures seeking an outlet of some sort.[33] Thus, an animal denied expression of its aggressive instinct might displace it onto another target. This model had some validity when talking about fixed action patterns, but when we move on to more complex behaviors, then the hydraulic model has serious shortcomings. Lorenz, for example, talked as if instincts were not only hydraulic, with an internal power, but pre-programmed, inevitable, fixed at birth, and existing for the good of the species.[34] Calling some behaviors "hard-wired" just puts a modern (electronic) gloss on these outmoded ideas, not a single one of which would be espoused by any modern behavioral scientist. This is not to disparage Lorenz and other early pioneers, but to acknowledge that we have moved on, largely thanks to being able to stand on their giant shoulders.

Behavioral scientists today use a model based on computation, which is very different from saying that the brain is a computer. Computation is just the performing of functions (computers are just a physical way to do this), and those myriad functions did not arise by accident; the brain evolved to solve particular kinds of problems. This is not unique to human brains. All brains, indeed, all nervous systems, exist to enable the organism to make useful moves in its *umwelt*, by evaluating threats and opportunities.[35] Plants do not need brains, because their static strategies preclude reacting to threats and opportunities, in these kinds of ways. Animals, like sea squirts, that get to a life stage where all the meaningful threats and opportunities have been resolved, have been known to eat their own brains. Even the common shrew reduces—by absorption—costly brain matter by 25% during hibernation, only to regrow it in the spring. Brains are expensive real estate and need to continually justify their upkeep by solving problems.

The information-processing model has replaced the hydraulic one throughout neuroscience, psychology, and behavioral biology. Viewed this way, instincts[36] are somewhat specialized problem-solving engines, sensitive to context. Some of these instincts might interact with one another, but many are walled off from their fellows (our self-serving biases in reasoning are just one obvious instance of this). Some people might call these engines "modules" or they might call them *open* (as opposed to *closed*) programs,[37] but one thing is beyond dispute: The brain evolved to solve fitness problems. Any model that implicitly or explicitly denies this can be safely jettisoned for the same reason that any creationist (or otherwise magical) theory can be jettisoned. By way of example, there can be no selection on "general purpose intelligence" because there are no general-purpose problems to solve. There is such a thing as general intelligence, but this really means being simultaneously good at a range of different tasks. It is for this reason that accounts of culture, unglued from biology, are unsatisfying.

Every brain that was not, to some extent, "organized in advance of experience"[38] would be ruthlessly outcompeted by those that were, so we can also assume that regular patterns of threats and opportunities would result in evolved responses to such patterns that maximized fitness. This is, indeed, exactly what we do find across taxa, and the study of it is called Life History Theory.

10.3 LIFE HISTORY THEORY AND HUMAN SEXUALITY

Imagine arriving at the airport car hire desk and considering the options. You are told that you have only two choices. A flashy, sporty number that will allow you to go fast, or a rather more boring sedan, which will reliably get you to your destination, but will be far less exciting. Which do you choose? If you are sensible, then you will want to ask some further questions. Questions such as "What are the roads like? Are there potholes? Are the bridges reliable? What is the gap between petrol stations?" You might also ask, "How corrupt are the local police, and which car will attract their attention?"

In short, a sensible car hire decision relies on knowing what the road ahead is like. For the last fifty or so years, biologists have become increasingly aware that natural selection has built in—in various ingenious ways that I will discuss in a moment—a stage of life, in many organisms, that amounts to asking the question, "What is the road ahead likely to be?" These mechanisms fall under the description of Life History Theory, a mid-level biological set of interlocked explanations and predictions that marry together Tinbergen's *adaptationist, mechanistic*, and *ontogenic* types of biological questions.

The field is vast but, to a first approximation, the major axis of variance concerns fast or slow life histories. In brief, various early experiences calibrate the growing organism for the likely road ahead. Fast life histories[39] are ones that maximize fitness in conditions of stress and risk, whereas slow ones maximize fitness in conditions of relative predictability. The calibration events typically (at least in humans) center on parenting input, and it is very tempting to think of certain forms of parenting—those that push the offspring down a fast path—as bad. It is important to realize that, from an ethological perspective, such parenting has evolved to calibrate the offspring to likely patterns of risk in its future.[40] If, as a society, we wish to *re*calibrate such individuals, then we can—and we should—recognize that an ounce of prevention would prevent a pound of cure here.

10.4   LIFE HISTORY AND SEXUALITY

We are used to thinking of human sexuality in terms of labels and identity but, biologically, sexuality is a strategy for maximizing genetic transfer, and strategic options can change throughout the lifetime. We have known for some time that female sexuality is more fluid than male. This is not the place to offer a detailed exegesis of the various inter-related development and (possible) adaptive pathways that lead to the much greater variance in female than male sexuality.[41] That said, it seems likely that masculinity as a trait is a balanced polymorphism; alloparenting—i.e., parenting shared amongst women with little or no male involvement—is, and has been throughout human history, a common strategic option;[42] much selection is sexually antagonistic; and finally, life history speed is mediated hormonally.

Exclusive same-sex preference in women is typically measured at 1% or less. However, openness to same-sex pairings is much higher with, for example, French and UK women reporting a rate of 9–12% having same-sex attractions. Other cultures have reported even higher figures, such as nearly one-quarter of New Zealand women reporting at least one same-sex attraction.[43] The famous Kinsey report in the 1950s, where same-sex attraction was much more taboo, reported that only 2% of women self-described as lesbian, but fully 13% reported having experienced an orgasm with another woman. There is also considerable modern evidence that female sexuality is more fluid than that of males.[44]

Humans are not the only animals that sometimes exhibit lesbian preferences. Other species include, famously, bonobos and gulls. Lesbian gulls seem to pair bond when male quality or investment is low,[45] whereas bonobos are thoroughly matriarchal and bisexual.[46] Humans are different from either. The possibility of stable same-sex relationships between women sets the stage for societies with degrees of sexual separatism.

10.5   INTIMATE AND ALOOF

And yet small marvel 't is if one, thro' woman's wile
Befooled shall be oft-times, and brought to sorrow sore,
For so was he betrayed, Adam, our sire, of yore,
And Solomon full oft! Delilah swift did bring
Samson unto his fate; and David too, the king,
By Bathsheba ensnared, grief to his lot must fall—
Since women these beguiled 't were profit great withal
And one might love them well, and yet believe them not!

— (Gawain, in Gawain and the Green Knight)

When you put a girl on a pedestal, don't be surprised when she looks down on you from that pedestal.
— (Anonymous redditor on r/redpill, retrieved 12/10/2018)

What is assumed to be normal, in one's attitudes toward the opposite sex, can vary considerably between cultures. One key cultural variation concerns us for the story of female sexual selection in general, and it manifests in approaches to understanding female orgasm. Specifically, this is the question of how, or indeed whether, the men and women of that culture typically live together. This variable was documented by the esteemed husband-and-wife anthropology team John and Beatrice Whiting in 1975.[47]

Having surveyed many cultures in the anthropological record, the Whitings drew attention to a key variable of intersexual relations, along which societies could be distinguished. They named the twin poles of this axis *intimate* and *aloof*. These were not all-or-nothing divisions; they varied in degree, but the differences between them explain a good deal about why studying a key human trait—like female orgasm—can present a moving target for anthropology or cross-cultural studies in general.

Simplifying somewhat, in *intimate* societies, both sexes—when partnered—live with one another, share the daily work, and child-care (although there is always some sex-based division of labor) but they (more or less) get along. Husbands tend to be present at the birth of their children. These societies tend to be more peaceful than the other sort of society—the *aloof* ones. In highly aloof societies, such as the Chukchee,[48] the Yurok,[49] and the Pomo,[50] the sexes live apart. Sometimes they live far apart—in separate huts in the same village, like the *Haus Tambaran* of New Guinea,[51] or the routine sexual separations found throughout North Africa and the Middle East. Sometimes, the aloof men and women even live in separate villages, where the boys might leave their mothers at puberty and start developing suspicion of, even contempt for, the opposite sex. Partners in aloof cultures do not routinely sleep in the same bed. There can even be specialized male and female languages in aloof cultures, which are taboo for the opposite sex to speak.

One version of the aloof arrangement is sometimes called a "female gardening ecology." This is an anthropological euphemism for, "The women do all the real work, while the men spend all their efforts showing off to one another and flying into magnificent rages at threats to their honor." The sexes in aloof societies can have importantly separate creation myths—often detailing how one sex tricked or stole important stuff like fire or magic from the other. The elders of both sexes tell stories about the other, in order put other members of their own sex off from fraternizing with the enemy.

Mating occurs (of course, otherwise there would be nowhere for little aloof society members to come from) but, in aloof settings, sexual intercourse can be somewhat clandestine. At extreme ends, this means the policing of others' behaviors—often with myths and slurs (or, sometimes honest warnings) about the opposite sex.[52] Young males may be encouraged by their elders to rape, or otherwise sexually molest women, if they can get away with it.[53] Other possibilities include the older males telling the younger ones that women are all wicked witches who will try to steal the young men's potency. The older males (selfless creatures that they are) offer to step in and protect the youngsters from this evil. This can involve homosexual initiation rituals as with, for example, the Sambia of Papua New Guinea, where the young men are believed to have to ingest sperm to grow. Women, who have their own internal growth drive connected to menstrual blood, need to be kept away from the boys for years to avoid pollution.[54] At the same time, the older aloof women are typically telling the younger ones that all men are beasts and rapists (with some justification, it must be admitted) and that female-only living is the only viable way to go.

Of course, manipulating the minds of the younger members is key to maintaining aloof societies. Heaven forbid that the young men and women became friendly with one another, or the jig would be up. Lest we think that such divisions are confined to remote cultures, it should be noted that militaristic societies from the Spartans[55] to the British Empire public school system,[56] with its foundational single-sex "sink or swim" boarding school ethic, deliberately engineered to generate an officer class,[57] have all contained large elements of sexual aloofness and sexual separatism. In Sweden, in the present day, there exist *Riksorganisationen för kvinnojourer och tjejjourer i Sverige*

(ROKS), which teach young, often abused, women that the world is run by a conspiracy of male predators (some of whom dress as police) who, with the full backing of regular society, kidnap and torture women for fun. Only by staying within the confines of the ROKS can they be safe. Recent all-female protests in South Korea, called the 4B movement, prompted by abhorrent male behaviors and lax legal responses, also have strong female separatist elements.[58]

These are somewhat extreme examples, but most of us can easily see some of the psychological components for aloof societies around us every day—at least, those of us with access to the internet can.[59] The (potential) cognitive and emotional toolkit for such sexual separatism therefore seems likely to be a cross-cultural universal, whether it actually gets activated. This matters for the study of female orgasm, because societies that lean towards the aloof will tend towards a narrow set of views about female (hetero) sexual pleasure. In case it is not already obvious, aloof cultures will tend to deny female orgasm's existence or, at least, downplay its importance unless (and this is a very important caveat) the women have considerable independent economic clout. This needs to be born in mind when we hear traveler's tales concerning the "non-existence of female orgasm among the *Inventovolk*, demonstrating that it is a cultural construct." We should not be too hasty. Before we rush to such conclusions, we need to be very clear about who has been asked these questions, and how these questions have been asked.

10.5.1   A Pair of Aloof Cultures and the Lessons to Be Drawn from Them

What happens to sex in those cultures where the men and women treat each other with suspicion or worse? I will consider two in some detail, the second much more than the first, because it appears— at first sight—to be a culture where female orgasm does not exist at all. The first culture is one where, despite sexual separatism, men and women are close to being equal in power. The second is one where women are very much socially subservient. It will turn out that female orgasm occurs in both, but the experience is framed in importantly different ways.

10.5.1.1   Mangaia

Mangaia, called *Au'au* by the locals, is the oldest island in the Pacific Ocean. Part of the Polynesian archipelago, it looks like what most of us would think of, if asked to imagine an island paradise. The exact sort of place that I, aged about ten, used to read about with rapt attention in adventure books by Sir Arthur Grimble.[60] A visit to their lovely tourist site[61] tells you about their gorgeous beaches, stunning caves, and curiously large doughnuts.

There are, however, things about a culture that do not make it onto tourist guides or the sort of adventure books suitable for ten-year-olds. They are the sort of thing that anthropologists and sex researchers like to document. One such fact is that the males of this island traditionally surgically adjust their penises, by the use of incision. More surprisingly, the surgically adjusted young males are also instructed by experts in generating female pleasure during sex. The experts—older women, typically into their thirties, who drill the younger men in sexual techniques—teach these young men the importance of cunnilingus, nipple stimulation, and the surety that any female partners have had several orgasms before being allowed to ejaculate inside them. If a young man, so prepared, fails to bring his Mangaian partner to orgasm, she will typically abandon him—and tell others why she has done so. Informal versions of these sorts of arrangements are not unknown in western societies, but with the Mangaians—as well as other Melanesian islanders reported on by the anthropologist Davenport—it is well established. Mangaians state that female orgasm is an essential learned skill and that this process occurs through the efforts of a good man.[62]

Symons, who we remember from chapter five as the original champion of the by-product account of female orgasm, interprets this as following:

> These pacific people are the exception, not the rule. Davenport (1977, p. 149) summarizes the ethno-graphic literature thus: In most of the societies for which there are data, it is reported that men take the

initiative and, without extended foreplay, proceed vigorously towards climax without much regard for achieving synchrony with the woman's orgasm. Again and again, there are reports that coitus is primarily completed in terms of the man's passions and pleasures, with scant attention paid to the woman's response. If women do experience orgasm, they do so passively.[63]

Noted 1970s sex researcher Mary Jane Sherfey, in her *Nature and Evolution of Female Sexuality*, proposed that the sexual milieu of these islanders spoke to an insatiable female sexual appetite that had been suppressed (by males) to control female sexuality and enable agriculture. There is something to what she argues, to be sure. However, it should be noted that much of the suppression of female sexuality comes from other women—a perspective that is obvious from the vantage point of understanding that, biologically, your bitterest rivals are often members of your own sex, those who are in direct competition with you for the best mates, for example. In addition, any simple division of cultures into "pre-agrarian, and therefore pro-sex" versus "post-agrarian and therefore anti-sex" is too one-dimensional to do justice to the facts. There are simply too many counterexamples. For example, the Karnataka temples, which have advice for maximizing sexual pleasure for men and women literally carved into the walls, were created in a highly agrarian culture.

Symons considers the Sherfey argument—sexual insatiability—as the other side of the coin to Desmond Morris' pair-bond argument concerning female orgasm.[64] Sherfey was, as he noted, "a sexual radical for whom paradise is an endless, orgiastic sexual indulgence, while Morris *et al.* are sexual conservatives who favor a paradise of intimate, sexually intense monogamy."[65] Symons considers that neither option is viable, and I would agree. But there is another perspective possible—that the choosiness of female orgasm relates to the choosiness of female sexuality in general, and this choosiness is refracted through the differing sexual strategies that females can adopt. In addition, as I hope my quick sketch of Mangaian sexual norms has made clear, the idea of the women experiencing orgasm "passively" is a mischaracterization.

The Mangaian women have the economic and social clout to be able to choose (including choosing to leave) partners. Structures such as marriage (with virginity enforced beforehand and financial dependence afterwards), the spreading of misinformation about female sexual pleasure—even to the point of guilt and shame—and extreme genital mutilation all take away from the ability to leave if the sex is unsatisfactory. They speak to parental control of female sexuality, mediated through a partner whom she is artificially motivated not to be able to defect from, to the detriment of female choice. These sort of strictures, and worse, are the norm in the second set of aloof cultures I want to consider.[66]

10.5.1.2 Sudan

Female genital mutilation/cutting (FGM/C) involves some or all of the external genitalia of women, often while they are still young girls, being removed, frequently in conditions of low sterility. Renowned anthropologist Hanny Lightfoot-Klein[67] describes the most extreme, "full pharonic," as being a "scraping clean" of the whole of the external genitalia, followed by infibulation (sewing up) so as to leave a mere pinhole. Under these conditions, penetrative sex is reported as being horribly painful and difficult, if not sometimes impossible.[68] In addition to the trauma of the operation itself, there are frequent later complications. Urination can take two hours or more. Menstrual fluid can become backed up and lead to infection. Often a husband will try to penetrate his wife over the course of up to six months before both give up due to the pain and difficulty. Complications, both physical and psychological, are common and severe, in contradiction to the claims of some cultural relativists.[69] Typical medical complications include "infections, tetanus, hemorrhage and fever"[70] and psychological effects both before and after the operation include, but are not limited to, "phobic behaviour . . . extreme anxiety . . . emotional volatility . . . emotional withdrawal."[71] The report of these negative health outcomes has been overwhelmingly supported by recent meta-analyses of all the available data.[72]

10.5.2   Severe FGM/C and Sexual Response

It is easy to see that the more severe the FGM/C, the more damaged the woman's sexual response would be. What is surprising is that in cases where the entirety of the external genitalia are "scraped clean," that there is any sexual desire or sexual response in the women at all. However, this does appear to be the case.[73] Despite what one observer describes as "the more extensive operations,"[74] many women in question seemed capable of experiencing sexual pleasure and, in some cases, able to give "vivid descriptions of orgasm."[75] While the glans of the clitoris is removed in the severest FGM/C, as we saw in chapter six, other sensitive tissue remains, including the remainder of the clitoris, hidden beneath the surface, and other sensitive internal tissue such as the G-spot, as well as other nerves both in the anterior vaginal wall and elsewhere, which are untouched.

While the woman's capacity for self-stimulation following FGM/C must be greatly reduced, it is quite possible that she would still be stimulated, in these areas and in the remainder—i.e., internal majority of—the clitoris, by penetrative intercourse. This stands in stark contradiction to by-product accounts of female orgasm, which rely on the notion that penises are poorly suited to stimulation of female sexual organs. Even more strikingly, these reports cannot be explained away by reference to desires to please investigators, because for Sudanese women the display of *any* sexual pleasure to their men or others is culturally taboo. Other, more modern investigations support the Light-foot-Klein accounts in a population of Somali women who have also received FGM/C.

The practice of the "full pharonic," which is used throughout Sudan, is probably the most severe FGM/C practiced on the planet. The whole practice needs to be seen in the context of:

> Arabic Islamic culture . . . [F]amily honor in Sudanese society is defined in greatest measure by the sexual purity of its women. Because of this, a modesty code is rigorously imposed on them, which generally includes female seclusion, veiling of the face or head in public, child betrothal, the virginity test of brides, definitive transfer of sexual rights at marriage, the early remarriage of divorcees and widows, as well as genital mutilation.[76]

The descriptions of the practice, and consequences that follow, are heavily based on those given by Lightfoot-Klein, with some additions from other sources. I am sharing them in some detail because my experience is that they are not widely known in the field, and they deserve to be so for a whole host of reasons. Certainly, squeamishness should not be the sole arbiter of truth.[77]

The mildest FGM/C is that of the mild *Sunna*,[78] where a pricking, slitting, or removal of the clitoral hood (*prepuce*) occurs. This appears to be the practice advocated by the prophet Mohammed in the Koran.[79] Then there are more severe practices where partial or total excision of the clitoris occurs through *clitoridectomy*, where there is removal of the external part of the clitoris along with removal of some or all of the labia minora. The most severe practice is the full *pharonic* circumcision. In this, the external part of the clitoris is totally removed, along with the labia minora and the inner layers of the labia majora. These raw edges are then stuck together with cat gut or thorns. A small sliver of wood or straw is inserted into the vagina to provide a passage for urine and menstrual blood. The young girl's legs are tied together for about forty days while the heavy scar tissue forms. The result is a look where the female appears "scraped clean." The operation is typically carried out on girls six to nine years old without anesthetic. They are held down by female relatives, and the person carrying out the procedure is typically a midwife or grandmother, who holds high status in the patrilineal system.

> [Grandmothers] have influence and authority over their daughters-in-law . . . as well as their own daughters . . . [a]re as respected as fathers. . . . [T]hey are most often the initiators of the infibulation ceremonies . . . [a]nd must be considered the chief perpetrators."[80]

After the area has healed up, there is nothing that could count as a clitoris, as this term is used in the by-product account of female orgasm.

Lightfoot-Klein interviewed over 300 women who had undergone this particular, most severe practice, sometimes using an interpreter, sometimes speaking the native language. She asked her participants several questions about erogenous sensitivity. Typical questions included: "What parts of your body are the most sensitive?," "What about your scar?," and "What about inside?" She found that this "sequence was generally acceptable and elicited the desired information."[81]

Lightfoot-Klein discovered that the women were reluctant to admit to sexual desire at first, but that this was because of social expectations to the contrary rather that any actual lack of desire. Taboos on expressing sexual desire are circumvented in this culture by women's "using smoke," a practice involving a woman squatting over a small fire of sandalwood with her cloak tented about her. The resultant, very noticeable, smell sends out a clear signal for some distance that a woman is sexually receptive. "That is what she does . . . that is what we *all* do when we want intercourse with our husbands. When he smells that odor, he knows exactly what it means." Other activities that females used to show interest to their males without breaking the taboo about females' showing sexual desire include banging around the pots and pans at night. These kinds of practices carry the same sort of plausible deniability as coming up for some coffee might have in western cultures, allowing an expression of private sexual desire, while maintaining public probity.[82]

These practices of advertising female sexual desire show that event the severest clitoridectomy does not, in all cases, as some commentators have averred (and one might reasonably expect), entirely remove such desire, surprising as that may seem.[83] However, such severe mutilation would, presumably, remove or seriously curtail the capacity of the woman to easily satisfy any desire in the absence of the man, for example, by means of masturbation or from another woman's touch. A recent, WHO-sponsored meta-analysis of all the relevant data so far showed, unsurprisingly, that FGM/C reduces sexual pleasure in women and is associated with poor health outcomes.[84] However, the experiences of women who have undergone FGM/C provide a natural experiment of what occurs if women are completely without, what is routinely thought of, albeit incorrectly, as the clitoris.

Recall what I said earlier about aloof and intimate societies? This would be a good example of how aloofness between the sexes plays out in practice. Women in the Sudanese culture (in complete contradiction to the Mangaians) are expected to be passive during sex. The only sexual position that is culturally sanctioned is with the man on top, and the woman is expected to remain still and make no noise. One respondent described it to Lightfoot-Klein in the following way:

> Custom in Sudan dictates that the woman acts completely uninterested, he tells me, even if she strongly desires sex. Each partner has to play an assigned role. "She acts the part of the rape victim, and he acts the part of the rapist. Everything proceeds quite normally after that."[85]

These women therefore provide a strong test-case of the by-product account's claim that only the externally visible part of the clitoris is responsible for sexual pleasure[86] and that penetration does not typically stimulate this area in women. To summarize, and emphasize, the women who have undergone a full pharonic have no clitoris as the by-product advocates would understand it to be and, furthermore, they have no reason to lie about sexual pleasure to please males or anyone else. To the contrary. Admission of sexual pleasure in these women is a societal taboo and had to be coaxed out with sensitive questioning. Here is one such account: Lightfoot-Klein had asked one respondent, aged about 40, who had had a full pharonic circumcision, whether she was able to enjoy sexual intercourse:

> The woman's eyes opened wide, and she gave an incredulous guffaw. What followed was a most amazing performance. She doubled up with laughter, began slapping her thighs, and finally fell off her seat onto the floor, where she continued to rock with explosive and uncontrollable laughter. . . "[Y]ou must be completely *mad* to ask . . . a question like that! . . . *A body is a body*, and no circumcision can change that! No matter what they cut away from you—they cannot change that!"[87]

Many, probably most, women find sexual intercourse to be painful after the full pharonic, but some women *do* enjoy sex despite it. They report the phenomenology of orgasm and in ways remarkably similar to those I reported in chapter eight. Given that expression of sexual desire, or satisfaction, for women is taboo in Sudan, the remarkable consistency between the reports, tending towards *saturation*, is good evidence of reliability. There is no reason to believe that these accounts result from collusion, given that Sudanese women do not talk to one another about sex, producing socially desirable answers (these answers are the opposite of socially desirable) or lying to please menfolk (who were not present during the interviews).

It is worthy of note that most of the women in question reported their orgasmic experiences as from being with a second partner, one that they had had the opportunity to choose, rather than one that had been arranged for them. In other words, these experiences were often in the context of second marriages. This suggests that women, when given choice, choose those that are likely to please them the most. This finding fits with findings in our own studies (chapter eight) that older women tended to know better what they wanted from lovers and to report more satisfying orgasms.

Finally, these accounts I share here are a selection from hundreds of possible such accounts. They make no claim to be representative of *all* women who have undergone FGM/C, but this is irrelevant to the point at issue. They provide interesting data that are officially impossible on the by-product account of human female orgasm, while also providing intriguing clues to the complex phenomenology of all human female orgasms. The themes crop up with sufficient regularity to put one in mind of the goal of *saturation* in qualitative research:[88]

History #3: "I feel as if I am unconscious and shaking. It is almost unbearably sweet in my whole body, and if my baby fell out of the bed, I could not pick it up."[89]

History #7: She has a strong orgasm with him 30% of the time . . . The strongest sensation is experienced at the contact of his penis with her cervix, and her orgasm, when it occurs, is precipitated by his ejaculation. She has strong vaginal pulsations, and feels as if she were under sedation.[90]

History #9: When asked if she enjoys sex, she laughs and retorts, "Is there any woman who doesn't?" She is able to have very strong orgasm close to 100% of the time. She has little sensation in what remains of the sex organs, except in the scar area. . . At orgasm she feels as if she were under sedation, and she experiences strong vaginal contractions. This is followed by complete relaxation.[91]

History #10: The area of her scar is pleasurably sensitive to an extreme degree. However, the strongest sensation is inside her vagina. She has orgasm 90% of the time and becomes "completely unconscious." She laughs happily as she describes this. She lubricates very much, feels a pleasurable shock, and then relaxes completely.[92]

History #11: She has strongly erogenous sensation in the region of the scar, as well as in the walls of the vagina and her cervix. She has orgasm about half of the time. Intercourse lasts for one hour, and it is so important to him that she reaches orgasm that she pretends when she is unable to do so. . . When she is receptive, she can indicate this by the use of sandalwood smoke and fragrant oil.[93]

All of these accounts emphasize the fact that much sensation is inside the vagina, that vaginal contractions occur during orgasm, and that there are feelings attendant on orgasm akin to floating and fainting and, frequently, of total safety. These accounts are consistent with what western respondents described in some, but not all, of their orgasms, as I detailed in chapters seven and eight, and are all consistent with the action of oxytocin.[94] Finally, it feels worth emphasizing that these accounts stand in stark contradiction to the claim that only stimulation of the glans of the clitoris can create orgasm in women. Finally, the explanation offered by social constructionists,[95] that female orgasm is therefore a cultural construct, cannot be taken seriously, even on its own terms. These Sudanese women are not even supposed, by the lights of their culture, to enjoy sex at all, so there is nowhere from which the motivation for such a cultural construction could come.

10.6　CONCLUSIONS

Comparing extreme cultural variations allows us to test some key hypotheses about female orgasm. Specifically: First, does it fit the pattern of a by-product, claiming that the clitoris is a small, mainly external organ, unsuitable to be stimulated by penetrative intercourse? The answer, from studying the responses of those who have suffered full clitoridectomies, dramatically falsifies this hypothesis. Second, given that women in cultures that tend towards sexual separatism can both demand and experience orgasms, does this allow us to choose between the mate choice and sire choice functional alternatives for female orgasm? In intimate societies, the general pattern seems to be that female sexuality is more likely to be celebrated, and the promotion of women's orgasms forms a central part of that (along with male pleasure). In aloof societies, the pattern is more complex. Where women have political and social clout—as with the Mangaians—they demand sexual pleasure, even if pair-bonding along more familiar heterosexual lines is not the goal. This alone tells us something interesting and undermines the idea that female orgasm *always* functions as only pair-bonding. In those aloof cultures, where the women lack political clout—such as the Sudanese groups studied by Lightfoot-Klein—the move is to remove (or attempt to remove) female pleasure entirely. This also carries with it implications that female orgasm is intimately connected to female choice, and both parental and paternal control try to circumvent this choice. Scholars have sometimes implied that studying female orgasm functionally implies that we must choose between mate choice (usually simplified to sperm selection) and sire choice (usually simplified to pair-bonding). But there may be no need to choose one over the other. The action of oxytocin produces (potentially) *both* effects, and this may be one of the reasons that the nature and functions of female orgasm have been tough to pin down until now. It looks more likely that it can do *both*, with emphases changing in different ecologies. The variety of women's orgasmic experiences—as explored in chapter eight—is fully consistent with this position.

NOTES

1　As mentioned in chapter two. Anand, M. R. (1958). *Kama Kala: Some notes on the philosophical basis of Hindu erotic sculpture.* London: Skilton.
2　See Stearns, S. C. (1976). Life-history tactics: A review of the ideas. *The Quarterly Review of Biology,* *51*(1), 3–47, for a foundational document in the field. However, things have developed considerably in the last fifty years. In general, life history theory marries together insights from developmental and evolutionary biology to show how organisms have evolved to respond to frequent patterns of threat and opportunity in their phylogeny.
3　Henrich, J., Heine, S. J., & Norenzayan, A. (2010). The weirdest people in the world? *Behavioral and Brain Sciences,* *33*(2–3), 61–83. Similar points could be made about animal samples that are also, effectively, convenience samples in many instances. Webster, M. M., & Rutz, C. (2020). How STRANGE are your study animals? *Nature,* *582*(7812), 337–340. The entire premise of this book is that ease of sampling has bedevilled sex research, and that we should enthusiastically embrace *in*convenience sampling.
4　This mismatch becomes particularly acute in the case of medicine. See Gurven, M. D., & Lieberman, D. E. (2020). WEIRD bodies: Mismatch, medicine and missing diversity. *Evolution and Human Behavior,* *41*(5), 330–340, for a discussion of how various ideological squeamishnesses are preventing effective health care across populations.
5　A recent study suggested that the Catholic church undermined many of the WEIRD properties in Europe by circumventing kin-based institutions, leading to more nuclear families. This is an interesting idea, but it exists alongside the division of intimate/aloof to be discussed here, not as an alternative to it. Schulz, J., Bahrami-Rad, D., Beauchamp, J., & Henrich, J. (2018). *The origins of WEIRD psychology.* Available at SSRN 3201031.
6　It is, of course, possible to take this notion too far. The foundational model of personality psychology, the five-factor model, has been found to be stable across WEIRD and non-WEIRD cultures. See Doğruyol, B., Alper, S., & Yilmaz, O. (2019). The five-factor model of the moral foundations theory is stable across WEIRD and non-WEIRD cultures. *Personality and Individual Differences,* *151*, 109547. Sex differences in personality have also found to be robust across culture in large (tens of thousands of *n*) studies. See, e.g., Kaiser, T., Del Giudice, M., & Booth, T. (2020). Global sex differences in personality: Replication with an open online dataset. *Journal of Personality,* *88*(3), 415–429.

It is also worth noting that the mere act of describing behaviors from the outside can make them seem more exotic than they really are, as beautifully parodied in Miner, H. (1956). Body ritual among the Nacirema. *American Anthropologist, 58*(3), 503–507.

7 For example, in a series of experiments that puzzled nobody except economists, it has been found that every culture studied prefers fairer splits of treats, and will take a hit rather than be treated unfairly (rather than simply maximizing returns as simple economic models predicted). Oosterbeek, H., Sloof, R., & Van De Kuilen, G. (2004). Cultural differences in ultimatum game experiments: Evidence from a meta-analysis. *Experimental Economics, 7*, 171–188.

8 See Kimura, M. (1994). *Population genetics, molecular evolution, and the neutral theory: Selected papers.* University of Chicago Press, for an authoritative account.

9 E.g., Gould, S. J. (2000). The spice of life. *Leader to Leader, 15*, 14–19.

10 See Pinker, S. (2003). *The blank slate: The modern denial of human nature.* Penguin., for a detailed account of the history of this idea.

11 Anderson, W. S. (Ed.). (1997). *Ovid's metamorphoses.* University of Oklahoma Press. The myth of Pygmalion is in Book Ten.

12 This aphorism is usually attributed to British statistician George E. P. Box.

13 For example, we have known that the universe is expanding for some time. However, it has been found more recently that this expansion is accelerating, and no one is quite sure why. The "reason" given is *dark energy*, which is a detailed and mathematically sophisticated description of our ignorance. Carroll, S. M. (2001). *Dark energy and the preposterous universe* (arXiv preprint astro-ph/0107571).

14 Selection occurs where an allele, which performs a useful function, tends towards fixation at a locus. There are various ways in which this can happen: (1) purifying, which weeds out deleterious mutations; (2) stabilizing, where the species norm tends to increase fitness; and (3) directional, where average values differ from the optimal phenotype. Stated and measured female partner preferences sometimes follow one or the other model, but at least *some* follow directional selection. For example, taller men tend to have more partners than average (directional) but fewer children if they tend towards extremes of height (stabilizing). Nettle, D. (2002). Height and reproductive success in a cohort of British men. *Human Nature, 13*(4), 473–491. There are also interesting phenomena such as sexually antagonistic and frequency dependent selection. I do not have space to get into too much detail about what kinds of selection might be at play, partly because many kinds can only be viewed in retrospect. The point is to follow up from the realization that humans do not mate *randomly* to explore what the non-random patterns, linked to female orgasm, might be.

15 Harrow, J., Frankish, A., Gonzalez, J. M., Tapanari, E., Diekhans, M., Kokocinski, F., . . . Hubbard, T. J. (2012). GENCODE: The reference human genome annotation for the ENCODE project. *Genome Research, 22*(9), 1760–1774.

16 Counting these differences rapidly gets technical, and I have no space to get into the technicalities here. See Prüfer, K., Munch, K., Hellmann, I., Akagi, K., Miller, J. R., Walenz, B., . . . Pääbo, S. (2012). The bonobo genome compared with the chimpanzee and human genomes. *Nature, 486*(7404), 527–531, for an authoritative discussion. The point is to simply make an order-of-magnitude observation.

17 Not that this stopped some scholars from trying to teach chimpazees and other creatures human language throughout the 1970s and beyond. For thorough accounts of the failure of these attempts, see Terrace, H. S. (2019). *Why chimpanzees can't learn language and only humans can.* Columbia University Press, and Pinker, S. (1995). *The language instinct: The new science of language and mind* (Vol. 7529). Penguin.

18 Famously, the genome of onions carries more information than the genome of humans, as easily demonstrated in high school laboratories. This could prompt one to think that most DNA has no particular function—is junk. However, another interpretation is possible. Namely, that regulatory functions are at least as important as direct coding. For an example of this, see Yoon, J. H., Abdelmohsen, K., Kim, J., Yang, X., Martindale, J. L., Tominaga-Yamanaka, K., . . . Gorospe, M. (2013). Scaffold function of long non-coding RNA HOTAIR in protein ubiquitination. *Nature Communications, 4*(1), 2939.

19 They also seem to be under (mutual) selection, at least when we are looking for long-term mates. Miller, G. F. (2007). Sexual selection for moral virtues. *The Quarterly Review of Biology, 82*(2), 97–125.

20 Curry, O. S., Mullins, D. A., & Whitehouse, H. (2019). Is it good to cooperate? Testing the theory of morality-as-cooperation in 60 societies. *Current Anthropology, 60*(1), 47–69.

21 For example, the peoples of the north do not "have 50 words for snow," contrary to popular belief. Martin, L., & Pullman, G. K. (1991). *The great Eskimo vocabulary hoax.* For a detailed exploration of why humans *want* to believe, in the teeth of the evidence (and the evident fact of the possiblity of translation), that separate languages produce incommensurable ways of seeing the world, see McWhorter, J. H. (2014). *The language hoax: Why the world looks the same in any language.* Oxford University Press.

22 Gould, S. J. (1995). A task for paleobiology at the threshold of majority. *Paleobiology*, *21*(1), 1–14 emphasized that behavior does not fossilize; however, it *is* preserved in something that lasts as long as rocks do.

23 Tinbergen, N. (2005). On aims and methods of ethology. *Animal Biology*, *55*(4), 297–321.

24 Richard Dawkins first introduced the term "meme" in Dawkins, R. (2006). *The selfish gene*. (Original work published 1976). The idea was taken up by Dennett, D. C. (1993). *Consciousness explained*. Penguin, and Blackmore, S. (2000). *The meme machine* (Vol. 25). Oxford Paperbacks. Recently, Daniel Dennett has mounted a robust defense of the concept in Dennett, D. C. (2017). *From bacteria to Bach and back: The evolution of minds*. W.W. Norton.

25 Of course, they could be indirectly coded. For instance, children go through a stage of being fascinated by fire. This could easily be part of an inherited toolkit that makes the ability to make fire somewhat canalized in humans so that they do not have to learn it from scratch in each generation. Kolko, D. J. (Ed.). (2002). *Handbook on firesetting in children and youth*. Elsevier. This could be an example of the Baldwin effect—of selection for learning. Baldwin, J. M. (1896). A new factor in evolution (continued). *The American Naturalist*, *30*(355), 536–553.

    Fire making is a good trick, that can be lost if the population falls below a certain level, as perhaps happened in Tasmania following massacre and disease. However, see also Gott, B. (2002). Fire-making in Tasmania: Absence of evidence is not evidence of absence. *Current Anthropology*, *43*(4), 650–656. Whether this example is true, it is a reminder that much human cognitive capacity is distributed through the population rather than isolated in individuals.

26 The key foundational book of this classical version of evolutionary psychology is Barkow, J. H., Cosmides, L., & Tooby, J. (Eds.). (1995). *The adapted mind: Evolutionary psychology and the generation of culture*. Oxford University Press.

27 Ryan, C., & Jethá, C. (2010). *Sex at dawn: The prehistoric origins of modern sexuality*. Harper Collins. This book generated considerable controversy at the time. The general sense seemed to be that there simply *had* to be a single (and morally salient) answer to this question. However, it is not clear that such an answer is either available or desirable.

28 Buss, D. M., & Schmitt, D. P. (1993). Sexual strategies theory: An evolutionary perspective on human mating. *Psychological Review*, *100*(2), 204–232. Predicted patterns of female orgasm relating to socio-sexuality will be discussed later.

29 E.g., in some hard-scrabble Tibetan farms with matrilineal inheritance. Crook, J. H. (1994). Explaining Tibetan polyandry: Socio-cultural, demographic, and biological perspectives. In J. H. Crook & H. A. Osmaston (Eds.), *Himalayan Buddhist villages* (pp. 735–786). Shri Jainendra Press.

30 Mild polygny—e.g., high-prestige men with two wives—has been common throughout human hunter-gatherer history, with maybe five-sixths of traditional societies allowing for it. Marlowe, F. W. (2005). Hunter-gatherers and human evolution. *Evolutionary Anthropology: Issues, News, and Reviews*, *14*(2), 54–67. There are, of course, some extremes, with rulers such as Ishmael the bloody having harems in the hundreds.

31 See Bloch, R. H. (2009). *Medieval misogyny and the invention of western romantic love*. University of Chicago Press, for an account of how a functionally integrated system of endocrine mechanisms was, in reality, dreamed up by medieval troubadours.

32 Some 90% of humans across cultures marry at some point. Buss, D. M. (1985). Human mate selection: Opposites are sometimes said to attract, but in fact we are likely to marry someone who is similar to us in almost every variable. *American Scientist*, *73*(1), 47–51.

33 Incidentally, the idea that humans always use current technology to analogize cognition is a remarkably persistent, and demonstrably incorrect, notion that I examine in detail here: King, R. (2016). I can't get no (boolean) satisfaction: A reply to Barrett et al. (2015). *Frontiers in Psychology*, *7*, 1880. In brief, the functionalist ideas of George and Mary Boole that made computational technology possible, predated actual computers by more than 100 years, therefore they cannot be the "new technology" used to analogize cognition. Boole, G. (1911). *The laws of thought (1854)* (Vol. 2). Open Court Publishing Company.

34 Lorenz, K. Z. (1966). A. The psychobiological approach: Methods and results-evolution of ritualization in the biological and cultural spheres. *Philosophical Transactions of the Royal Society of London. Series B, Biological Sciences*, *251*(772), 273–284.

35 The term *umwelt* referring to the totality of salient signs in an organism's environment was introduced in von Uexküll, J. (1926). *Theoretical biology*. K. Paul, Trench, Trubner & Company Limited.

36 Call them "modules" if preferred. Literally nothing of importance hangs on this.

37 This is the terminology used by Ernst Mayr in Mayr, E. (1974). Behavior programs and evolutionary strategies: Natural selection sometimes favors a genetically "closed" behavior program, sometimes an" open" one. *American Scientist*, *62*(6), 650–659.

38 The phrase was first used by AI researcher Gary Marcus.

39 When we say *fast*, this is literally true. Organisms on fast trajectories are literally aging quicker than ones on slow trajectories in terms of thir cell division. The fact that the speed at which someone is aging at the cellular level and their likelihood of cognitions that discount their futures are highly correlated is a fascinating prediction of LHT, and appears to be true. Hill, K. (1993). Life history theory and evolutionary anthropology. *Evolutionary Anthropology: Issues, News, and Reviews*, 2(3), 78–88.

40 See, for example, Belsky, J., Ruttle, P. L., Boyce, W. T., Armstrong, J. M., & Essex, M. J. (2015). Early adversity, elevated stress physiology, accelerated sexual maturation, and poor health in females. *Developmental Psychology*, 51, 816–822.

    Belsky, J., Schlomer, G. L., & Ellis, B. J. (2012). Beyond cumulative risk: Distinguishing harshness and unpredictability as determinants of parenting and early life history strategy. *Developmental Psychology*, 48(3), 662–673, for nuanced discussion of how parenting and assessment of local riskiness (and hence likely life opportunities and threats) are calculated with surprising accuracy by parents.

41 For anyone that desires this, see Luoto, S., Krams, I., & Rantala, M. J. (2019). A life history approach to the female sexual orientation spectrum: Evolution, development, causal mechanisms, and health. *Archives of Sexual Behavior*, 48, 1273–1308.

42 Hrdy, S. B. (2009). *Mothers and others: The evolutionary origins of mutual understanding.* Belknap Press of Harvard University Press, is the most authoritative work tracing the phylogeny of alloparenting. Since then, considerable work has been done on integrating the concept of alloparenting into more general theories of life history, and patterns of female-female pairings in humans, and other animals such as herring gulls. Young, L. C., & VanderWerf, E. A. (2014). Adaptive value of same-sex pairing in Laysan albatross. *Proceedings of the Royal Society of London B: Biological Sciences*, 281(1775), 20132473. https://doi.org/10.1098/rspb.2013.2473.

    Young, L. C., Zaun, B. J., & VanderWerf, E. A. (2008). Successful same-sex pairing in Laysan albatross. *Biology Letters*, 4, 323–325.

43 Luoto, S., Krams, I., & Rantala, M. J. (2019). A life history approach to the female sexual orientation spectrum: Evolution, development, causal mechanisms, and health. *Archives of Sexual Behavior*, 48, 1273–1308.

44 Diamond, L. M. (2008). *Sexual fluidity: Understanding women's love and desire.* Harvard University Press. See also Kuhle, B. X., & Radtke, S. (2013). Born both ways: The alloparenting hypothesis for sexual fluidity in women. *Evolutionary Psychology*, 11, 304–323, for a discussion of how sexual fluidity in women might have evolved to intersect with shared parenting.

45 Harris, R. F. (1996). Gender bender. *Current Biology*, 6(7), 765.

46 De Waal, F. B. (1995). Bonobo sex and society. *Scientific American*, 272(3), 82–88.

47 Whiting, J. W., & Whiting, B. B. (1975). Aloofness and intimacy of husbands and wives. *Ethos*, 3(2), 183–207. I am also particularly indebted to Henry Harpending who, shortly before his untimely death, drew my attention both to this paper and a deleted passage in his own, Cochran, G., & Harpending, H. (2009). *The 10,000 year explosion: How civilization accelerated human evolution.* Basic Books. This latter explored and developed similar territory.

48 The Chukchee (or Chukchi) people are nomadic hunter-gatherers living around the Bering Sea. Bogoras, W. (1909). *The Chukchee* (Vol. 11). EJ Brill Limited.

49 The Yurok are Native Americans who lived along the Californian coast. Waterman, T. T., & Kroeber, A. L. (1934). *Yurok marriages* (Vol. 35, No. 1). University of California Press.

50 The Pomo are another indigenous people of California. Aginsky, B. W. (1939). Population control in the Shanel (Pomo) tribe. *American Sociological Review*, 4(2), 209–216.

51 The "Haus Tamberan" is a male-only worship structure in Papua New Guinea, see Gardi, R. (1960). *Tambaran: An encounter with cultures in decline in New Guinea.* Constable.

52 See Playà, E., Vinicius, L., & Vasey, P. L. (2017). Need for alloparental care and attitudes toward homosexuals in 58 countries: Implications for the kin selection hypothesis. *Evolutionary Psychological Science*, 3(4), 345–352. https://doi.org/10.1007/s40806-017-0105-9 for how this can be mediated by attitudes toward homosexual behaviors, and how this can tie into the alloparenting pattern discussed earlier.

53 Derek Freeman documented this in Freeman, D. (1983). *Margaret Mead and Samoa: The making and unmaking of an anthropological myth.* Australian National University Press (in Mead, M., Sieben, A., & Straub, J. (1973). *Coming of age in Samoa.* Penguin). At least part of the difference seems to lie in the fact that Mead talked to young women (through translators) while Freeman talked to their fathers.

54 Herdt, G. H. (2017). Fetish and fantasy in Sambia initiation. In *Rituals of manhood* (pp. 44–98). Routledge.

55 Cartledge, P. (1981). Spartan wives: Liberation or licence? *The Classical Quarterly*, *31*(1), 84–105.

56 George Monbiot writes in the *Guardian* newspaper about this with considerable psychological insight here: Monbiot, G. (2019). Newspaper article about English public school system. *Guardian Newspaper.* www.theguardian.com/commentisfree/2019/nov/07/boarding-schools-boris-johnson-bullies last accessed 06/11/2019.

57 Nash, P. (1961). Training an elite: The prefect-fagging system in the English public school. *History of Education Quarterly*, *1*(1), 14–21.

58 The "4Bs" come from the Korean word "bi" meaning "no." They refer to *bihon* (no heterosexual marriage), *bichulsan* (no childbirth), *biyeonae* (no dating), and *bisekseu* (no heterosexual sexual relationships). See Anna Louie Sussman's (2023) article in The Cut: www.thecut.com/2023/03/4b-movement-feminism-south-korea.html last accessed 28/06/2023.

59 For more scholarly examination of male and female separatism, see Nathanson, P., & Young, K. K. (2001). *Spreading misandry: The teaching of contempt for men in popular culture*. McGill-Queen's Press-MQUP.

60 Grimble, A. F. (1952). *A pattern of islands*. J. Murray.

61 https://cookislands.travel/blog/top-10-reasons-visit-mangaia last accessed 25/11/2019.

62 Davenport, W. H. (1977). Sex in cross-cultural perspective. In *Human sexuality in four perspectives* (pp. 115–163). This part is quoted on p. 122.

63 Symons, D. (1979). *The evolution of human sexuality* (pp. 85–86). Oxford University Press.

64 Morris, D. (1994). *The naked ape*. Random House. His rather inelegantly termed "poleax" hypothesis—that orgasm caused women to fall asleep and therefore hang around the male who helped generate them—falls at an early hurdle. Orgasms do not typically make women sleepy, although they do men.

65 Symons (1979, p. 94).

66 Other mechanisms for subverting female choice include purdah, suttee, and foot binding, all of which curtail the woman's ability to get up and leave the relationship—even after the death of the spouse in some cases.

67 Lightfoot-Klein, H. (1984). *Prisoners of ritual: An odyssey into female genital circumcision in Africa*. Haworth Press, Inc. (Reprinted 1989). I have kept the initial date to help easily distinguish this book from a similarly titled journal article by the same author.

68 For detailed accounts, see Kassindja, F., & Bashir, L. M. (1999). *Do they hear you when you cry?* Dell Publishing, Lightfoot-Klein, H. (1989). The sexual experience and marital adjustment of genitally circumcised and infibulated females in the Sudan. *The Journal of Sex Research*, *26*(3), 375–392, 380; and Lightfoot-Klein, H. (1984). *Prisoners of ritual: An odyssey into female genital circumcision in Africa*. Haworth Press. The quotes from women who have received FGM/C come from the latter work unless otherwise noted.

69 Obermeyer, C. M. (1999). Female genital surgeries: The known, the unknown, and the unknowable. *Medical Anthropology Quarterly*, *13*(1), 79–106. It is a curious feature of cultural relativism that there is a tendency to conflate "unknowable" with "not talked about in polite society." Obermeyer makes the surprising claim that we go beyond our facts when suggesting that removing non-pathological body parts from non-consenting children might present ethical difficulties. Cultural relativism can, of course, give no account of how cultural values and practices change, given that any judgment of value within that culture will be viciously circular (if right and wrong can only be assessed relative to that culture, then how can a culture judge *itself*, and, hence, change?). In light of that, it might surprise readers to know that western (surgical) culture also promoted clitoridectomy until surprisingly recently. Isaac Baker-Brown was thrown out of the Royal College of Surgeons over the publication of his 1866 book, whose title summarizes his views nicely, *The Curability of Certain Forms of Insanity, Epilepsy, Catalepsy, and Hysteria in Females*. His so-called cure for these things was clitoridectomy. What is even less well known is that Baker-Brown was not expelled for *performing* these operations, but rather he was expelled for *publicizing* how often they were performed on young women who proved troublesome to their families. See Sheehan, E. (1981). Victorian clitoridectomy: Isaac Baker Brown and his harmless operative procedure. *Medical Anthropology Newsletter*, *12*(4), 9–15.

70 Lightfoot-Klein (1989, p. 13).

71 Lightfoot-Klein (1989, p. 60).

72 See Berg, R. C., Denison, E., & Fretheim, A. (2010). *Psychological, social and sexual consequences of female genital mutilation/cutting (FGM/C): A systematic review of quantitative studies. Report from Kunnskapssenteret nr 13–2010*. Nasjonalt kunnskapssenter for helsetjenesten, and Berg, R. C., & Denison, E. (2012). Does female genital mutilation/cutting (FGM/C) affect women's sexual functioning? A systematic review of the sexual consequences of FGM/C. *Sexuality Research and Social Policy*, *9*, 41–56 for authoritative and systematic analyses of these points.

73 For more details on this surprising finding, see Assaad, M. (1980). Female circumcision in Egypt: Social implications, current research, and prospects for change. *Studies in Family Planning*, *11*(1), 3–16; Berg, R. C., Denison, E., & Fretheim, A. (2010). *Psychological, social and sexual consequences of female genital mutilation/cutting (FGM/C): A systematic review of quantitative studies* (Report from Kunnskapssenteret nr 13–2010). Oslo: Nasjonalt kunnskapssenter for helsetjenesten; El Dareer, A. (1982). *Woman, why do you weep?* Zed Press; Gruenbaum, E. (1996). The cultural debate over female circumcision: The Sudanese are arguing this one out for themselves. *Medical Anthropological Quarterly*, *10*, 455–475; Khattab, H. (1996). *Women's perceptions of sexuality in rural Giza* (Monographs in Reproductive Health, No. 1). The Population Council; Lightfoot-Klein, H. (1984). *Prisoners of ritual: An odyssey into female genital circumcision in Africa*. Harrington Park Press. Reprinted 1989 by Haworth Press; Lightfoot-Klein, H. (1989). The sexual experience and marital adjustment of genitally circumcised and infibulated females in the Sudan. *Journal of ex Research*, *26*(3), 375–392; Obermeyer, C. M. (1999). Female genital surgeries: The known, the unknown, and the unknowable. *Medical Anthropology Quarterly*, *13*(1), 79–106.

74 Lightfoot-Klein (1989, p. 95).

75 The quote is from Obermeyer, C. M. (1999). Female genital surgeries: The known, the unknown, and the unknowable. *Medical Anthropology Quarterly*, *13*(1), 95.

76 Lightfoot-Klein (1989, p. 64).

77 Assaad, M. (1980). Female circumcision in Egypt: Social implications, current research, and prospects for change. *Studies in Family Planning*, *11*(1), 3–16; Berg, R. C., Denison, E., & Fretheim, A. (2010). *Psychological, social and sexual consequences of female genital mutilation/cutting (FGM/C): A systematic review of quantitative studies* (Report from Kunnskapssenteret nr 13–2010). Oslo: Nasjonalt kunnskapssenter for helsetjenesten; Catania, L., Abdulcadir, O., Puppo, V., Verde, J. B., Abdulcadir, J., & Abdulcadir, D. (2007). Pleasure and orgasm in women with female genital mutilation/cutting (FGM/C). *The Journal of Sexual Medicine*, *4*(6), 1666–1678;

El Dareer, A. (1982). *Woman, why do you weep?* Zed Press; Gruenbaum, E. (1996). The cultural debate over female circumcision: The Sudanese are arguing this one out for themselves. *Medical Anthropological Quarterly*, *10*, 455–475; Hayes, R. O. (1975). Female genital mutilation, fertility control, women's roles, and the patrilineage in modern Sudan: A functional analysis 1. *American Ethnologist*, *2*(4), 617–633.

Khattab, H. (1996). *Women's perceptions of sexuality in rural Giza. Monographs in Reproductive Health, 1*. The Population Council; Obermeyer, C. M. (1999). Female genital surgeries: The known, the unknown, and the unknowable. *Medical Anthropology Quarterly*, *13*(1), 79–106.

78 From the Arabic for *tradition*.

79 "[A] woman used to perform circumcision in Medina. Muhammad said to her, 'Do not cut severely as that is better for a woman and more desirable for a husband.'" This passage is in dispute. Umm 'Atiyyah; Abu Dawud, al-Bayhaq (Hadith 5229, 2007).

80 Hayes, R. O. (1975). Female genital mutilation, fertility control, women's roles, and the patrilineage in modern Sudan: A functional analysis 1. *American Ethnologist*, *2*(4), 632.

81 These quotes come from Lightfoot-Klein (1989, p. 23).

82 Pinker, S. (2007). *The stuff of thought: Language as a window into human nature*. Penguin.

83 Lax, R. F. (2000). Socially sanctioned violence against women: Female genital mutilation is its most brutal form. *Clinical Social Work Journal*, *28*(4), 403–412.

84 Berg, R. C., Denison, E., & Fretheim, A. (2010). *Psychological, social and sexual consequences of female genital mutilation/cutting (FGM/C): A systematic review of quantitative studies* (Report from Kunnskapssenteret nr 13–2010). Oslo: Nasjonalt kunnskapssenter for helsetjenesten.

85 Lightfoot-Klein (1984, p. 280).

86 See chapter six for a fuller description. The principal expositions of the view that only the glans of the clitoris is sensitive would be: Gould, S. J. (1987). Freudian slip. *Natural History*, *96*(2), 14–21; Lloyd, E. A. (2005). *The case of the female orgasm: Bias in the science of evolution*. Harvard University Press; Masters, W. H., & Johnson, V. E. (1966). *Human sexual response*. Little, Brown; Symons, D. (1979). *The evolution of human sexuality*. Oxford University Press; Wallen, K., & Lloyd, E. A. (2008). Clitoral variability compared with penile variability supports nonadaptation of female orgasm. *Evolution & Development*.

Wallen, K., & Lloyd, E. A. (2011). Female sexual arousal: Genital anatomy and orgasm in intercourse. *Hormones and Behavior*, *59*(5), 780–792. Notice that there is no middle ground here. Either the Sudanese women are lying, a frankly implausible view in the circumstances, or there is more that is sensitive than the glans of the clitoris.

87   Lightfoot-Klein (1984, pp. 25–26).

88   See Hood, J. C. (2007). Orthodoxy vs. power: The defining traits of grounded theory. In *The Sage handbook of grounded theory* (pp. 151–164) for a further, and more techincal, discussion of the concept of saturation in qualitative research. These data were not formally analyzed this way, but the concept is still a useful one.

89   Lightfoot-Klein, 1984, p. 250.

90   Lightfoot-Klein, 1984, p. 255.

91   Lightfoot-Klein, 1984, p. 257.

92   Lightfoot-Klein, 1984, pp. 258–259.

93   Lightfoot-Klein, 1984, p. 259.

94   These are detailed in chapter eight. The main references are Carter, C. S., Williams, J. R., Witt, D. M., & Insel, T. R. (1992). Oxytocin and social bonding. *Annual New York Academy of Science, 652*, 204–211.

Fisher, H. E., Aron, A., Mashek, D., Li, H., and Brown, L. L. (2002). Defining the brain systems of lust, romantic attraction, and attachment. *Archives of Sexual Behavior, 31*, 413–419.

Zak, P. J., Kurzban, R., & Matzner, W. T. (2005). Oxytocin is associated with human trustworthiness. *Hormones and Behaviour, 48*, 522–527.

Ayinde, B. A., Onwukaeme, D. N., & Nworgu, Z. A. M. (2006). Oxytocic effects of the water extract of *Musanga cecropioides* R. Brown (Moraceae) stem bark. *African Journal of Biotechnology, 5*, 1350–1354.

Russell, J. A., Leng, G., & J. Douglas, A. J. (2003). The magnocellular oxytocin system, the fount of maternity: Adaptations in pregnancy. *Frontiers in Neuroendocrinology, 24*, 27–61.

Blaicher, W., Gruber, D., Bieglmayer, C., Blaicher, A. M., Knogler, W., & Huber J. C. (1999). The role of oxytocin in relation to female sexual arousal. *Gynaecologic and Obstetric Investigation, 47*, 125–126.

Carmichael, M. S., Humbert, R., Dixen, J., Palmisano, G., Greenleaf, W., & Davidson, J. M. (1987). Plasma oxytocin increases in the human sexual response. *Journal of Clinical Endocrinology and Metabolism, 64*, 27–31.

Carmichael, M. S., Warburton, V. L., Dixen, J., & Davidson, J. M (1994). Relationships among cardiovascular, muscular and oxytocin responses during human sexual activity. *Archives of Sexual Behavior, 23*, 59–79.

95   E.g., Obermeyer, C. M. (1999). Female genital surgeries: The known, the unknown, and the unknowable. *Medical Anthropology Quarterly, 13*(1), 79–106.

11 The Battles and Truces of the Sexes

What you have to understand is that women want three things from a man. First, they want a man who makes them feel like a *woman*: someone dynamic, sexy, and dominant. Second, then want someone they can rely on: honest, trustworthy, and reliable. And third? They want those two men to never, *ever*, meet one another.

— (Anonymous, at her request, stand-up comic)

Love is rarely pure and never simple. The underlying neurological and evolutionary reasons for this are not hard to find. Key biological features of humans that matter for our story are the following: significant sexual dimorphism, concealed estrus, large obligate investment in offspring, and mixed mating strategies. What does all this mean, and imply? I discussed some features of our sexual dimorphism in chapter three, and I will elaborate further here. What about the other features? Unlike most primates, who advertise their fertility,[1] women do not do this, and our offspring demand a lot of investment.[2] This, at least sometimes, drives us to pair-bond with each other—what we normally call love—but also allows for more mixed mating strategies, where either sex may somewhat exploit the other, and we have evolved strategies to counter exploitation. Across the board, the likelihood of exploitative strategies—deception, coercion, manipulation, terrorization, intimidation, or force, from both men and women—increase when one is on fast life history trajectories.[3] This final chapter will explore why human mating is such a rich, multi-level, and diverse affair, and where female orgasm fits into this complex picture.

11.1 WHAT IS THIS THING YOUR SPECIES CALLS "LOVE"?

Humans have evolved three somewhat separate neurochemical systems involved in that deceptively simple word *love* and, while these systems can line up simultaneously in one relationship, they can also pull in somewhat different directions. These systems, and the strategic options that result from them, need to be understood for a complete picture of female orgasm to be possible. This is because, as I hope has become increasingly obvious as the book has progressed, female orgasm is complex in nature, unlike male orgasm, which is relatively straightforward. Similarly, human love lives are as likely to involve conflict as they are cooperation, but the good news is that, despite what some readings of science may be taken to imply, love is not an illusion. On the contrary, we can demonstrate its existence(s) objectively. However, that revelation comes at a price, which is to realize that it is not a simple matter.

It must be admitted that the history of behavioral science, when it comes to analyzing love, has not been entirely encouraging. It has been previously proposed (roughly) that we fall in love with others because said others have something we lack,[4] or that we match up in terms of similar features,[5] that we take the plunge because of a somewhat dry cost/benefit analysis,[6] or that we (in complicated ways) relive our opposite-sex parent relationship in adult life.[7]

It is not that these theories do not, each in their own way, measure something of value. However, these dominant models in the field of psychology—roughly simplifying to complementarity theory, similarity theory, equity theory, and Freud's Oedipus and Electra models—have all suffered from being largely content-free, simplistic, single-variable models that lack both a sound functional foundation[8] and *consilience*. Consilience is the minimal condition that a scientific theory fits (not

DOI: 10.1201/9781003372356-11

necessarily be fully re-describable in other terms) with other things we know from other fields, especially ones that are more basic.[9] In this case we want our models to be at least compatible with what we know from behavioral, evolutionary, and neuroscience, in ways, for example, that a blank slate theory cannot achieve.

Can we do better than this? I believe that we can, and that an understanding of female orgasm can aid us in this search. One thing common to all these models is that they lack an appreciation of the fact that all humans—men and women—have mixed mating strategies, not one.

11.2   LOVE SYSTEMS

The three key underlying systems that generate desire and affection are, respectively (and simplifying considerably), one based on sex hormones—estrogen and testosterone—one based on dopamine, and one based on oxytocin, or its male-centered counterpart vasopressin. The sex hormonal system[10] drives unfocused sexual desire, and is much stronger in men than in women;[11] the dopaminergic one focuses that desire on a particular individual, and the oxytocin system generates long-term bonding. This adult pair-bonding oxytocin system seems to have developed phylogenetically from the more fundamental mother/child bonding system, and became subsequently repurposed by natural selection.[12] We also know that when we show someone a picture of their beloved in a brain scanner, then their dopaminergic circuits—the self-same ones that fire up if cocaine is snorted—light up like a Christmas tree. This element of love is a bit like being addicted.[13]

These three systems correspond to well-documented and exhaustively researched relationship, and experiential, patterns first detailed by Robert Sternberg in his triangular theory of love.[14] This classic formulation, which also has solid cross-cultural support,[15] separates the singular word "love" into components of passion, intimacy, and commitment. One can be in the throes of any single one, or be dominated by two, or more rarely, have all three. However, once it is appreciated that these are separate systems, and can pull in somewhat different directions, it becomes obvious why human love lives can be so fraught.

Passion on its own is the classic love at first sight, hormonal desire in the absence of either commitment or intimacy. On the other hand, if passion dies down, we can find ourselves in companionate love, where there are both intimacy and commitment, but lacking the same erotic spark. Long-term relationships can slide into commitment alone where futures are shared, but not bedrooms. This lack of desire—or, more usually, lack of feeling *desired*, as noted relationship therapist Ester Perel eloquently explains, is often the proximate motivation for affairs.[16] Other combinations are also possible, such as a holiday romance combining both passion and intimacy, but without commitment, or the whirlwind courtship of passion and commitment, without really getting to know one another in intimate ways. Consummate love, where intimacy, passion, and commitment are all present, is possibly the rarest of all, and most authorities agree that it takes some work—perhaps aided by institutions and rituals—to achieve.

Where does female orgasm fit into all this? Briefly, female orgasm is associated with all three. Sexual activity is more likely to be initiated by women during peak estrus, when hormones are at their height,[17] orgasms generate short-term rewards, and satisfying ones can generate feelings of commitment. The orgasm-linked reward system might work by utilizing the dopaminergic circuits of the frontal convexity.[18] There is some dispute over exactly how this mechanism works, but there seems to be no doubt that a reward system is being operated. For example, one effect of masturbatory orgasm is to reduce behavioral inhibition, and the regions of the brain associated with this have been directly tested.[19] In terms of long-term commitment, the magnocellular oxytocin system is being invoked.[20]

Therefore, partnered orgasms function emotionally, at least, to help focus women on a particular individual, continue to have sexual relations with them, and generally promote at least the possibility of commitment: At least, commitment for long enough to grow a child to approximately the point where it has realistically put childhood disease behind, which ancestrally would have been about five years.[21] Remember, until recently, every sexual encounter carried the potential for pregnancy for a woman, and although the use of the contraceptive pill has altered that in a variety of ways—such as generating some increased sociosexuality (especially in the presence of alcohol)—our evolved psychologies have not always kept up, leaving women much less happy than men are with one-night stands, even with the risk of pregnancy largely removed.[22] This, as much as pregnancy rates (perhaps surprisingly, the introduction of the pill *increased* these), can skew our data. Additionally, there is some evidence that the pill (which alters hormone levels to fool the body into behaving as if pregnant in some ways) may even alter partner preferences.[23]

When we factor this into our mixed mating strategies, the picture becomes even more complex. Both sexes have both long- and short-term mating strategies (LTM and STM, respectively), as outlined in chapter nine. Some data suggest that women had more orgasms in the context of affairs with high-value men—measured in terms of, for example, symmetry—especially during times of peak fertility.[24] However, not all of the ovulatory data on affairs have been replicated, and we also know that women's orgasm rates in one-night stands are generally comparatively low.[25] This is not to say that a gene harvesting strategy, outside of the primary mate-ship, can never work to the woman's (really the genes') advantage, but how to make sense of these apparently conflicting data?

Here is one way: Recent re-analyzing of the data suggest that female STM—characterized as gene farming behind the back of a committed partner—is real enough, but probably more of a rarity than we once believed. A seemingly more common women's strategy is a "try before you buy" one—of testing potential partner replacements. As the streetwise saying goes, "she may not be seeing someone else, but she knows who it would be if you were not around." This strategy of moving from one partner to another only when that move is secure has been referred to as *monkey branching*, and it fits the observed data that women report high levels (70% plus) of being in love with these alternate partners.[26] I will say something about each strategy in turn, while acknowledging that there is some degree of overlap between them.

11.3 EXTRA-PAIR PATERNITY

In early sexual strategies theory,[27] women's STM strategy proper was partly predicated on the perceived amount of extra-pair paternities. While it now seems unlikely that the "sperm wars" proposed by Baker and Bellis[28] take place often enough to drive adaptations, and the hunt for the predicted "killer sperm" which would attack rival sperm in the woman's reproductive tract, has turned out to be a dead end, there is still enough extra-pair paternity to drive countervailing adaptations in men.[29] The key one of these is sexual jealousy, which can manifest in individual rage, or institutionalized practices focusing on control of female sexuality.

It is a striking fact that a few of us do not have the biological father that we think we do. How many? Well, that is a disputed number. But whatever figure you find debated, you never find one that has estimates of lower than 1%. Let me try to say something about why this might matter. I work in a psychology department that, until recently, had the largest, and best equipped, driving simulator in Ireland. Some of my colleagues are therefore steeped in the various statistics and lore of driving, and by osmosis, so am I. In that atmosphere, I learned that—for example—my lifetime chance of dying in a vehicular accident (including being a pedestrian or cyclist hit by vehicles) is roundabout 1%. That is quite a lot. Many readers likely already know of someone who died this way. You can bet that people who study numbers like that do things like wear reflective clothing when biking home; they do not text (or God forbid, drink) while driving. They take a bit of extra care.

The reason I say this is that this 1% figure crops up repeatedly in discussions of extra-pair paternity. Some might believe that a lifetime risk of 1% (or thereabouts) is not sufficient to drive any sort of adaptive change across our species. I will come to the technical reasons why this thought is wrong in a minute. But forget evolution, let me just try to pump your intuitions for a second. Does knowing that you have a comparable lifetime risk of auto-related death have absolutely *no* impact on you? You would not, for example, ask the dealer whether your next car had airbags? You would not put bright lights on your bicycle? You would just breeze through red lights without a second thought?

Humans are conscious beings, able to reflect on their behaviors and make corrections. Some of the time, at least. Natural selection is a blind process, but it is more than capable of seeing 1% lifetime differences, if these are persistent and consistent across populations. Back in the 1920s, when the first statisticians began to mathematize Darwinism, Haldane demonstrated that a mere 1% advantage per generation could take you from an allele[30] that exists in 0.001% of the population to 99.9% of the population in 4,000 generations. Four thousand generations are about 100,000 years—give or take the length of time that humans have been recognizably anatomically and psychologically human.[31] So, a mutation that gave you a 1% edge, and cropped up just once in 100,000 humans (a reasonable population estimate for the time) could be in everyone (near enough) by now.

Lacking such an edge can have even more striking effects. Zubrow[32] showed that if you had two populations living alongside one another, a 1% difference in mortality rates would be all that was needed to drive that population to extinction in thirty generations. Thirty generations are barely noticeable at the time scales evolutionists consider. I can still read books written thirty generations ago. Changes like this could be going on right now, and not be detectable at the scales of magnification that humans are comfortable with. It is even possible to demonstrate that a population of mouse-sized critters could evolve to be elephant-sized ones in 12,000 generations and—this is the critical point—those changes would not even be detectable as a 1% difference in each generation (averaged across individuals, of course).[33]

I hope I have convinced you that 1% matters. At least, at the scales we are considering. If a 1% chance of death can matter, then a 1% chance of pouring your lifetime care and resources into someone who is not your genetic offspring can certainly matter. At least, it can matter to the genes that underpin any such behaviors. Any edge that they can get to see or guard against, such as a potential genetic dead-end, could easily get selected. And being cuckolded is, in genetic terms, potentially worse than death. Many species, under some circumstances, can increase their fitness by sacrificing themselves for their offspring. It might not be true that a praying mantis dad is happy, exactly, to be munched by the mother of his four-hundred-odd offspring, but his genes certainly can be selected to generate this sacrificial behavior.[34]

In any case, the 1% figure is unlikely to be evenly distributed across the human population. Extra-pair paternity figures as high as 16% have been quoted, but these come from Child Support Agency numbers, where couples already have good reason to believe that one has been unfaithful.[35] Other have quoted figures ranging from 2% to 10%.[36]

And what are the effects of such uncertainty? They can be many. Jealousy—sometimes to the point of irrational rage. On a more subtle level, parents may simply not feel as invested in children who are not their own. A somewhat disturbing survey in the US in the 1970s supported the suspicions aroused by anyone who has read some children's fairy tales: In one large survey, only 53% of stepfathers, and 25% of stepmothers, reported any parental feelings towards their stepchildren. Even more disturbing than that, the chance of death goes up by a factor of seventy when one is a stepchild.[37] The absolute chance of death is still not great, but the increase in mortality is steep. None of this should be taken to indicate that stepparents cannot take care of children. Indeed, they can. But it would be irresponsible not to know this is working somewhat against the natural grain of human nature and, in high-stress situations—such as unstable socioeconomic circumstances—this is an additional risk.[38]

11.4 MATE-SWITCHING RATHER THAN DUAL MATING?

A less risky strategy than a full-blown affair is to keep a potential partner simmering on the back burner. Modern technology has made this a great deal easier, of course, with the possibility of connecting to people literally anywhere.[39] Many of the remarks pertaining to the dual mating hypothesis apply to this strategy too, but several researchers have noted that many women tend to find one-night stands more emotionally taxing than most men do, and they are more likely to report dissatisfactions with their primary relationship when having affairs. Thus, the lining up of a potential back-up through sexual trialing is a more accurate description of the data.[40] This would fit well with the "try before you buy" model of female orgasm that has been emerging from the data in the book thus far. If gene harvesting was the sole strategic goal, then falling in love with your affair partner would be counter-productive. However, over 70% of women report doing exactly this.[41]

11.5   ORGASM TIMING

The existence of a fertility-related function to women's orgasms, implies that we would expect to find that they occur more frequently during estrus—e.g., peak fertility. Although the data on high-value males being preferred as partners at ovulation is disputed, the timing of orgasm with peak fertility is well documented, irrespective of sexual orientation,[42] and it should be noted that the amount of overall sex, and associated attractive behaviors, also increases during this time.[43]

A interesting study by Zietsch and Santtila in 2013 found that self-reported orgasm correlated with the number of babies that the women had. This correlation was stronger in fraternal twins than identical ones, which implies that environmental differences superseded genetic ones (in the women). Surprisingly, the authors took this to imply that orgasm did not promote conception, but this is to make exactly the assumption that has been one of the major themes of the book—of not treating female orgasm as a facultative mechanism.[44] It would have been more interesting if, instead of the genetic makeup of the *women*, they had studied the genetic makeup of their *partners*. After all, this is a large part of what constitutes the *environment* of the orgasm here. We know from other studies[45] that something like 70% of the variance in coital orgasm frequency is due to environmental effects, and the obvious environmental effect to have tested here would have been the compatibility (or lack thereof) with their partner.[46]

In many traditional cultures, women are sequestered during their periods, and considered taboo in some way. This, of course, has the effect of narrowing down the window of control of their fertility. In the absence of these practices, men can often not be sure when women are at their most fertile, and they can thus be potentially exploited for resources for a child who is not theirs. This sets the stage for a variety of potential conflicts. Men compete with one another, socially and physically, largely for access to women, while women compete for access to *their* preferred partners, at the potential expense of other women.

11.6   MALE-MALE AGGRESSION

Did you know that you can watch the last sword duel on YouTube? It took place in 1967, which is very nearly in my lifetime.[47] Affairs of honor, where men were expected to defend threats to their status with potentially lethal force, are not some arcane part of our distant, dimly remembered, past; they are recent history. Across the planet, men still compete in ritual combat, with credible threats of damage or death, with much greater frequency than do women.[48] Not far from where I am typing this, Traveler men still regularly compete in bare-knuckle fights over threats to their status.[49]

At the start of the book, I drew attention to the marvelous Zallinger picture of the Evolution of Man but cautioned that it tended to downplay the less violent, and more female-centric, aspects of sexual selection on males. I would be remiss if I did not say *something* about that sort of selection, however. Mainstream biological thought is firmly of the opinion that selection has (and still does) fall most heavily on the males of our species. This is technically known as the Bateman gradient (outlined in chapter two).[50] While all this is true, a word of caution needs to be said. When calculating these things, it is typically *spousal* numbers that are used and, as we have seen, extra-pair copulations are a feature of humans. Paternity is therefore typically assumed rather than directly measured. Mate quality is typically not measured either. Additionally, we are a mutually sexually selecting species with strong evidence for both past, and recent, sexual selection on the females of our species too.[51]

Male-male competition is more obviously aggressive than female-female, and the bulk of studies into aggression have focused on it—either men competing for status in various social hierarchies, which itself brings attractiveness in its wake, or men getting together and attacking other men en masse. Historically, the latter was strongly associated with grabbing women as prisoners and concubines and, unfortunately, that tendency has not been fully eradicated even in the modern era. All this noisy violence has tended to distract from the fact that women compete with one another too, but compete they do. In this section I will review some of the data on male intrasexual aggression, and what can be said about its influence on the sort of creatures we find ourselves to be. But, the picture would be incomplete without considering female intrasexual aggression alongside it.

A robust finding across both archaeology and studying of contemporary hunter-gatherer groups (our best windows into our evolved pasts) is a sexual division of labor. Very broadly, men have gathered the bulk of the protein—in the form of big game, often cooperatively hunted—while women have accrued the bulk of the calories—usually in the form of gathering of plants, with occasional opportunistic small game added. This view has recently been challenged, although the challenge itself has been subject to rigorous scrutiny.[52]

Whatever the outcomes of this assessment, there are unarguably a raft of adaptations that fit men for combat and/or hunting big game. For example (on the average, of course) their broader shoulders[53] and longer limbs make them better at throwing.[54] This could be an adaptation to throwing objects at other men in battle, but it could equally be an adaptation to millennia of throwing spears at big game. Hands are better suited to holding than punching, and good evidence for this is that combat sports that focus on striking, rather than grappling, require extensive protection of the delicate hand bones. This set of adaptations also speaks to requirements to have relationships with kin, or coalitions, to teach you how to make tools, or to make them for you.[55]

It is impossible to know for sure, but when, for example, one combines these broad shoulders with a greater ability to catch (e.g., *block*) thrown objects,[56] then combat seems at least part of the story, as antelope do not throw spears back. This conclusion is supported by the fact that men, in general, have a greater interest in developing combat-related skills than do women.[57]

In general, men are taller, heavier, and with greater upper body strength.[58] The bulk of these differences hit at puberty, where men go through a stage that cements their forms of competitiveness.[59] They have thicker jaws[60] and skin that cuts less easily. They have faster basal metabolic rates,[61] higher muscle-to-fat ratios, and greater ability with rotational alignment[62] and reaction times in general.[63] They have both larger hearts, and higher systolic pressure, which fits well with their larger lung capacity.[64] Their blood has a higher ratio of hemoglobin to carry oxygen more effectively.[65] They have adaptations to running this whole system at maximum capacity, and dissipating the resultant heat, such as greater resistance to dehydration, larger circulating blood volume, and greater capacity to sweat.[66] Of course, all this means that they tend to die younger, even when controlling for all the risk-taking that testosterone also encourages.[67]

In mammals, male-male competition tends to reveal itself in slower growth to maturity of males compared to females, later sexual maturity, larger size, and more behavioral aggression than the females.

In extreme cases, this has meant that the most socially powerful men, e.g., emperors, cornered the sexual market. As Laura Betzig documents, in the first six great civilizations, Mesopotamia, Egypt, Aztec, Inca, India, and China, the emperor had a harem of hundreds, and killed any men who interfered. Genetic analysis confirms that this, at times, created bottlenecks in human genetics where the effective population of women was seventeen times that of men.[68] Genghis Khan's DNA (or that of his brothers) is still detectable in roughly 5% of the Asian male population.[69]

Men are also responsible for the bulk of physical aggression. Boys practice at it by engaging in rough-and-tumble play in much greater amounts than do girls, from two years old onwards, and they do this across all cultures measured, although the manner of such expression can be locally malleable.[70] The figures for violent crime show female rates at about one-tenth of male ones across thirty-one countries.[71]

At least some of these things are directly, or indirectly, attractive to women. For example, upper body strength predicts reproductive success in Hadza hunter-gatherers.[72] However, is this because upper body strength makes you more likely to be successful in hunting, and hence attractive, or does the mere appearance of likely success drive female sexual selection? It may well be both, and it would be strange indeed if women did not pick up on likely markers of the sorts of success that mattered in both ancestral and modern environments, independently of said success.[73] And, there is independent evidence that they do just this, with, for example, greater muscularity and grip strength predicting higher numbers of sexual partners in a modern environment, even when not carrying back an antelope. This pattern is reflected in both self and other-perceived levels of attractiveness too.[74]

Women also pick up on cues to likely success in these arenas in other ways. For example, waist to shoulder width ratio increases male attractiveness to heterosexual women, and they are particularly attuned to it.[75] Height is an oft-repeated desiderata for women, and it is notable that average male height is only roughly 7% greater than women's. Sometimes this is taken to indicate reduced competition for mates over evolutionary time, but this is not the whole story. There is a reason why combat sports have weight categories, not height ones. While height may be a rough-and-ready first approximation to size differences, muscle differences between men and women are much greater, and muscle difference conveys much greater combat advantage than reach. Men have 65% more lean muscle than women, and this is concentrated in the arms, where they have 72% more strength. These differences amount to three standard deviations (or more) from female strength, on average.[76]

Wilson and Daly document how male-male aggression occurs typically over status threats which, in an ancestral environment, would be key to reproductive success. If you allow yourself to be put down in lawless environments, then you will be kept there, as any man who remembers the boyhood schoolyard can remember.[77]

How does all this square with my assertion, in chapter nine, that Darwin was right when he said that the power to charm women has been more important than the power to defeat other men in battle? For one thing, the direct things that women select for are somewhat important in understanding our shape and dispositions, and women are not shy about letting men know what qualities they find attractive.[78]

At least as important as these direct selection effects is the fact that all of these adaptations allow men to compete with one another, and if there is one thing that stands above all other attractive male qualities, it is *confidence*.[79] Nothing gives confidence better than winning in whatever local hierarchies you operate, and there are typically multiple ones on offer. Women do not even have to observe such winning directly to pick up on confidence and reputation. To a first approximation, all these qualities mentioned allow for confidence as a (more or less) honest signal. Why honest? Because other men are often trying to knock it down. Often, but not always, because men also need to maintain coalitions, which are a better protection against other men than the sheer brute force of one individual, which is rarely up to the task. We are more like chimpanzees than gorillas in that respect. Confidence is therefore perhaps a more complex human analogue to the bower bird's bower—a more or less honest signal to the female of the species—partly because other males are often trying to undermine it.[80]

In a meta-analysis of more than 150,000 participants, across ninety-one studies, looking at 431 effect sizes, Lidborg's team found that the markers that male puberty (e.g., the adolescent burst of testosterone) enhances, specifically voice pitch, height, digit ratios, muscularity, and strength, all predicted mating success. In addition, where measurement was possible (which is tricky in populations that use birth control), muscularity and strength also predicted reproductive success. Surprisingly, facial masculinity did not predict either.[81]

What can we make of these findings? The split we seem to be faced with is that either male qualities are being selected for because they enhance male competitive success, *or* they are directly attractive to women because they index qualities like immunocompetence. However, the sense that this is a forced choice may be false, as I argued in chapter three. In the limit, whether *we* as humans appreciate the goodness of genes is irrelevant, because only differential gene representation matters here.

As anthropologist, and primatologist, Joyce Benenson has pointed out, these sex-differentiated qualities have sometimes been framed as denigrating to women, but this would be a mistake. As she emphasizes, qualities like "lower pain threshold" can be framed as "greater sensitivity to pain," and pain evolved to protect you, every bit as much as armor or a poison sting did.[82] The sci-fi fantasy of the super soldier who "does not feel pain" is not an insightful one. Such a soldier would die for want of being able to avoid damage. How do we know this? Because humans who suffer from being pain free have all sorts of medical issues and often die young.[83]

11.7 PRESTIGE AND DOMINANCE

A man can be short, fat and going bald but if he has fire women will desire him.

— (Mae West)

Does this mixed bag of selective effects contradict what I have been saying about female sexual selection, acting through differentially occurring orgasms, having had an impact on male behavior and morphology? No, because although contests might not even be witnessed by women, *winning* is. For men, but not women, status—which in the west covaries very closely with income—is crucially linked to fertility, when you control for childless men.[84]

And now that the use of violence is quite regulated in modern society, we can expect female choice factors to be even more important than previously thought, and said selection to be even more acute. In addition, confidence, especially that displayed in front of other men, is a good proxy for the potential for violence, as in ages up until now such displays could have been met with violent responses. One example of this is voice pitch, where humans exhibit considerable sexual dimorphism, with deeper pitches being generally regarded as both sexually attractive and dominant, and with men adjusting their vocal pitches downwards in the presence of attractive women.[85]

11.8 ALPHA MALES?

It serves me right for putting all my eggs in one bastard.

— Dorothy Parker

With evident ruefulness, noted primatologist Frans de Waal has said that he feels somewhat responsible for the introduction of the term *alpha male* into everyday discourse.[86] Originally used to refer to the (currently) dominant wolf in a pack, the term bled over into primate research to mean something like "male leader" and thence into everyday talk about the battle of the sexes to mean "god's gift to women." Whether the women agree with such estimations is more open to interpretation, but some index of male dominance, status, or prestige makes its way onto most lists of desiderata from heterosexual women. What can be said for it? Humans do not have alpha males in the way that wolves do, and apparently even these change from day to day. As de Waal is at pains to point out, chimpanzees and humans are not physically strong enough to dominate a group without considerable political support. It is painfully obvious that, as the saying goes, "And any man who must say 'I am king' is no true king at all."[87]

Confidence is something of an honest signal, although women have to learn to distinguish it from its various semblances such as youthful arrogance, for the simple reason that other men in the environment will try to undermine it. They do this by means of direct physical competition or indirect attacks on someone's reputation for toughness or reproductive viability. In addition, confidence can be faked. Indeed, the vexed question of whether women really like "bad boys" can be partly explained by their need to explore the varieties of real and fake confidence that men project in their social spheres.

11.9   FEMALE-FEMALE POLITICS

When women kiss it always reminds one of prize fighters shaking hands.

— H. L. Mencken

I trust I debunked notions of sexually coy women in chapter two. Now it is time to say something about some other supposed feminine qualities—their supposed lack of interest in personal status, and their alleged relative political passivity. Once again, subtlety and cunning have been mistaken for disinterest, or lack of ability. This matters for discussing female orgasm because, as we have seen in previous chapters, deception over, or interference in, choice mechanisms is ubiquitous among primates,[88] and, given the fact that female orgasm is just such a choice mechanism, we should expect that to occur here. Some accounts of women's politics imply that they are inherently egalitarian and cooperative, in stark contrast to men's noisy and competitive politics. However, as Joyce Benenson points out, egalitarianism can be a very effective strategy for bringing your opponents down. For example, she showed that women report feeling bad when others succeed, in ways that men typically don't, on twenty-two different social items. Benenson calls the typically female forms of competition "safe, subtle and solitary."[89] While Anne Campbell also documents the varieties of indirect aggression that women typically favor.[90]

Modern women are descended from gatherers who engaged in scramble competition,[91] and another feature of our ancestry was the division of formal marriage into matrilocal and patrilocal arrangements. Obviously, I cannot do full justice to the complexities here, but the second type of arrangement left ancestral women having to negotiate complex webs of alliances built around non-kin, rendering them exquisitely sensitive to signs of loyalty and betrayal.[92]

One upshot of all this is that romantic love is perhaps more likely to be generated where couples do not have the pressure of kin around them, because it can act as a glue where customs like non-returnable bride prices are not the norm.[93]

As Sarah Blaffer Hrdy puts it, females "cooperate selfishly, and there is a perpetual undercurrent of competition."[94] In intrasexual competition where denigrating rivals through strategic gossip, men focus on their rivals' supposed heterosexual incapacities. Women focus on the supposed infidelity of rivals—colloquially called "slut shaming," or by casting doubt on loyalty or trustworthiness.[95] This sort of tactic can disrupt a rival's social networks that, in ancestral times, would have been crucial to her, and her offspring's, survival. Trust is a major part of our social capital since so much of what we do relies on people (more or less) being able to be relied upon.

Boys start with rough-and-tumble play from an early age and balance interpersonal violence with the potential need for coalitional aggression. Notoriously, boys can have fierce physical contests and then be firm friends, consistent with their ability to form hierarchical pacts. But female aggression is often not like this, and more closely fits the pattern of patrilocal rather than matrilocal marriage. If girls are married off to groups of non-kin, then they would have to acquire less overt means of establishing, and maintaining, alliances and getting one over on rivals, and this is indeed what they do with tactics of gossip, excluding from the group, and attacking their appearance or reliability.[96] These attacks are seen as highly effective by victims[97] and can be compared to the behavior of those dominant female baboons who harass subordinate ones in ways that cause hormonal fluctuations that lessen reproductive viability.[98]

11.10   GOSSIP

Reputation, reputation, reputation! O, I have lost my reputation! I have lost the immortal part of myself, and what remains is bestial.

— (Cassio in Shakespeare's *Othello*, Act 2 Scene 3)

Sticks and stones may break your bones, but bones heal. Reputations, however, often do not. People have taken their own lives over nothing more than their beliefs about what others thought of them.

For hypersocial beings like us, there is almost nothing more harmful than having one's reputation sullied.[99] If intimacy is the gradual sharing of potentially harmful revelations to increase trust and connectedness, gossip is its dark side: The ability to use insider information (or made-up information) to destroy that social capital. The degree of intrasexual competition in a group predicts both positive attitudes toward spreading gossip, and its frequency. In addition, women gossip a great deal more than men and tend to focus on appearance, which is vital information in terms of reproductive status and viability—and other social information. Men tend to gossip about achievement, which also, as I hope I showed above, conveys vital reproductive information. And it works. Adolescent girls who use relational aggression towards other girls ended up being more popular with boys of their own age.[100]

11.11 BEYOND GOSSIP

I do not believe in using women in combat, because females are too fierce.

— Margaret Mead

A lot of cultures involve, and have involved, a low level of polygyny. It would be a mistake to think that the women in such a set-up were always allies, although this is more likely to happen if they are related. A new wife lowers the resources that the husband can bring to the other wives, and thus she will often be the target of aggression, which is usually verbal, but can be physical.[101] It can even get more extreme than fighting. In a good example of the tension between the cooperative, and conflict-ridden, natures of reproduction, consider the Dogon people of Mali. They are polygynous, like many cultures. From a genetic point of view, the males who get multiple wives benefit from multiple chances to reproduce. While the co-wives get the benefit of a high-value mate, the trade-off is that he is splitting resources among other wives and children. Unless, that is, those other children happen to have some sort of mishap. Strassman documented hundreds of cases of children being poisoned by co-wives in cases that can tie up the Mali courts for years. These mishaps are somewhat selective in their application. Boys, who stand to inherit land from their fathers, are two and a half times more likely to be poisoned than are their sisters.[102]

11.12 THEORIES OF FEMALE SEXUAL CONTROL

In an influential paper, Roy Baumeister, and Jean Twenge, noted the considerable degree of suppression of female sexuality across time and space and considered some possible explanations for it.[103] One set of attempted controls center around the asymmetry of human mating—males can never be totally certain of paternity.[104] Plenty of species (such as lions and gorillas) engage in infanticide in cases where another male's offspring is present and, in humans, while male-driven infanticide is rare, it is not totally unknown, and neglect leading to the death of infants is, sadly, a demonstrable risk factor.[105] The culturally specific attempts to remove female desire discussed in the previous chapter show clear elements of this, although as a detailed examination of the practice of FGM shows, it is not simply the case that men impose this on women. Indeed, as Lightfoot-Klein documented, many of the men she surveyed wished their partners were intact.

Any simplistic *men versus women* story does not do justice to the observed facts or to simple logic. For instance, while men in relationships might benefit from their partners being chaste, men *not* in relationships (and given our history of mild polygyny that always means more men than women) would definitely not so benefit. As we saw in chapter three, many men are open to low-cost sexual encounters and are, at least in most western cultures, more likely to say that their partner's pleasure is more important than their own.[106]

What about the idea that women can exert coercive control over other women? Baumeister and Twenge claim that this idea lacks prima facie plausibility, given that "men have held superior political and social power throughout most of history."[107] In respect of this, it needs to be noted that while the people with power may have been usually men, it does not follow that most men have had power. Historically, power—at least political power—has been concentrated in the hands of the few. There is also more than one type of power. They ask the question: "Why would women want to suppress female sexuality? Sex is undoubtedly a major potential source of pleasure and fulfilment in life, and for women to stifle their own sexuality."[108] Alas, this statement reveals a level of biological naivete. From a gene's-eye perspective, the other members of your sex can be your most bitter enemies. You can also form coalitions, but only someone who has not paid attention to the vast female repertoire of intrasexual competitive behaviors—including, but not limited to, reputational denigration, impugning fertility, and harassment occasioning counter-fertility producing hormonal fluctuations—could possibly have missed this. True, men often miss female-female competition, but behavioral scientists are meant to dig a little deeper rather than believe the surface propaganda. And it is not only your direct sexual rivals who have an interest in your sexuality which does not always fully align with your interests. Your parents must also be factored in.

11.13   PARENT/OFFSPRING CONFLICT

Trivers' seminal (1974) paper reminds us that the gene's-eye view makes some counter-intuitive predictions, which all turn out to be true. Offspring are the parents' genetic payload, but they do not completely align in terms of interests because each one only has 50% of the transmitted genes. Thus, we expect conflict over resource allocation, especially at key periods, with, for example, the offspring demanding more than the parent might be willing to give because they can hedge their bets. As Trivers warns us, conflict in some species, including the human species, is expected to extend to the adult reproductive role of the offspring: Under certain conditions parents are even expected to attempt to mold an offspring, "against its better interests, into a permanent non-reproductive."[109] What would (ultimately) motivate (female) parents to reduce (if not completely remove) the sexual choices of their daughters only? This is a conflict over mate choice. What are the options? One is that life history–driven trade-offs between genetic quality, and parental investment, would lead to children preferring high genetic-quality mates, at the expense of parental investment. Since this cost would involve the parents, they would be motivated to exert control over it by, for example, interfering with the offspring's ability to easily make certain selections.[110] Another option is that female children exercising their mate choice will make compromises that the parents see as less desirable.[111] However, it is not clear that the latter option makes predictions that are easily distinguishable from the first.

Control of female sexuality is expected to peak at adolescence, when hormonal increases provoke sexual behavior. A large amount of the top-down control exerted seems to be by mothers on daughters,[112] with mothers typically being the source of sexual information, which is, of course, open to manipulation.[113]

A good example of this sort of parental control was outlined in the previous chapter when discussing FGM/C. It is tempting to place FGM/C under the "male control" banner, but any such conclusion would be simplistic. Recall that Lightfoot-Klein reported that the Sudanese men she spoke to preferred the women who visibly enjoyed sex, supporting the idea that male partners are naturally attuned to signs of female orgasm. This suspicion is confirmed when we investigate men who had multiple wives, some who had received FGM/C, and some who had not. Shandall concluded that "something other than men's sexual satisfaction must be at stake in continuing the practice."[114] Surprisingly, perhaps, it is the women of the group who defend the continued practice of FGM (Boddy, 1989, 1998).[115] Some scholars (Hicks, 1996)[116] even reported that men's attempts to reduce the severity of the practices were thwarted by the women.

Parents are not rivals for sexual attention, however, peers are. When it comes to peers, it seems that other women exert the most suppressive forces, usually in the reputational forms noted previously. It should be becoming obvious from the foregoing that humans, as sophisticated eusocial beings, routinely cloak their desires behind plausible deceptions. Thus, we must address the topic of signaling.

11.14 SIGNALS AND NOISES

Economist Robin Hanson tells us that *everything* is signaling.[117] What is the *real* reason that you engage in conversation, support certain political causes, and educate your children? The "elephant in the room," as he puts it, is that the real reason for doing all these things is to send out a signal—I am a good ally, conscientious, the bearer of good genes. It should not be controversial that some of our motives are opaque to us. However, the "everything is signaling" approach cannot be literally true for the same reason that "everybody lies" could not be true in chapter eight. While it is salutary to realize that there is often a (more or less) hidden signaling aspect to many of our behaviors, this should not be taken to be the sole *real* reason, any more than any of Tinbergen's other factors, in understanding a trait.

That said, signals and the ability to separate salient ones from random noise are very important. We know that men are highly sensitive to female orgasm sounds and behaviors, to the extent that women often feel pressured (or can be paid) to simulate same. Most women have faked orgasm at least once, with those endorsing "anti-feminist values" more likely to fake than those who do not, with those who endorsed what the authors described as endorsing more feminine roles were *less* likely to fake.[118] Women are well aware that men feel inadequate if their partner does not orgasm.[119]

We are now in a position to explain these otherwise curious facts. When asked directly about what their and their partners' orgasms meant to them, both men and women reported the following themes:

1) The purpose and end of sex
2) "It's more about my partner's orgasm"
3) The ultimate pleasure
4) Not a simple physiological response
5) Faking is not uncommon

These (mostly non gendered) themes illustrate the complex and sometimes contradictory meanings around orgasm, and reveal meaning to be dependent on situation and context.[120]

11.15 "YOU SAY THAT AS IF IT'S A BAD THING"

There is, therefore, cooperation amidst all the heterosexual conflict. However, there is often a curious asymmetry in the way that this is reported. For example, Chadwick and van Anders tell us that "female orgasm functions as a masculinity marker for men."[121] At the same time, a large, and representative, meta-analysis of five separate studies of Finnish women admonishes us that "women care more about their partner's orgasms than they do their own."[122] But, strip away the barely covert moralizing (this sort of analysis is never applied to lesbian sex, for example) from these accounts and what remains? *Each* person caring about their partner's pleasure during sex. One might have hoped that this was something to be encouraged. Interestingly, in the Finnish study, the key determinants of women's sexual satisfaction were not, as the researchers had predicted, multiple lifetime partners, or masturbatory technique. Instead, they were the "ability to concentrate, mutual sexual initiations, and partner's good sexual techniques. A relationship that felt good and worked well emotionally, and where sex was approached openly and appreciatively."[123] And this is where we came in. Human sex is non-accidentally intimate in nature, not merely mechanical.

11.16 INTIMACY

> You know you love someone when you cannot put into words how they make you feel.
>
> —Margaret Mead

Notoriously, some men may not hang around much after ejaculating, but many do, and as we have seen, the ability to generate oxytocin over time is physiologically linked to reproductively important functions such as sperm transport. We humans are not alone in this, technically called "post-copulatory activity." For example:

> A survey using 131 randomly-chosen species and conservative behavioral criteria showed that copulatory courtship occurred in just 80% of the species . . . Male insects and spiders use virtually all parts of their bodies as they tap, slam, squeeze, bite, lick, rub, shake, gently rock, or twist the female, coat her with liquid, cover her eyes with semitransparent colored plates, wrap her symbolically in weak silken lines, feed her, wave at her, and sing to her . . . Similar behavior also occurs in other groups of animals. . . . Unless one makes the unlikely supposition that this behavior, which is often energetic, persistent, and stereotyped, represents selectively neutral "mistakes" or incidental movements by the males . . . courtship at this late stage would seem to function to induce favorable female responses.[124]

Orgasm is not the end of sex for humans, any more than bare copulation is the end of sex for animals much simpler than ourselves. Intimacy, eroticism, and mutual connection are not mere decoration. They are intrinsic to human sexual behavior in its deepest expression, and thus it is ironic that they have been treated as mere sidelines in much (although not all) sex research. These things are a commonplace of sex therapists, of course, who have to deal with more than mere mechanical issues in their work.

It has sometimes taken scholars a while to catch up with common wisdom. It has been perfectly natural to focus on genital action, and easy to measure proxies for sexual desire such as urges for masturbation and sexual fantasies. However, this approach ignores vast elements of human sexual desires and sexual behavior—at least sexual behavior that equates to long-term, rather than short-term, mating. What has been left out includes what one scholar has, reasonably enough, called the ignoring of "major components of women's sexual satisfaction: trust, intimacy, the ability to be vulnerable, respect, communication, affection, and pleasure from sexual touching."[125] Given that women, both during pregnancy and beyond, are likely to want both conscious, and unconscious, assurances of support, provisioning, and commitment, it is utterly unsurprising that these are associated with arousal consistent with long-term mating strategies. At the risk of seeming overly blunt, humans can make themselves orgasm very easily (unless prevented by mutilation, or imposed guilt). Yet we seek out the sharing of sexual release with one another, and use this as a tester, and confirmer, of commitment. This should not surprise us. Even other primates form consortships—the monopolizing of a partner.[126] These are comparatively short-term ones, by human standards. But the "wham, bam, thank you ma'am" approach is, not only, not the sole strategic option for humans, but it is also not the best one to secure either paternity certainty (from the man's perspective) or commitment to the relationship (from the woman's). This is also fully consistent with the drop in sexual desire attendant on various stressors, or in response to taking SSRIs.[127] Basson argues strongly and cogently for a different model for women's sexual desire and arousal than that of men's.

This intimacy expresses itself, in sexual encounters, with a tendency to mirror our own bodies in the touch and gaze of our partners.[128] They touch those areas that they prefer to be touched, and rate the intensity of such touch and gaze as strong with the partner's body as their own. We show neurological signs of mutual *entrainment* with partners[129] that we get to know well and unconsciously (but this can be made conscious) mold each other to mutual pleasure.[130] Men tended to show higher levels of alignment than women, but both tend to treat our partner's bodies as extensions of their own.[131]

11.17   CONCLUSION: SEX IS CAPTIVATING

Foreplay should begin the moment the previous orgasm ends.

—Sex Therapist Emily Morse[132]

Humans routinely make having a partner the focus of their lives. Sex is obviously a part of this, to the extent that its absence is tough to do without. People can substitute—by pornography,[133] or sex workers, for example—but these are typically seen as substitutes.[134] We can substitute for other needs as well—these days, often using online services. You can share finances with a housemate, play with gaming partners, care for a pet. You can vent your anger online, or get your physical vanity validated on image-based social media. But these are all a form of psycho-social junk food. They lack substance. Even, online meetings can be shown to not stimulate the same brain regions as face-to-face ones.[135] Only the presence of another human being, one that you feel strongly about, and feel this to reciprocated, not only fulfills these functions but allows you, as a mature adult, to fully believe in them. We find it easy to believe that some people give up when their soulmate dies, and there is a name for this, psychogenic death. And, notoriously, humans do not stop loving one another even after death. Instead, it is contempt that is the opposite of love, not indifference, and contempt is a mechanism for dissolving the bond between humans, and almost always impossible to come back from.

Assumptions, as I have said a few times in this book, are the things we do not know we are making. We know (or think we know) all about male orgasm. How it feels, what it does, *why* it does what it does, and how to measure its (fairly obvious) associative effects. It was only natural to think that when female orgasm did not fit the same mold, then it would be seen as deviant, or deficient in some way. But once it is appreciated that female orgasm tracks specifically female sexual selection patterns and needs, and not male ones, then a different picture emerges. Recall where we have got to so far. Female orgasm is intensely related to mood. It is not homogenous, like male orgasm, but heterogenous. Some female orgasms are associated with sperm retention, although this has often been difficult to establish as a single event in the laboratory. However, women's refraction patterns are utterly unlike male ones, resulting in the possibility of multiple orgasms, closely connected in time, and when women are allowed to call the sexual shots, they tend to favor this sort of activity. All of these increase oxytocin levels, which in turn is demonstrably associated with a range of fertility-centered effects including, but not limited to, a peristaltic pumping action delivering sperm to the dominant follicle over a matter of an hour. These orgasms can be relatively demanding to bring about. They can be cryptic, or even faked, in ways that male orgasm cannot. They are associated with signals that are highly motivating to male (and female) partners, and to the woman who has them, even allowing her to manipulate male partners if she wishes.

A mixture of things: Unthinking androcentrism, general squeamishness at acknowledging our animal natures, the unavoidably cryptic nature of female choice mechanisms, all coupled with the fact that humans are actively running interference on one another's sexual strategies, while promoting their own, have all conspired to create a perfect fog around female orgasm. I hope that fog has now been lifted somewhat.

Researchers into female orgasm, naturally enough, searched for an event that was as easily pinned down as male orgasm and, when this was not forthcoming, some resorted to trying to explain it away. But, as I hope this book has shown, they have been looking at it in the wrong way. We know—and have known for decades—that female orgasm covaries strongly with oxytocin, and we know that oxytocin creates measurable effects including, but not limited to, increased trust (e.g., pair-bonding) and uterine peristalsis (e.g., sperm selection). If modify the question as to what female orgasm—in the sense of the outward events—is *for,* to add the aspect of what the various events around female orgasm *signify*, then an answer becomes clearer. We may not have to choose between the mate choice, and sire choice, hypotheses, because the evidence is that both occur. We do not have to find a single event, such as insuck, associated with each and every female orgasm (although it does come about with some) because the sensations, emotions, and signals associated with female orgasm are the outward sign that oxytocin action is occurring. And, it is *this* that has the fitness benefits of its own. Some of us, if I am right, have fallen for the oldest fallacy in science: Mistaking the model for reality. When we observe human mating out of captivity, we get to see it in its fullest expression.

And it seems that when it comes to orgasm, as with other activities, women can multi-task.

NOTES

1 Typically, with things like enlarged and reddened sexual regions. See, e.g., Zinner, D. P., Nunn, C. L., van Schaik, C. P., & Kappeler, P. M. (2004). Sexual selection and exaggerated sexual swellings of female primates. In *Sexual Selection in Primates: New and Comparative Perspectives*, 71–89.

2 Geary, D. C. (2000). Evolution and proximate expression of human paternal investment. *Psychological Bulletin, 126*(1), 55.

3 Reynolds, J. J., & McCrea, S. M. (2016). Life history theory and exploitative strategies. *Evolutionary Psychology, 14*(3).

4 Winch, R. F. (1958). *Mate-selection; a study of complementary needs*. Harper.

5 Assortative mating is definitely a feature of the human mating landscape, as explored here: Thiessen, D., & Gregg, B. (1980). Human assortative mating and genetic equilibrium: An evolutionary perspective. *Ethology and Sociobiology, 1*(2), 111–140.

6 Berscheid, E., & Walster, E. (1974). A little bit about love. *Foundations of Interpersonal Attraction, 1,* 356–381.

7 Freud, S. (2017). *Three essays on the theory of sexuality: The 1905 edition*. Verso Books. To the extent that these theories developed into modern attachment and life history theory, which is open to debate, they have empirical support.

8 (apart from assortative mating, which has strong empirical support across taxa)

9 Wilson, E. O. (2000). *Consilience*. Alfred A. Knoff.

10 It is, of course, much more complex than I am stating here. There are, for example, many different androgens—literally "masculinizing hormones"—that get grouped together for convenience as "testosterone." Women also produce some of them, although in much smaller amounts. Bahrke, M. S., Yesalis, C. E., & Wright, J. E. (1990). Psychological and behavioural effects of endogenous testosterone levels and anabolic-androgenic steroids among males: A review. *Sports Medicine, 10,* 303–337.

11 FtM transsexuals who take testosterone as part of their transition report immediate (within half an hour) increases in sexual desire. Irwig, M. S. (2017). Testosterone therapy for transgender men. *The Lancet Diabetes & Endocrinology, 5*(4), 301–311.

   Hooven, C. (2021). *T: The story of testosterone, the hormone that dominates and divides us*. Henry Holt and Company.

12 And, to that extent, Freud might have had a point about the Oedipus complex, but got things exactly the wrong way round.

13 Fisher, H. E., Aron, A., & Brown, L. L. (2006). Romantic love: A mammalian brain system for mate choice. *Philosophical Transactions of the Royal Society B: Biological Sciences, 361*(1476), 2173–2186.

14 Sternberg, R. J. (1986). A triangular theory of love. *Psychological Review, 93*(2), 119.

15 Sorokowski, P., Sorokowska, A., Karwowski, M., Groyecka, A., Aavik, T., Akello, G., . . . Sternberg, R. J. (2021). Universality of the triangular theory of love: Adaptation and psychometric properties of the triangular love scale in 25 countries. *The Journal of Sex Research, 58*(1), 106–115.

16 Perel, E. (2006). *Mating in captivity* (p. 272). HarperCollins.

17 Matteo, S., & Rissman, E. F. (1984). Increased sexual activity during the midcycle portion of the human menstrual cycle. *Hormones and Behavior, 18*(3), 249–255.

18 Georgiadis, J. R., Kortekaas, R., Kuipers, R., Nieuwenburg, A., Pruim, J., Reinders, A. S., & Holstege, G. (2006). Regional cerebral blood flow changes associated with clitorally induced orgasm in healthy women. *European Journal of Neuroscience, 24*(11), 3305–3316.

   However, see Prause, N. (2011). The human female orgasm: Critical evaluations of proposed psychological sequelae. *Sexual and Relationship Therapy, 26*(4), 315–328, for an alternative view. No one suggests that orgasm does not act as some form of reinforcer, however.

19 Georgiadis, J. R., Kortekaas, R., Kuipers, R., Nieuwenburg, A., Pruim, J., Reinders, A. S., & Holstege, G. (2006). Regional cerebral blood flow changes associated with clitorally induced orgasm in healthy women. *European Journal of Neuroscience, 24*(11), 3305–3316.

20 The effects of oxytocin are discussed at length in chapters seven and eight. Some key papers include:

   Blaicher, W., Gruber, D., Bieglmayer, C., Blaicher, A. M., Knogler, W., & Huber J. C. (1999). The role of oxytocin in relation to female sexual arousal. *Gynaecologic and Obstetric Investigation, 47,* 125–126.

   Campbell, A. (2008). Attachment, aggression and affiliation: The role of oxytocin in female social behavior. *Biological Psychology, 77*(1), 1–10.

   Carmichael, M. S., Humbert, R., Dixen, J., Palmisano, G., Greenleaf, W., & Davidson, J. M. (1987). Plasma oxytocin increases in the human sexual response. *Journal of Clinical Endocrinology and Metabolism, 64,* 27–31.

Carmichael, M. S., Warburton, V. L., Dixen, J., & Davidson, J. M (1994). Relationships among cardiovascular, muscular and oxytocin responses during human sexual activity. *Archives of Sexual Behavior, 23*, 59–79.

Insel, T. R. (2010). The challenge of translation in social neuroscience: A review of oxytocin, vasopressin, and affiliative behavior. *Neuron, 65*(6), 768–779.

21 Fisher, H. (2016). *Anatomy of love: A natural history of mating, marriage, and why we stray (completely revised and updated with a new introduction)*. W.W. Norton.

22 Campbell, A. (2008). The morning after the night before: Affective reactions to one-night stands among mated and unmated women and men. *Human Nature, 19*(2), 157–173.

23 Hill, S. (2019). *This is your brain on birth control: The surprising science of women, hormones, and the law of unintended consequences*. Penguin.

24 Thornhill, R., & Gangestad, S. W. (1996). The evolution of human sexuality. *Trends in Ecology & Evolution, 11*(2), 98–102; Gangestad, S. W., & Thornhill, R. (1997). The evolutionary psychology of extrapair sex: The role of fluctuating asymmetry. *Evolution and Human Behavior, 18*(2), 69–88; Gangestad, S. W., & Thornhill, R. (1998). Menstrual cycle variation in women's preferences for the scent of symmetrical men. *Proceedings of the Royal Society of London. Series B: Biological Sciences, 265*(1399), 927–933; Shackelford, T. K., Weekes, V. A., LeBlanc, G. J., Bleske, A. L., Euler, H. A., & Hoier, S. (2000). Female coital orgasm and male attractiveness. *Human Nature, 11*, 299–306.

25 See, for example, Eschler, L. (2004). The physiology of the female orgasm as a proximate mechanism. *Sexualities, Evolution & Gender, 6*(2–3), 171–194.

Hevesi, K., Horvath, Z., Sal, D., Miklos, E., & Rowland, D. L. (2021). Faking orgasm: Relationship to orgasmic problems and relationship type in heterosexual women. *Sexual Medicine, 9*(5), 100419.

26 Buss, D. M. (2016). *The evolution of desire: Strategies of human mating*. Hachette UK.

27 See the extensive notes in chapter nine, beginning with Buss, D. M., & Schmitt, D. P. (1993). Sexual strategies theory: An evolutionary perspective on human mating. *Psychological Review, 100*(2), 204–232.

28 Baker, R. (1996). *Sperm wars*. Basic Books.

29 Baker, R. R., & Bellis, M. A. (1989). Number of sperm in human ejaculates varies in accordance with sperm competition theory. *Animal Behaviour, 37*, 867–869. They appeared to show that double-mating (which they rather luridly titled a "sperm war") occurred in around 0.1% of all copulations. However, a prediction of these findings—that sperm would adapt to compete directly with one another—has not been replicated.

Moore, H. D. M., Martin, M., & Birkhead, T. R. (1999). No evidence for killer sperm or other selective interactions between human spermatozoa in ejaculates of different males *in vitro*. *Proceedings of the Royal Society of London B, 266*, 2343–2350.

That said, being cued to multi-male pairings *does* interestingly produce more male ejaculate. Kilgallon, S. J., & Simmons, L. W. (2005). Image content influences men's semen quality. *Biology Letters, 1*(3), 253–255.

Pound, N. (2002). Male interest in visual cues of sperm competition risk. *Evolution and Human Behavior, 23*(6), 443–466.

30 Gene variant.

31 Haldane, J. B. S. (1927). A mathematical theory of natural and artificial selection, part V: Selection and mutation. In *Mathematical proceedings of the Cambridge philosophical society* (Vol. 23, No. 7, pp. 838–844). Cambridge University Press.

32 Zubrow, E. (1989). The demographic modelling of Neanderthal extinction. In *The human revolution* (pp. 212–231). Edinburgh University Press.

33 Stebbins, G. L. (1982). *Darwin to DNA, molecules to humanity*. Demonstrated the mouse to elephant evolution.

34 Birkhead, T. R., Lee, K. E., & Young, P. (1988). Sexual cannibalism in the praying mantis Hierodula membranacea. *Behaviour*, 112–118.

35 James, W. H. (1993). The incidence of superfecundation and of double paternity in the general population. *AMG Acta geneticae medicae et gemellologiae: Twin Research, 42*(3–4), 257–262. This does exist but is very rare.

36 Anderson, K. G. (2006). How well does paternity confidence match actual paternity? *Current Anthropology, 47*, 513–520.

37 *Source:* National Safety Council. www.nsc.org/learn/safety-knowledge/Pages/injury-facts.aspx last accessed 29/09/2017.

38 Duberman, L. (1975). *The reconstituted family: A study of remarried couples and their children.* Burnham Incorporated. Reported on stepparent feelings:

Daly, M., & Wilson, M. (1988). Evolutionary social psychology and family homicide. They show that the Cinderella effect can be all too real. Daly and Wilson's work disturbed so many people that attempts were made to silence it.

39 Dibble, J. L., Drouin, M., Aune, K. S., & Boller, R. R. (2015). Simmering on the back burner: Communication with and disclosure of relationship alternatives. *Communication Quarterly, 63*, 329–344.

Dibble, J. L., & Drouin, M. (2014). Using modern technology to keep in touch with back burners: An investment model analysis. *Computers in Human Behavior, 34*, 96–100.

40 Buss, D. M., & Schmitt, D. P. (2019). Mate preferences and their behavioral manifestations. *Annual Review of Psychology, 70*, 77–110.

41 Buss, D. M., & Schmitt, D. P. (2019). Mate preferences and their behavioral manifestations. *Annual Review of Psychology, 70*, 77–110.

42 Matteo, S., & Rissman, E. F. (1984). Increased sexual activity during the midcycle portion of the human menstrual cycle. *Hormones and Behavior, 18*(3), 249–255.

43 Miller, G., Tybur, J. M., & Jordan, B. D. (2007). Ovulatory cycle effects on tip earnings by lap dancers: Economic evidence for human estrus? *Evolution and Human Behavior, 28*(6), 375–381.

Gangestad, S. W., Thornhill, R., & Garver, C. E. (2002). Changes in women's sexual interests and their partner's mate–retention tactics across the menstrual cycle: Evidence for shifting conflicts of interest. *Proceedings of the Royal Society of London. Series B: Biological Sciences, 269*(1494), 975–982.

Schleifenbaum, L., Driebe, J. C., Gerlach, T. M., Penke, L., & Arslan, R. C. (2021). Women feel more attractive before ovulation: Evidence from a large-scale online diary study. *Evolutionary Human Sciences, 3*, e47.

44 Zietsch, B. P., & Santtila, P. (2013). No direct relationship between human female orgasm rate and number of offspring. *Animal Behaviour, 86*(2), 253–255.

45 For instance, Dawood, K., Kirk, K. M., Bailey, J. M., Andrews, P. W., & Martin, N. G. (2005). Genetic and environmental influences on the frequency of orgasm in women. *Twin Research and Human Genetics, 8*(1), 27–33. Also see Dunn, K. M., Cherkas, L. F., & Spector, T. D. (2005). Genetic influences on variation in female orgasmic function: A twin study. *Biology Letters, 1*(3), 260–263.

46 Garver-Apgar, C. E., Gangestad, S. W., Thornhill, R., Miller, R. D., & Olp, J. J. (2006). Major histocompatibility complex alleles, sexual responsivity, and unfaithfulness in romantic couples. *Psychological Science, 17*(10), 830–835.

47 The last duel (with epees) took place between politicians in France in 1967. Viewable at: Schlager7 [Youtube Video]. (1967). *Last sword duel.* www.youtube.com/watch?v=e68nuAcSuWQ last accessed 19/08/2023.

48 Buss, D. M., & Shackelford, T. K. (1997). Human aggression in evolutionary psychological perspective. *Clinical Psychology Review, 17*(6), 605–619.

Chagnon, N. A. (1968). *The fierce people.* Holt, Rinehart & Winston. This is, by far, the best anthropological account of ritual male/male combat and its distinction from lethal warfare in non-state society.

49 King, R., & O'Riordan, C. (2019). Near the knuckle: How evolutionary logic helps explain Irish Traveller bare-knuckle contests. *Human Nature, 30*(3), 272–298.

50 This is technically the angle of the slope of the linear regression of reproductive success on mating success (Arnold & Duvall, 1994). It can also be called the sexual selection gradient for this reason (Janicke et al., 2016).

51 For past sexual selection on females, see Campbell, 2013; Low, 1979, Low et al., 1987; Schmitt & Buss, 1996. For current sexual selection on females, see Borgerhoff Mulder & Ross, 2019).

52 See Anderson, A., Chilczuk, S., Nelson, K., Ruther, R., & Wall-Scheffler, C. (2023). The myth of man the hunter: Women's contribution to the hunt across ethnographic contexts. *PLoS One, 18*(6), e0287101 for claims that women participate in hunting as much as men. Responses to the paper have suggested that the coding techniques biased towards societies where women did any hunting at all, rather than look at the volume of hunting that they did, did not correct for small versus big game hunting, included non-foragers as foragers, and lumped in ritual hunting festivals with actual hunting. See www.vivekvenkataraman. com/blog last accessed 11/07/2023 for a detailed breakdown of the methodology of the paper.

53 Syed, U. A. M., Davis, D. E., Ko, J. W., Lee, B. K., Huttman, D., Seidl, A., . . . Abboud, J. A. (2017). Quantitative anatomical differences in the shoulder. *Orthopedics, 40*(3), 155–160.

54 Jardine, R., & Martin, N. G. (1983). Spatial ability and throwing accuracy. *Behavior Genetics, 13*, 331–340.

55 See Morgan, M. H., & Carrier, D. R. (2013). Protective buttressing of the human fist and the evolution of hominin hands. *Journal of Experimental Biology*, *216*(2), 236–244, for an alternative view and my response to them here:
King, R. (2013). Fists of furry: At what point did human fists part company with the rest of the hominid lineage? *Journal of Experimental Biology*, *216*(12), 2361–2361.

56 Watson, N. V., & Kimura, D. (1989). Right-hand superiority for throwing but not for intercepting. *Neuropsychologia*, *27*(11–12), 1399–1414.

57 Deaner, R. O., Balish, S. M., & Lombardo, M. P. (2016). Sex differences in sports interest and motivation: An evolutionary perspective. *Evolutionary Behavioral Sciences*, *10*(2), 73.

58 Alexander, R. D., Hoogland, J. L., Howard, R. D., Noonan, K. M., & Sherman, P. W. (1979). Sexual dimorphism and breeding systems in pinnipeds, ungulates, primates and humans. In N. A. Chagnon & W. Irons (Eds.), *Evolutionary biology and human social behavior: An anthropological perspective* (pp. 402–435). Duxbury Press.

59 Loomba-Albrecht, L. A., & Styne, D. M. (2009). Effect of puberty on body composition. *Current Opinion in Endocrinology, Diabetes and Obesity*, *16*(1), 10–15.

60 Humphrey, L. T., Dean, M. C., & Stringer, C. B. (1999). Morphological variation in great ape and modern human mandibles. *The Journal of Anatomy*, *195*(4), 491–513.

61 Garn, S. M., Clark, L. C., & Harper, R. V. (1953). The sex difference in the basal metabolic rate. *Child Development*, 215–224.

62 Voyer, D. (1995). Effect of practice on laterality in a mental rotation task. *Brain and Cognition*, *29*(3), 326–335.

63 Der, G., & Deary, I. J. (2006). Age and sex differences in reaction time in adulthood: Results from the United Kingdom health and lifestyle survey. *Psychology and Aging*, *21*(1), 62.

64 Tanner, J. M. (1986). 1 normal growth and techniques of growth assessment. *Clinics in Endocrinology and Metabolism*, *15*(3), 411–451.

65 Waalen, J., Felitti, V., & Beutler, E. (2002). Haemoglobin and ferritin concentrations in men and women: Cross sectional study. *BMJ*, *325*(7356), 137.

66 Burse, R. L. (1979). Sex differences in human thermoregulatory response to heat and cold stress. *Human Factors*, *21*(6), 687–699.

67 Courtenay, W. H. (2003). Key determinants of the health and well-being of men and boys. *International Journal of Mens Health*, *2*, 1–30.

68 Betzig, L. (1993). Sex, succession, and stratification in the first six civilizations: How, powerful men reproduced, passed power on to their sons, and used power to defend their wealth, women, and children. In L. Ellis (Ed.), *Social stratification and socioeconomic inequality, Volume 1: A comparative biosocial analysis* (pp. 37–74). Praeger.

69 For example, Abilev, S., Malyarchuk, B., Derenko, M., Wozniak, M., Grzybowski, T., & Zakharov, I. (2012). The Y-chromosome C3* star-cluster attributed to Genghis Khan's descendants is present at high frequency in the Kerey clan from Kazakhstan. *Human Biology*, *84*(1), 79–89.

70 Whiting, B., & Edwards, C. P. (1973). A cross-cultural analysis of sex differences in the behavior of children aged three through 11. *The Journal of Social Psychology*, *91*(2), 171–188.

71 Simon, R. J., & Baxter, S. (1989). Gender and violent crime. In N. A. Weiner & M. E. Wolfgang (Eds.), *Violent crime, violent criminals* (pp. 171–197). SAGE.

72 Apicella, C. L. (2014). Upper-body strength predicts hunting reputation and reproductive success in Hadza hunter–gatherers. *Evolution and Human Behavior*, *35*(6), 508–518.

73 Atkinson, J., Pipitone, R. N., Sorokowska, A., Sorokowski, P., Mberira, M., Bartels, A., & Gallup, G. G., Jr. (2012). Voice and handgrip strength predict reproductive success in a group of indigenous African females. *PLoS One*. https://doi.org/10.1371/journal.pone.0041811.

74 Lassek, W. D., & Gaulin, S. J. (2009). Costs and benefits of fat-free muscle mass in men: Relationship to mating success, dietary requirements, and native immunity. *Evolution and Human Behavior*, *30*(5), 322–328.

75 Pazhoohi, F., Garza, R., Doyle, J. F., Macedo, A. F., & Arantes, J. (2019). Sex differences for preferences of shoulder to hip ratio in men and women: An eye tracking study. *Evolutionary Psychological Science*, *5*, 405–415.

76 Lassek, W. D., & Gaulin, S. J. (2022). Substantial but misunderstood human sexual dimorphism results mainly from sexual selection on males and natural selection on females. *Frontiers in Psychology*, *13*, 859931.

77 Wilson, M., & Daly, M. (1985). Competitiveness, risk taking, and violence: The young male syndrome. *Ethology and Sociobiology, 6*(1), 59–73.

78 Sell, A., Lukazsweski, A. W., & Townsley, M. (2017). Cues of upper body strength account for most of the variance in men's bodily attractiveness. *Proceedings of the Royal Society B: Biological Sciences, 284*(1869), 20171819. doi: 10.1098/rspb.2017.1819.

79 Bale, C., & Archer, J. (2013). Self-perceived attractiveness, romantic desirability and self-esteem: A mating sociometer perspective. *Evolutionary Psychology, 11*(1).

80 Marshall, A. J. (1954). Bower-birds. *Biological Reviews, 29*(1), 1–45.

81 Lidborg, L. H., Cross, C. P., & Boothroyd, L. G. (2020). *Masculinity matters (but mostly if you're muscular): A meta-analysis of the relationships between sexually dimorphic traits in men and mating/ reproductive success.* BioRxiv.

82 Benenson, J. F., Webb, C. E., & Wrangham, R. W. (2022). Self-protection as an adaptive female strategy. *Behavioral and Brain Sciences, 45*, e128.

83 Recently the gene variant responsible for this mutation in something like 1 per 1 million humans has been identified and explored via knockout and gain of function research in mice. People with this muta- tion often damage and scald themselves while young, as do mutant mice. Leipold, E., Liebmann, L., Korenke, G. C., Heinrich, T., Gießelmann, S., Baets, J., . . . Kurth, I. (2013). A de novo gain-of-function mutation in SCN11A causes loss of pain perception. *Nature Genetics, 45*(11), 1399–1404.

84 Fieder, M., & Huber, S. (2007). The effect of sex and childlessness on the association between status and reproductive output in modern society. *Evolution and Human Behavior, 28*, 392–398.

85 Puts, D. A., Hill, A. K., Bailey, D. H., Walker, R. S., Rendall, D., Wheatley, J. R., . . . Ramos-Fernandez, G. (2016). Sexual selection on male vocal fundamental frequency in humans and other anthropoids. *Proceedings of the Royal Society B: Biological Sciences, 283*(1829), 20152830.

86 De Waal, F. (2022). *Different: What apes can teach us about gender.* Granta Books.

87 Martin, G. R. (2003). *A storm of swords, Part 2.* Blood and Gold.

88 For example, in baboons, dominant females regularly harass ones lower in the hierarchy, and this has been directly linked to lower fertility in the latter. Smuts, B., & Nicolson, N. (1989). Reproduction in wild female olive baboons. *American Journal of Primatology, 19*(4), 229–246.

89 Benenson, J. F., & Abadzi, H. (2020). Contest versus scramble competition: Sex differences in the quest for status. *Current Opinion in Psychology, 33*, 62–68. The quote is from p. 62.

90 Campbell, A. (1999). Staying alive: Evolution, culture, and women's intrasexual aggression. *Behavioral and Brain Sciences, 22*(2), 203–214.

91 Benenson, J. F., & Abadzi, H. (2020). Contest versus scramble competition: Sex differences in the quest for status. *Current Opinion in Psychology, 33*, 62–68.

92 Murdock, G. P. (1967). *Ethnographic atlas.* Pittsburgh University Press.

93 Rosenblatt, P. C. (1967). Marital residence and the functions of romantic love. *Ethnology, 6*(4), 471–480.

94 Hrdy, S. B. (2009). *The woman that never evolved.* Harvard University Press. The quote is from p. 100.

95 Owens, L., Shute, R., & Slee, P. (2000a). "Guess what I just heard!": Indirect aggression among teenage girls in Australia. *Aggressive Behavior, 26*, 67–83.
   Owens, L., Shute, R., & Slee, P. (2000b). "I'm in and you're out . . . :" Explanations for teenage girls' indirect aggression. *Psychology, Evolution, and Gender, 2*, 19–46.

96 Hess, N. H., & Hagen, E. H. (2006). Sex differences in indirect aggression: Psychological evidence from young adults. *Evolution and Human Behavior, 27*(3), 231–245.

97 Galen, B. R., & Underwood, M. K. (1997). A developmental investigation of social aggression among children. *Developmental Psychology, 33*, 589–600.

98 Dunbar, R. I. (1980). Determinants and evolutionary consequences of dominance among female gelada baboons. *Behavioral Ecology and Sociobiology, 7*, 253–265.

99 Dunbar, R. I. M. (1996). Groups, gossip, and the evolution of language. In A. Schmitt, K. Atzwanger, K. Grammer, & K. Schäfer (Eds.), *New aspects of human ethology* (pp. 77–89). Springer.
   Dunbar, R. I. (2004). Gossip in evolutionary perspective. *Review of General Psychology, 8*(2), 100– 110. https://doi.org/10.1037/1089–2680.8.2.100.

100 Smith, R. L., Rose, A. J., & Schwartz-Mette, R. A. (2010). Relational and overt aggression in child- hood and adolescence: Clarifying mean-level gender differences and associations with peer acceptance. *Social Development, 19*, 243–269.

101 Burbank, V. K. (1987). Female aggression in cross-cultural perspective. *Behavior Science Research, 21*, 70–100. This study compared 137 different cultures.

102 Strassmann, B. I. (1997). Polygyny as a risk factor for child mortality among the Dogon. *Current Anthropology, 38*, 688–695.

103 Baumeister, R. F., & Twenge, J. M. (2002). Cultural suppression of female sexuality. *Review of General Psychology, 6*(2), 166–203.

104 Buss, D. M. (1994). The strategies of human mating. *American Scientist, 82*(3), 238–249.

105 Chagnon, N. A. (1968). *The fierce people.* New York: Holt, Rinehart & Winston.

106 Janus, S. S., & Janus, C. L. (1993). *The Janus report on sexual behavior.* Wiley.

107 Baumeister, R. F., & Twenge, J. M. (2002). Cultural suppression of female sexuality. *Review of General Psychology, 6*(2), 166–203. The quote comes from page 170.

108 Baumeister, R. F., & Twenge, J. M. (2002). Cultural suppression of female sexuality. *Review of General Psychology, 6*(2), 166–203. Quote is from p. 170.

109 Trivers, R. L. (1974). Parent-offspring conflict. *American Zoologist, 14*(1), 249–264. The quote is from p. 249.

110 van den Berg, P., Fawcett, T. W., Buunk, A. P., & Weissing, F. J. (2013). The evolution of parent–offspring conflict over mate choice. *Evolution and Human Behavior, 34,* 405–411.

111 Apostolou, M. (2014). *Sexual selection under parental choice: The evolution of human mating behaviour.* Psychology Press.

112 Libby, R. W., Gray, L., & White, M. (1978). A test and reformulation of reference group and role correlates of premarital sexual permissiveness theory. *Journal of Marriage and the Family.*

113 Nolin, M. J., & Petersen, K. K. (1992). Gender differences in parent-child communication about sexuality: An exploratory study. *Journal of Adolescent Research, 7*(1), 59–79.

114 Shandall, A. A. (1967). Circumcision and infibulation of females. *Sudan Medical Journal, 5,* 178–212. The quote comes from p. 93 and is cited in Baumeister, R. F., & Twenge, J. M. (2002). p. 183.

115 Boddy, J. (1989). *Wombs and alien spirits: Women, men and the Zar cult in northern Sudan.* University of Wisconsin Press. Boddy, J. (1998). Violence embodied? Circumcision, gender, politics, and cultural aesthetics. In R. E. Dobash & R. P. Dobash (Eds.), *Rethinking violence against women* (pp. 77–110). SAGE.

116 Hicks, E. K. (1996). *Infibulation: Female mutilation in Islamic northeastern Africa.* Transaction.

117 Simler, K., & Hanson, R. (2017). *The elephant in the brain: Hidden motives in everyday life.* Oxford University Press.

118 Harris, E. A., Hornsey, M. J., Larsen, H. F., & Barlow, F. K. (2019). Beliefs about gender predict faking orgasm in heterosexual women. *Archives of Sexual Behavior, 48,* 2419–2433.

119 Salisbury, C. M., & Fisher, W. A. (2014). "Did you come?" A qualitative exploration of gender differences in beliefs, experiences, and concerns regarding female orgasm occurrence during heterosexual sexual interactions. *The Journal of Sex Research, 51*(6), 616–631.

120 Opperman, E., Braun, V., Clarke, V., & Rogers, C. (2014). "It feels so good it almost hurts": Young adults' experiences of orgasm and sexual pleasure. *The Journal of Sex Research, 51*(5), 503–515. The quote comes from p. 503.

121 Chadwick, S. B., & van Anders, S. M. (2017). Do women's orgasms function as a masculinity achievement for men? *The Journal of Sex Research, 54*(9), 1141–1152.

122 Kontula, O., & Miettinen, A. (2016). Determinants of female sexual orgasms. *Socioaffective Neuroscience & Psychology, 6*(1), 31624. This analysis involved over 12000 women from 18 to 81 years old. The quote comes from the abstract summary.

123 Kontula, O., & Miettinen, A. (2016). Determinants of female sexual orgasms. *Socioaffective Neuroscience & Psychology, 6*(1), 31624. p. 1.

124 Eberhard, W. G. (2009). Postcopulatory sexual selection: Darwin's omission and its consequences. *Proceedings of the National Academy of Sciences, 106*(supplement_1), 10025–10032.

125 Basson, R. (2000). The female sexual response: A different model. *Journal of Sex & Marital Therapy, 26*(1), 51–65. The quote is from p. 52.

126 Manson, J. H. (1997). Primate consortships: A critical review. *Current Anthropology, 38*(3), 353–374.

127 Kavoussi, R. (1996, November). Clinical profile/safety and efficacy data. In *Wellbutrin advisory panel meeting.*

128 Maister, L., Fotopoulou, A., Turnbull, O., & Tsakiris, M. (2020). The erogenous mirror: Intersubjective and multisensory maps of sexual arousal in men and women. *Archives of Sexual Behavior,* 1–15.

129 Safron, A. (2016). What is orgasm? A model of sexual trance and climax via rhythmic entrainment. *Socioaffective Neuroscience & Psychology, 6*(1), 31763.

130 Fleischman, D. S. (2016b). An evolutionary behaviorist perspective on orgasm. *Socioaffective Neuroscience & Psychology, 6*(1), 32130.

131 Tsakiris, M., Maister, L., Fotopoulou, A., & Turnbull, O. (2018). The erogenous mirror: Intersubjective and multisensory maps of sexual arousal.

132 As said on the Chris Williamson Podcast, *Why Is Everyone Having Such Bad Sex.* Modern Wisdom
    645. Last accessed 11/08/2023.
133 Or erotica, which is pornography that you approve of.
134 Which is not to argue against their existence, often times a substitute is better than nothing.
135 Zhao, N., Zhang, X., Noah, J. A., Tiede, M., & Hirsch, J. (2023). Separable processes for live "In-person"
    and live "Zoom-like" faces. *Imaging Neuroscience.* https://doi.org/10.1162/imag_a_00027.

References

Abilev, S., Malyarchuk, B., Derenko, M., Wozniak, M., Grzybowski, T., & Zakharov, I. (2012). The Y-chromosome C3* star-cluster attributed to Genghis Khan's descendants is present at high frequency in the Kerey clan from Kazakhstan. *Human Biology*, *84*(1), 79–89.

Adams, D., & Carwardine, M. (2013). *Last chance to see*. Penguin Random House.

Addiego, F., Belzer, E. G., Jr., Comolli, J., Moger, W., Perry, J. D., & Whipple, B. (1981). Female ejaculation: A case study. *Journal of Sex Research*, *17*(1), 13–21.

Aginsky, B. W. (1939). Population control in the Shanel (Pomo) tribe. *American Sociological Review*, *4*(2), 209–216.

Ah-King, M. (2013). On anisogamy and the evolution of "sex-roles". *Trends in Ecology & Evolution*, *28*, 1–2.

Ainsworth, C. (2018). [Graphic] *Sex redefined: The idea of 2 sexes is overly simplistic*. https://blogs.scientific american.com/sa-visual/visualizing-sex-as-a-spectrum/ last accessed 24/10/2018.

Alexander, R. D., Hoogland, J. L, Howard, R. D., Noonan, K. M., & Sherman, P. W. (1979). Sexual dimorphism and breeding systems in pinnipeds, ungulates, primates and humans. In N. A. Chagnon & W. Irons (Eds.), *Evolutionary biology and human social behavior: An anthropological perspective* (pp. 402–435). Duxbury Press.

Allen, C., Cobey, K. D., Havlíček, J., & Roberts, S. C. (2016). The impact of artificial fragrances on the assessment of mate quality cues in body odor. *Evolution and Human Behavior*, *37*(6), 481–489.

Alonzo, S. H., & Sinervo, B. (2001). Mate choice games, context-dependent good genes, and genetic cycles in the side-blotched lizard, Uta stansburiana. *Behavioral Ecology and Sociobiology*, *49*(2–3), 176–186.

Alvergne, A., Jokela, M., & Lummaa, V. (2010). Personality and reproductive success in a high-fertility human population. *Proceedings of the National Academy of Sciences*, *107*(26), 11745–11750.

Alvergne, A., & Lummaa, V. (2010). Does the contraceptive pill alter mate choice in humans? *Trends in Ecology & Evolution*, *25*(3), 171–179.

Alwaal, A., Breyer, B. N., & Lue, T. F. (2015). Normal male sexual function: Emphasis on orgasm and ejaculation. *Fertility and Sterility*, *104*(5), 1051–1060.

Alzate, H. (1985). Vaginal eroticism and female orgasm: A current appraisal. *Journal of Sex & Marital Therapy*, *11*(4), 271–284.

Ammersbach, R. (1930). Sterilität und frigidität. *München. Medizinische Wochenschrift*, *77*, 225–227.

Anand, M. R. (1958). *Kama Kala: Some notes on the philosophical basis of Hindu erotic sculpture*. Skilton.

Anderson, A., Chilczuk, S., Nelson, K., Ruther, R., & Wall-Scheffler, C. (2023). The myth of man the hunter: Women's contribution to the hunt across ethnographic contexts. *PLoS One*, *18*(6), e0287101.

Anderson, K. G. (2006). How well does paternity confidence match actual paternity? *Current Anthropology*, *47*, 513–520.

Anderson, W. S. (Ed.). (1997). *Ovid's metamorphoses*. University of Oklahoma Press.

Angel, K. (2010). The history of 'female sexual dysfunction' as a mental disorder in the 20th century. *Current Opinion in Psychiatry*, *23*(6), 536.

Anon. (1968). Sub-umbra, or sport among the she-noodles. In *The Pearl: A journal of voluptuous reading, the underground magazine of Victorian England*. Grove. (Original work published 1879)

Apicella, C. L. (2014). Upper-body strength predicts hunting reputation and reproductive success in Hadza hunter–gatherers. *Evolution and Human Behavior*, *35*(6), 508–518.

Apostolou, M. (2014). *Sexual selection under parental choice: The evolution of human mating behaviour*. Psychology Press.

Apostolou, M., Sullman, M., Birkás, B., Błachnio, A., Bushina, E., Calvo, F., . . . Font-Mayolas, S. (2023). Mating performance and singlehood across 14 nations. *Evolutionary Psychology*, *21*(1), 14747049221150169.

Aristophanes. (431 BC/trans. 1973). *Lysistrata*. Penguin.

Aristotle. (1983). *Physics: Books I and II*. Oxford University Press.

Arnold, S. J., & Duvall, D. (1994). Animal mating systems: A synthesis based on selection theory. *The American Naturalist*, *143*(2), 317–348.

Arsan, E., & Kovič, B. (1967). *Emmanuelle* (Vol. 1). le Terrain vague.

Ashford, M., Timberman, S., Beverly, C., Lipman, A., & Verno, J. (2013). *Masters of sex* [TV Show]. Distributed by CBS Television, produced by Round Two and Timberman/Beverly productions.

Assaad, M. (1980). Female circumcision in Egypt: Social implications, current research, and prospects for change. *Studies in Family Planning, 11*(1), 3–16.

Atkinson, J., Pipitone, R. N., Sorokowska, A., Sorokowski, P., Mberira, M., Bartels, A., & Gallup, G. G., Jr. (2012). Voice and handgrip strength predict reproductive success in a group of indigenous African females. *PLoS One.* https://doi.org/10.1371/journal.pone.0041811.

Ayinde, B. A., Onwukaeme, D. N., & Nworgu, Z. A. M. (2006). Oxytocic effects of the water extract of *Musanga cecropioides* R. Brown (Moraceae) stem bark. *African Journal of Biotechnology, 5*, 1350–1354.

Babbitt, D. D. (Ed.). (1936). *The Dhammapada*. Oxford University Press.

Bae, B. I., Jayaraman, D., & Walsh, C. A. (2015). Genetic changes shaping the human brain. *Developmental Cell, 32*(4), 423–434.

Bahrke, M. S., Yesalis, C. E., & Wright, J. E. (1990). Psychological and behavioural effects of endogenous testosterone levels and anabolic-androgenic steroids among males: A review. *Sports Medicine, 10*, 303–337.

Bailey, M. (1991, July 7). Cupboard love: Sex secrets of the British Museum. *The Observer*, p. 7.

Baker, R. (1996). *Sperm wars*. Basic Books.

Baker, R., & Bellis, M. A. (1989). Number of sperm in human ejaculates varies in accordance with sperm competition theory. *Animal Behaviour, 37*, 867–869.

Baker, R., & Bellis, M. A. (1993a). Human sperm competition: Ejaculate adjustment by males and the function of masturbation. *Animal Behaviour, 46*, 861–885.

Baker, R., & Bellis, M. A. (1993b). Human sperm competition: Ejaculate manipulation by females and a function for the female orgasm. *Animal Behaviour, 46*, 887–909.

Baldinger-Melich, P., Urquijo Castro, M. F., Seiger, R., Ruef, A., Dwyer, D. B., Kranz, G. S., . . . Koutsouleris, N. (2020). Sex matters: A multivariate pattern analysis of sex-and gender-related neuroanatomical differences in cis-and transgender individuals using structural magnetic resonance imaging. *Cerebral Cortex, 30*(3), 1345–1356.

Baldwin, J. M. (1896). A new factor in evolution (continued). *The American Naturalist, 30*(355), 536–553.

Bale, C., & Archer, J. (2013). Self-perceived attractiveness, romantic desirability and self-esteem: A mating sociometer perspective. *Evolutionary Psychology, 11*(1).

Baranowski, A. M., & Hecht, H. (2015). Gender differences in and similarities in receptivity to casual sex invitations: Effects of location and risk perception. *Archives of Sexual Behavior, 44*, 2257–2265.

Barbach, L. G. (1975). *For yourself: The fulfilment of female sexuality*. Doubleday.

Barkow, J. H., Cosmides, L., & Tooby, J. (Eds.). (1995). *The adapted mind: Evolutionary psychology and the generation of culture*. Oxford University Press.

Barrett, L. F. (2019). Blog post *Zombie ideas*. www.psychologicalscience.org/observer/zombie-ideas last accessed 22/08/2023.

Basson, R. (2000). The female sexual response: A different model. *Journal of Sex &Marital Therapy, 26*(1), 51–65.

Bateman, A. J. (1948). Intra-sexual selection in Drosophila. *Heredity, 2*(Pt. 3), 349–368.

Baumeister, R. F., & Twenge, J. M. (2002). Cultural suppression of female sexuality. *Review of General Psychology, 6*(2), 166–203.

Becker, J. B., McClellan, M. L., & Reed, B. G. (2017). Sex differences, gender and addiction. *Journal of Neuroscience Research, 95*(1–2), 136–147.

Bell, A. P., & Weinberg, M. S. (1978). *Homosexualities: A study of diversity among men and women*. Simon and Shuster.

Belsky, J., Ruttle, P. L., Boyce, W. T., Armstrong, J. M., & Essex, M. J. (2015). Early adversity, elevated stress physiology, accelerated sexual maturation, and poor health in females. *Developmental Psychology, 51*, 816–822.

Belsky, J., Schlomer, G. L., & Ellis, B. J. (2012). Beyond cumulative risk: Distinguishing harshness and unpredictability as determinants of parenting and early life history strategy. *Developmental Psychology, 48*(3), 662–673.

Belyaev, D. K., Plyusnina, I. Z., & Trut, L. N. (1985). Domestication in the silver fox (*Vulpes fulvus*): Changes in physiological boundaries of the sensitive period of primary socialization. *Applied Animal Behaviour Science, 13*(4), 359–370.

Belyaev, D. K., & Trut, L. N. (1964). Behaviour and reproductive function of animals. II. Correlated changes under breeding for tameness. *Bulletin of the Moscow Society of Naturalists Biological Series (in Russian), 69*(5), 5–14.

Belyaev, D. K., & Trut, L. N. (1982). Accelerating evolution. *Science in the USSR, 5*, 24–29, 60–64.

Benenson, J. F., & Abadzi, H. (2020). Contest versus scramble competition: Sex differences in the quest for status. *Current Opinion in Psychology, 33*, 62–68.

Benenson, J. F., Webb, C. E., & Wrangham, R. W. (2022). Self-protection as an adaptive female strategy. *Behavioral and Brain Sciences, 45,* e128.

Bensman, L., Hatfield, E., & Doumas, L. A. (2011). *Two people just make it better: The psychological differences between partnered orgasms and solitary orgasms* [Doctoral dissertation, University of Hawai'i at Mānoa].

Bentler, P. M., & Peeler, W. H. (1979). Models of female orgasm. *Archives of Sexual Behavior, 8,* 405–423.

Berg, R. C., & Denison, E. (2012). Does female genital mutilation/cutting (FGM/C) affect women's sexual functioning? A systematic review of the sexual consequences of FGM/C. *Sexuality Research and Social Policy, 9,* 41–56.

Berg, R. C., Denison, E., & Fretheim, A. (2010). *Psychological, social and sexual consequences of female genital mutilation/cutting (FGM/C): A systematic review of quantitative studies* (Report from Kunnskapssenteret nr 13–2010). Oslo: Nasjonalt kunnskapssenter for helsetjenesten,

Berger, P. L., & Luckmann, T. (1991). *The social construction of reality: A treatise in the sociology of knowledge* (No. 10). Penguin UK.

Berglund, A., Rosenqvist, G., & Svensson, I. (1986). Reversed sex roles and parental energy investment in zygotes of two pipefish (Syngnathidae) species. *Marine Ecology Progress Series, 29,* 209–215.

Berscheid, E., & Walster, E. (1974). A little bit about love. *Foundations of Interpersonal Attraction, 1,* 356–381.

Betzig, L. (1993). Sex, succession, and stratification in the first six civilizations: How, powerful men reproduced, passed power on to their sons, and used power to defend their wealth, women, and children. In L. Ellis (Ed.), *Social stratification and socioeconomic inequality, Volume 1: A comparative biosocial analysis* (pp. 37–74). Praeger.

Birdwhistell, R. L. (1970). Masculinity and femininity as display. In *Kinesics and context: Essays on body motion* (pp. 39–46).

Birkhead, T. R. (2000). *Promiscuity: An evolutionary history of sperm competition.* Harvard University Press.

Birkhead, T. R., Lee, K. E., & Young, P. (1988). Sexual cannibalism in the praying mantis Hierodula membranacea. *Behaviour,* 112–118.

Birkhead, T. R., & Møller, A. P. (1998). *Sperm competition and sexual selection.* Academic Press.

Blackmore, S. (2000). *The meme machine.* Oxford Paperbacks.

Blaicher, W., Gruber, D., Bieglmayer, C., Blaicher, A. M., Knogler, W., & Huber J. C. (1999). The role of oxytocin in relation to female sexual arousal. *Gynaecologic and Obstetric Investigation, 47,* 125–126.

Blanchard, R. (1989). The concept of auto-gynephilia and the typology of male gender dysphoria. *Journal of Nervous and Mental Disease, 177*(10), 616–623.

Bloch, R. H. (2009). *Medieval misogyny and the invention of western romantic love.* University of Chicago Press.

Boddy, J. (1989). *Wombs and alien spirits: Women, men and the Zar cult in northern Sudan.* University of Wisconsin Press.

Boddy, J. (1998). Violence embodied? Circumcision, gender, politics, and cultural aesthetics. In R. E. Dobash & R. P. Dobash (Eds.), *Rethinking violence against women* (pp. 77–110). SAGE.

Bogoras, W. (1909). *The Chukchee* (Vol. 11). EJ Brill Limited.

Boole, G. (1911). *The laws of thought (1854)* (Vol. 2). Open Court Publishing Company.

Borgerhoff Mulder, M., & Ross, C. T. (2019). Unpacking mating success and testing Bateman's principles in a human population. *Proceedings of the Royal Society B, 286*(1908), 20191516.

Braun, M., Wassmer, G., Klotz, T., Reifenrath, B., Mathers, M., & Engelmann, U. (2000). Epidemiology of erectile dysfunction: Results of the 'cologne male survey'. *International Journal of Impotence Research, 12*(6), 305.

Briggs, Z. (2012). Fifty shades of grey increases sex toy sales. Interview with Jacqueline Gold, CEO of Anne Summers. *Cosmopolitan Magazine.* www.cosmopolitan.com/uk/entertainment/a16634/fifty-shades-of-grey-increases-sex-toy-sales/ last accessed 28/04/2018.

Brindle, M., & Opie, C. (2016). Postcopulatory sexual selection influences baculum evolution in primates and carnivores. *Proceedings of the Royal Society B: Biological Sciences, 283*(1844)

Brisson, L. (1976). *Le mythe de Tirésias: essai d'analyse structurale.* Brill.

Brody, S., & Costa, R. M. (2008). Vaginal orgasm is associated with less use of immature psychological defense mechanisms. *The Journal of Sexual Medicine, 5*(5), 1167–1176.

Brooks, A. M. M. R. (2017). Eyes wide shut: The importance of eyes in infant gaze-following and understanding other minds. In *Gaze-following* (pp. 229–254). Psychology Press.

Browne, E. G. (2001). *Islamic medicine. Fitzpatrick lectures delivered at the Royal College of Physicians in 1919–1920* (Reprint). Goodword Books.

Brownmiller, S. (2013). *Against our will: Men, women and rape.* Open Road Media (Original work published 1975).

Buffum, J. (1986). Pharmacosexology update: Prescription drugs and sexual function. *Journal of Psychoactive Drugs, 18*(2), 97–106.

Burbank, V. K. (1987). Female aggression in cross-cultural perspective. *Behavior Science Research, 21,* 70–100.

Burns, J. (2016). *Forbes report on the sex toy industry.* www.forbes.com/sites/janetwburns/2016/07/15/adult-expo-founders-talk-15b-sex-toy-industry-after-20-years-in-the-fray/#2329ff675bb9 last accessed 9/9/2018.

Burse, R. L. (1979). Sex differences in human thermoregulatory response to heat and cold stress. *Human Factors, 21*(6), 687–699.

Buss, D. M. (1985). Human mate selection: Opposites are sometimes said to attract, but in fact we are likely to marry someone who is similar to us in almost every variable. *American Scientist, 73*(1), 47–51.

Buss, D. M. (1994). The strategies of human mating. *American Scientist, 82*(3), 238–249.

Buss, D. M. (2007). The evolution of human mating. *Acta Psychologica Sinica, 39*(3), 502–512.

Buss, D. M. (2016). *The evolution of desire: Strategies of human mating.* Hachette UK.

Buss, D. M., Goetz, C., Duntley, J. D., Asao, K., & Conroy-Beam, D. (2017). The mate switching hypothesis. *Personality and Individual Differences, 104,* 143–149.

Buss, D. M., & Haselton, M. (2005). The evolution of jealousy. *Trends in Cognitive Sciences, 9*(11), 506.

Buss, D. M., Larsen, R. J., Westen, D., & Semmelroth, J. (1992). Sex differences in jealousy: Evolution, physiology, and psychology. *Psychological Science, 3*(4), 251–256.

Buss, D. M., & Schmitt, D. P. (1993). Sexual strategies theory: An evolutionary perspective on human mating. *Psychological Review, 100*(2), 204–232.

Buss, D. M., & Schmitt, D. P. (2019). Mate preferences and their behavioral manifestations. *Annual Review of Psychology, 70,* 77–110.

Buss, D. M., & Shackelford, T. K. (1997). Human aggression in evolutionary psychological perspective. *Clinical Psychology Review, 17*(6), 605–619.

Buston, P. (2003). Social hierarchies: Size and growth modification in clownfish. *Nature, 424*(6945), 145.

Butler, C. A. (1976). New data about female sexual response. *Journal of Sex & Marital Therapy, 2*(1), 40–46.

Butler, J. (2004). *Undoing gender.* Routledge.

Byers, E. S., Goldsmith, K. M., & Miller, A. (2016). If given the choice, would you choose to be a man or a woman? *The Canadian Journal of Human Sexuality, 25*(2), 148–157.

Cahill, L. (2014). Fundamental sex difference in human brain architecture. *Proceedings of the National Academy of Sciences, 111*(2), 577–578.

Cahill, L., & Hall, E. D. (2017). Is it time to resurrect "lazaroids"?. *Journal of Neuroscience Research, 95* (1–2), 17–20.

Campbell, A. (1999). Staying alive: Evolution, culture, and women's intrasexual aggression. *Behavioral and Brain Sciences, 22*(2), 203–214.

Campbell, A. (2008a). The morning after the night before: Affective reactions to one-night stands among mated and unmated women and men. *Human Nature, 19*(2), 157–173.

Campbell, A. (2008b). Attachment, aggression and affiliation: The role of oxytocin in female social behavior. *Biological Psychology, 77*(1), 1–10.

Campbell, A. (2013). The evolutionary psychology of women's aggression. *Philosophical Transactions of the Royal Society B: Biological Sciences, 368*(1631), 20130078.

Carmichael, M. S., Humbert, R., Dixen, J., Palmisano, G., Greenleaf, W., & Davidson, J. M. (1987). Plasma oxytocin increases in the human sexual response. *Journal of Clinical Endocrinology and Metabolism, 64,* 27–31.

Carmichael, M. S., Warburton, V. L., Dixen, J., & Davidson, J. M (1994). Relationships among cardio-vascular, muscular and oxytocin responses during human sexual activity. *Archives of Sexual Behavior, 23,* 59–79.

Carr, A. F., & Ogren, L. H. (1960). The ecology and migrations of sea turtles. 4, The green turtle in the Caribbean Sea. *Bulletin of the AMNH, 121,* Article 1.

Carroll, L. (1917). *Through the looking glass: And what Alice found there.* Rand, McNally.

Carroll, S. M. (2001). *Dark energy and the preposterous universe* (arXiv preprint astro-ph/0107571).

Carter, C. S., Devries, A. C., & Getz, L. L. (1995). Physiological substrates of mammalian monogamy: The prairie vole model. *Neuroscience and Biobehavioral Reviews, 19,* 303–314.

Carter, C. S., Williams, J. R., Witt, D. M., & Insel, T. R. (1992). Oxytocin and social bonding. *Annual New York Academy of Science, 652,* 204–211.

Carter, M. (2013). *Interview: Anne Wood, the co-creator of the Teletubbies | Society.* Guardian.co.uk.

Cartledge, P. (1981). Spartan wives: Liberation or licence? *The Classical Quarterly, 31*(1), 84–105.

Catania, L., Abdulcadir, O., Puppo, V., Verde, J. B., Abdulcadir, J., & Abdulcadir, D. (2007). Pleasure and orgasm in women with female genital mutilation/cutting (FGM/C). *The Journal of Sexual Medicine, 4*(6), 1666–1678.

Cattell, R. B. (1946). *Description and measurement of personality.* Harrap.

Chadwick, S. B., Francisco, M., & van Anders, S. M. (2019). When orgasms do not equal pleasure: Accounts of "bad" orgasm experiences during consensual sexual encounters. *Archives of Sexual Behavior,* 1–25.

Chadwick, S. B., & van Anders, S. M. (2017). Do women's orgasms function as a masculinity achievement for men? *The Journal of Sex Research, 54*(9), 1141–1152.

Chagnon, N. A. (1968). *The fierce people.* Holt, Rinehart & Winston.

Chalker, R. (2000). *The clitoral truth: The secret world at your fingertips.* Seven Stories Press.

Chan, C. B., & Ye, K. (2017). Sex differences in brain-derived neurotrophic factor signaling and functions. *Journal of Neuroscience Research, 95*(1–2), 328–335.

Clark, R. D., & Hatfield, E. (1989). Gender differences in receptivity to sexual offers. *Journal of Psychology & Human Sexuality, 2*(1), 39–55.

Cochran, G., & Harpending, H. (2009). *The 10,000 year explosion: How civilization accelerated human evolution.* Basic Books.

Conan Doyle, A. (1893). The naval treaty. In *The memoirs of Sherlock Holmes* (pp. 238–282). George Newnes.

Conley, T. D. (2011). Perceived proposer personality characteristics and gender differences in acceptance of casual sex offers. *Journal of Personality and Social Psychology, 100*(2), 309.

Corl, A., Davis, A. R., Kuchta, S. R., Comendant, T., & Sinervo, B. (2010). Alternative mating strategies and the evolution of sexual size dimorphism in the side-blotched lizard, Uta stansburiana: A population-level comparative analysis. *Evolution: International Journal of Organic Evolution, 64*(1), 79–96.

Costa, R. M., Miller, G. F., & Brody, S. (2012). Women who prefer longer penises are more likely to have vaginal orgasms (but not clitoral orgasms): Implications for an evolutionary theory of vaginal orgasm. *The Journal of Sexual Medicine, 9*(12), 3079–3088.

Courtenay, W. H. (2003). Key determinants of the health and well-being of men and boys. *International Journal of Mens Health, 2,* 1–30.

Cowperthwaite, G. (2013). *Blackfish* [TV Documentary]. Magnolia Home Entertainment.

Crick, F. (1970). Central dogma of molecular biology. *Nature, 227*(5258), 561.

Cronin, H. (1993). *The ant and the peacock: Altruism and sexual selection from Darwin to today.* Cambridge University Press.

Crook, J. H. (1994). Explaining Tibetan polyandry: Socio-cultural, demographic, and biological perspectives. In J. H. Crook & H. A. Osmaston (Eds.), *Himalayan Buddhist villages* (pp. 735–786). Shri Jainendra Press.

Cross, S. J., Linker, K. E., & Leslie, F. M. (2017). Sex-dependent effects of nicotine on the developing brain. *Journal of Neuroscience Research, 95*(1–2), 422–436.

Cueva-Rolón, R., Sansone, G., Bianca, R., Gómez, L. E., Beyer, C., Whipple, B., & Komisaruk, B. R. (1996). Vagotomy blocks responses to vaginocervical stimulation after genitospinal neurectomy in rats. *Physiology & Behavior, 60*(1), 19–24.

Curry, O. S., Mullins, D. A., & Whitehouse, H. (2019). Is it good to cooperate? Testing the theory of morality-as-cooperation in 60 societies. *Current Anthropology, 60*(1), 47–69.

Cusack, C. M. (2012). Obscene squirting: If the government thinks it's urine, then they've got another thing coming. *Texas Journal of Women & Law, 22,* 45.

Custodio, A. M., Santos, F. C., Campos, S. G., Vilamaior, P. S. L., Oliveira, S. M., Goes, R. M., & Taboga, S. R. (2010). Disorders related with ageing in the gerbil female prostate (Skene's paraurethral glands). *International Journal of Experimental Pathology, 91*(2), 132–143.

Dachtler, J., & Fox, K. (2017). Do cortical plasticity mechanisms differ between males and females?. *Journal of Neuroscience Research, 95*(1–2), 518–526.

Daly, M., & Wilson, M. (1988). Evolutionary social psychology and family homicide. *Science, 242*(4878), 519–524.

Dartnell, L. (2014). *The knowledge: How to rebuild our world from scratch.* Random House.

Darwin, C. (1859). *The origin of species.* Murray.

Darwin, C. (1888). *The descent of man and selection in relation to sex* (Vol. 1). Murray.

Davenport, W. H. (1977). Sex in cross-cultural perspective. In F. A. Beach (Ed.), *Human sexuality in four perspectives* (pp. 115–163). Johns Hopkins University.

Davies, N. B. (1985). Cooperation and conflict among dunnocks, *Prunella modularis*, in a variable mating system. *Animal Behaviour, 33*, 628–648.

Davis, N. (2016). Mystery of the female orgasm may be solved . . . Scientists believe they can explain the evolutionary reason for women's sexual response. *The Guardian*. www.theguardian.com/society/2016/aug/01/mystery-of-the-female-orgasm-may-be-solved last accessed 25/08/2023.

Dawkins, R. (2006). *The selfish gene: With a new introduction by the author*. Oxford University Press. (Original work published 1976).

Dawood, K., Kirk, K. M., Bailey, J. M., Andrews, P. W., & Martin, N. G. (2005). Genetic and environmental influences on the frequency of orgasm in women. *Twin Research and Human Genetics, 8*(1), 27–33.

Deaner, R. O., Balish, S. M., & Lombardo, M. P. (2016). Sex differences in sports interest and motivation: An evolutionary perspective. *Evolutionary Behavioral Sciences, 10*(2), 73.

DeBruine, L. M. (2014). Women's preferences for male facial features. In *Evolutionary perspectives on human sexual psychology and behavior* (pp. 261–275). Springer.

De Graaf, R. (1672). *De Mulierum Organis Generationi Inservenientibus*. Lugduni Batavorum ex officina Hackiana.

Del Giudice, M., & Gangestad, S. W. (2023). No evidence against the greater male variability hypothesis: A commentary on Harrison et al.'s (2022) meta-analysis of animal personality. *Evolutionary Psychological Science*, 1–8.

Dempsey, M., King, R., & Nagy, A. (2018). A pot of gold at the end of the rainbow? A spectrum of attitudes to assisted reproductive technologies in Ireland. *Journal of Reproductive and Infant Psychology, 36*(1), 59–66.

Dennett, D. C. (1993). *Consciousness explained*. Penguin.

Dennett, D. C. (1995). *Darwin's dangerous idea*. Penguin.

Dennett, D. C. (2001). Are we explaining consciousness yet? *Cognition, 79*(1–2), 221–237.

Dennett, D. C. (2003). Who's on first? Heterophenomenology explained. *Journal of Consciousness Studies, 10*(9–10), 19–30.

Dennett, D. C. (2017). *From bacteria to Bach and back: The evolution of minds*. W.W. Norton.

Der, G., & Deary, I. J. (2006). Age and sex differences in reaction time in adulthood: Results from the United Kingdom health and lifestyle survey. *Psychology and Aging, 21*(1), 62.

De Waal, F. (2022). *Different: What apes can teach us about gender*. Granta Books.

De Waal, F. B. (1995). Bonobo sex and society. *Scientific American, 272*(3), 82–88.

Di Noto, P. M., Newman, L., Wall, S., & Einstein, G. (2013). The hermunculus: What is known about the representation of the female body in the brain? *Cerebral Cortex, 23*(5), 1005–1013.

Diamond, L. M. (2008). *Sexual fluidity: Understanding women's love and desire*. Harvard University Press.

Diamond, M. (1965). A critical evaluation of the ontogeny of human sexual behavior. *The Quarterly Review of Biology, 40*(2), 147–175.

Diamond, M. (2006). Biased-interaction theory of psychosexual development: "How does one know if one is male or female?" *Sex Roles, 55*(9–10), 589–600.

Diamond, M. (n.d.). *Online comment*. www.hawaii.edu/PCSS/ last accessed 24/10/2018.

Diamond, M., Jozifkova, E., & Weiss, P. (2011). Pornography and sex crimes in the Czech Republic. *Archives of Sexual Behavior, 40*(5), 1037–1043.

Diamond, M., & Sigmundson, H. K. (1997). Sex reassignment at birth: Long-term review and clinical implications. *Archives of Pediatrics & Adolescent Medicine, 151*(3), 298–304.

Dibble, J. L., & Drouin, M. (2014). Using modern technology to keep in touch with back burners: An investment model analysis. *Computers in Human Behavior, 34*, 96–100.

Dibble, J. L., Drouin, M., Aune, K. S., & Boller, R. R. (2015). Simmering on the back burner: Communication with and disclosure of relationship alternatives. *Communication Quarterly, 63*, 329–344.

Dick, P. K. (1985). *I hope I shall arrive soon*. Doubleday Books.

Dickins, T. E., & Barton, R. A. (2013). Reciprocal causation and the proximate–ultimate distinction. *Biology & Philosophy, 28*, 747–756.

Dickinson, R. L. (1949). *Human sex anatomy: A topographical hand atlas*. Williams & Wilkins.

Discover Magazine. (non attributed). On popular reporting of the idea that female ejaculate is urine. http://blogs.discovermagazine.com/seriouslyscience/2018/08/22/7025/ last accessed 08/03/2019.

Dixson, A. F. (1999). *Primate sexuality: Comparative studies of the prosimians, monkeys, apes, and human beings*. Oxford University Press.

Dixson, A. F. (2012). *Primate sexuality: Comparative studies of the prosimians, monkeys, apes, and human beings* (2nd ed.). Oxford University Press.

Doğruyol, B., Alper, S., & Yilmaz, O. (2019). The five-factor model of the moral foundations theory is stable across WEIRD and non-WEIRD cultures. *Personality and Individual Differences, 151,* 109547.

Dos Santos, F. C. A., & Taboga, S. R. (2006). Female prostate: A review about the biological repercussions of this gland in humans and rodents. *Animal Reproduction, 3*(1), 3–18.

Dos Santos, F. C. A., Carvalho, H. F., Góes, R. M., & Taboga, S. R. (2003). Structure, histochemistry, and ultrastructure of the epithelium and stroma in the gerbil (Meriones unguiculatus) female prostate. *Tissue and Cell, 35*(6), 447–457.

Dotson, A. L., & Offner, H. (2017). Sex differences in the immune response to experimental stroke: Implications for translational research. *Journal of Neuroscience Research, 95*(1–2), 437–446.

Doyle, H. H., & Murphy, A. Z. (2017). Sex differences in innate immunity and its impact on opioid pharmacology. *Journal of Neuroscience Research, 95*(1–2), 487–499.

Dreger, A. (2016). *Galileo's middle finger: Heretics, activists, and one scholar's search for justice.* Penguin Books.

Dreger, A. D. (1998). "Ambiguous sex"—or ambivalent medicine? Ethical issues in the treatment of intersexuality. *Hastings Center Report, 28*(3), 24–35.

Dreger, A. D. (1998). *Hermaphrodites and the medical invention of sex.* Harvard University Press.

Dreger, A. D. (1999). *Intersex in the age of ethics.* University Publishing Group.

Duberman, L. (1975). *The reconstituted family: A study of remarried couples and their children.* Burnham Incorporated.

Dugatkin, L. A., & Trut, L. (2017). *How to Tame a Fox (and build a dog): Visionary scientists and a Siberian tale of jump-started evolution.* University of Chicago Press.

Dunbar, R. I. (1980). Determinants and evolutionary consequences of dominance among female gelada baboons. *Behavioral Ecology and Sociobiology, 7,* 253–265.

Dunbar, R. I. (2004). Gossip in evolutionary perspective. *Review of General Psychology, 8*(2), 100–110.

Dunbar, R. I. M. (1996). Groups, gossip, and the evolution of language. In A. Schmitt, K. Atzwanger, K. Grammer, & K. Schäfer (Eds.), *New aspects of human ethology* (pp. 77–89). Springer.

Dunn, K. M., Cherkas, L. F., & Spector, T. D. (2005). Genetic influences on variation in female orgasmic function: A twin study. *Biology Letters, 1*(3), 260–263.

Eagly, A. H. (2018). The shaping of science by ideology: How feminism inspired, led, and constrained scientific understanding of sex and gender. *Journal of Social Issues, 74*(4), 871–888.

Eagly, A. H., & Wood, W. (2013). The nature–nurture debates: 25 years of challenges in understanding the psychology of gender. *Perspectives on Psychological Science, 8*(3), 340–357.

EAMES Productions. (1977). *Powers of ten* [Film]. www.youtube.com/watch?v=0fKBhvDjuy0 last accessed 20/12/2022.

Eberhard, W. G. (1985). *Sexual selection and animal genitalia.* Harvard University Press.

Eberhard, W. G. (2009). Postcopulatory sexual selection: Darwin's omission and its consequences. *Proceedings of the National Academy of Sciences, 106*(Suppl. 1), 10025–10032.

Ebert, D., & Hamilton, W. D. (1996). Sex against virulence: The coevolution of parasitic diseases. *Trends in Ecology & Evolution, 11*(2), 79–82.

Eibl-Eibesfeldt, I. (1971). *Love and hate: On the natural history of basic behaviour patterns* (Trans. from the German by Geoffrey Strachan). Methuen.

El Dareer, A. (1982). *Woman, why do you weep?* Zed Press.

Electronic Regulations. (2014). [UK Statute Law]. 2014 legislation banning (among other things) portrayals of urination, bondage and abusive language. www.legislation.gov.uk/uksi/2014/2916/pdfs/uksi_20142916_en.pdf last accessed 15/03/2019.

Eliot, L. (2019). The gendered brain: The new neuroscience that shatters the myth of the female brain. *Nature, 566*(7745), 453–454.

Ellis, B. J., & Symons, D. (1990). Sex differences in sexual fantasy: An evolutionary psychological approach. *Journal of Sex Research, 27*(4), 527–555.

Ellis, H. D. (1933). *The psychology of sex: A manual for students.* Heinemann.

Emlen, S. T., & Wrege, P. H. (2004). Size dimorphism, intrasexual competition, and sexual section in Wattled Jacana (Jacana jacana), a sex-role-reversed shorebird in Panama. *The Auk,* 391–403.

Eschler, L. (2004). The physiology of the female orgasm as a proximate mechanism. *Sexualities, Evolution & Gender, 6*(2–3), 171–194.

Evans, E. I. (1933). The transport of spermatozoa in the dog. *American Journal of Physiology, 105,* 287–293.

Fallopius, G. (1561). *Observationes anatomicae, ad Petrum Mannam.* Venetiis (work).

Fausto-Sterling, A. (2000). *Sexing the body: Gender politics and the construction of sexuality.* Basic Books.

Ferguson, E. E. (2006). Judicial authority and popular justice: Crimes of passion in fin-de-siècle Paris. *Journal of Social History*, 293–315.

Ferguson, J. K. (1941). A study of the motility of the intact uterus at term. *Surgical Gynecology and Obstetrics, 73*, 359–366.

Fernald, R. D. (1997). The evolution of eyes. *Brain, Behavior and Evolution, 50*(4), 253–259.

Fernandes, E. M. (2009). *The swinging paradigm: An evaluation of the marital and sexual satisfaction of swingers*. Union Institute and University.

Festinger, L. (1962). *A theory of cognitive dissonance* (Vol. 2). Stanford University Press.

Fieder, M., & Huber, S. (2007). The effect of sex and childlessness on the association between status and reproductive output in modern society. *Evolution and Human Behavior, 28*, 392–398.

Fieder, M., Huber, S., Bookstein, F. L., Iber, K., Schäfer, K., Winckler, G., & Wallner, B. (2005). Status and reproduction in humans: New evidence for the validity of evolutionary explanations on basis of a university sample. *Ethology, 111*(10), 940–950.

Figueredo, A. J., Sefcek, J. A., Vasquez, G., Brumbrach, B., King, J. E., & Jacobs, W. J. (2005). Evolutionary personality psychology. In D. M. Buss (Ed.), *Handbook of evolutionary psychology* (pp. 851–877). Wiley.

Fillod, O., & Richard, M. (2016). [Downloadable model of clitoris, life size for 3D printing]. www.thingiverse.com/thing:1876288 last accessed 25/08/2023.

Fine, C. (2017). *Testosterone rex: Unmaking the myths of our gendered minds*. Icon Books.

Fisher, H. (2016). *Anatomy of love: A natural history of mating, marriage, and why we stray (completely revised and updated with a new introduction)*. W.W. Norton.

Fisher, H. E., Aron, A., Mashek, D., Li, H., & Brown, L. L. (2002). Defining the brain systems of lust, romantic attraction, and attachment. *Archives of Sexual Behavior, 31*, 413–419.

Fisher, R. A. (1930). *The genetical theory of natural selection* (Mimicry). Dover.

Fisher, R. A. (1941). Average excess and average effect of a gene substitution. *Annals of Eugenics, 11*(1), 53–63.

Fisher, S. (1973). *The female orgasm*. Basic Books.

Flamini, M. A., Barbeito, C. G., Gimeno, E. J., & Portiansky, E. L. (2002). Morphological characterization of the female prostate (Skene's gland or paraurethral gland) of Lagostomus maximus maximus. *Annals of Anatomy-Anatomischer Anzeiger, 184*(4), 341–345.

Fleischman, D. S. (2016a). Sex differences in disease avoidance. In T. K. Shackelford & V. Weekes-Shackelford (Eds.), *Encyclopedia of evolutionary psychological science*. Springer.

Fleischman, D. S. (2016b). An evolutionary behaviorist perspective on orgasm. *Socioaffective Neuroscience & Psychology, 6*(1), 32130.

Foldes, P., & Buisson, O. (2009). REVIEWS: The clitoral complex: A dynamic sonographic study. *The Journal of Sexual Medicine, 6*(5), 1223–1231.

Fong, D. W., Kane, T. C., & Culver, D. C. (1995). Vestigialization and loss of nonfunctional characters. *Annual Review of Ecology and Systematics, 26*(1), 249–268.

Foucault, M. (1980). *Power/knowledge: Selected interviews and other writings, 1972–1977*. Pantheon.

Foucault, M. (2012). *Discipline and punish: The birth of the prison*. Vintage.

Fox, C. A. (1970). Reduction in the rise of systolic blood pressure during human coitus by the β-adrenergic blocking agent, propranolol. *Journal of Reproduction and Fertility, 22*(3), 587–590.

Fox, C. A. (1976). Some aspects and implications of coital physiology. *Journal of Sex and Marital Therapy, 2*, 205–213.

Fox, C. A., & Fox, B. (1969). Blood pressure and respiratory patterns during human coitus. *Journal of Reproduction and Fertility, 19*(3), 405–415.

Fox, C. A., & Fox, B. (1971). A comparative study of coital physiology, with special reference to the sexual climax. *Journal of Reproduction and Fertility, 24*, 319–336.

Fox, C. A., Meldrum, S. J., & Watson, B. W. (1973). Continuous measurement by radio-telemetry of vaginal pH during human coitus. *Journal of Reproduction and Fertility, 33*(1), 69–75.

Fox, C. A., Wolff, H. S., & Baker, J. A. (1970). Measurement of intra-vaginal and intra-uterine pressures during human coitus by radio-telemetry. *Journal of Reproduction and Fertility, 22*, 243–251.

Freeman, D. (1983). *Margaret Mead and Samoa: The making and unmaking of an anthropological myth*. Australian National University Press (in Mead, M., Sieben, A., & Straub, J. (1973). *Coming of age in Samoa*. Penguin).

Freud, A. (1936). *Das ich und die Abwehrmechanismen*. Wien Internationaler Psychoanalyt.

Freud, A. (1992). *The ego and the mechanisms of defence*. Karnac Books.

Freud, S. (1932). Female sexuality (Trans. J. Riviere). *International Journal of Psycho-Analysis, 13*, 281–297.

Freud, S. (2017). *Three essays on the theory of sexuality: The 1905 edition*. Verso Books.

Fry, S. (2010). *Interview for attitude* [Magazine]. Stream Publishing Ltd.

Gácsi, M., Miklósi, Á., Varga, O., Topál, J., & Csányi, V. (2004). Are readers of our face readers of our minds? Dogs (Canis familiaris) show situation-dependent recognition of human's attention. *Animal Cognition*, 7(3), 144–115.

Gaitskill, M. (2012). *Bad behavior: Stories*. Simon & Schuster.

Galen, B. R., & Underwood, M. K. (1997). A developmental investigation of social aggression among children. *Developmental Psychology*, *33*, 589–600.

Gallup, G. G., Jr., Ampel, B. C., Wedberg, N., & Pogosjan, A. (2014). Do orgasms give women feedback about mate choice? *Evolutionary Psychology*, *12*(5).

Gallup, G. G., Jr., & Frederick, D. A. (2010). The science of sex appeal: An evolutionary perspective. *Review of General Psychology*, *14*(3), 240.

Galton, F. (1908). *Local association for promoting eugenics*. Eugenics Education Society.

Gangestad, S. W., Garver-Apgar, C. E., Cousins, A. J., & Thornhill, R. (2014). Intersexual conflict across women's ovulatory cycle. *Evolution and Human Behavior*, *35*(4), 302–308.

Gangestad, S. W., Haselton, M. G., & Buss, D. M. (2006). Evolutionary foundations of cultural variation: Evoked culture and mate preferences. *Psychological Inquiry*, *17*(2), 75–95.

Gangestad, S. W., Thornhill, R., & Garver-Apgar, C. E. (2010). Fertility in the cycle predicts women's interest in sexual opportunism. *Evolution and Human Behavior*, *31*, 400–411.

Gangoly, O. C. (1957). *The art of the Chandelas* (Vol. 1). Rupa.

Gardi, R. (1960). *Tambaran: An encounter with cultures in decline in New Guinea*. Constable.

Garn, S. M., Clark, L. C., & Harper, R. V. (1953). The sex difference in the basal metabolic rate. *Child Development*, 215–224.

Garver-Apgar, C. E., Gangestad, S. W., Thornhill, R., Miller, R. D., & Olp, J. J. (2006). Major histocompatibility complex alleles, sexual responsivity, and unfaithfulness in romantic couples. *Psychological Science*, *17*(10), 830–835.

Geary, D. C. (2000). Evolution and proximate expression of human paternal investment. *Psychological Bulletin*, *126*(1), 55.

Gehring, W. J. (1996). The master control gene for morphogenesis and evolution of the eye. *Genes to Cells*, *1*(1), 11–15.

Gell-Mann, M. (1995). *The Quark and the Jaguar: Adventures in the simple and the complex*. Macmillan.

Genell, S. (1939). Experimental investigations of the muscular functions of the vagina and the uterus in the rat. *Acta Obstetricia et Gynecologica Scandinavica*, *19*, 113–175.

Georgiadis, J. R., Kortekaas, R., Kuipers, R., Nieuwenburg, A., Pruim, J., Simone Reinders, A. A. T., & Holstege, G. (2006). Regional cerebral blood flow changes associated with clitorally induced orgasm in healthy women. *European Journal of Neuroscience*, *24*, 3305–3316.

Georgiadis, J. R., Reinders, A. A. T., Paans, A. M. J., Renken, R., & Kortekaas, R. (2009). Men versus women on sexual brain function: Prominent differences during tactile genital stimulation, but not during orgasm. *Human Brain Mapping*, *30*, 3089–3101.

Gersick, A., & Kurzban, R. (2014). Covert sexual signaling: Human flirtation and implications for other social species. *Evolutionary Psychology*, *12*(3).

Gignac, G. E., Darbyshire, J., & Ooi, M. (2018). Some people are attracted sexually to intelligence: A psychometric evaluation of sapiosexuality. *Intelligence*, *66*(C), 98–111.

Gilbert, C. L., Jenkins, K., & Wathes, D. C. (1991). Pulsatile release of oxytocin into the circulation of the ewe during oestrus, mating and the early luteal phase. *Journal of Reproduction and Fertility*, *91*, 337–346.

Gildersleeve, K., Haselton, M. G., & Fales, M. R. (2014a). Do women's mate preferences change across the ovulatory cycle? A meta-analytic review. *Psychological Bulletin*, *140*, 1205–1259.

Gildersleeve, K., Haselton, M. G., & Fales, M. R. (2014b). Meta-analyses and p-curves support robust cycle shifts in women's mate preferences: Reply to Wood and Carden (2014) and Harris, Pashler, and Mickes (2014). *Psychological Bulletin*, *150*(5), 1272–1280.

Gillespie, D. O., Russell, A. F., & Lummaa, V. (2008). When fecundity does not equal fitness: Evidence of an offspring quantity versus quality trade-off in pre-industrial humans. *Proceedings Biological Science*, *22*(275), 713–722.

Gobinath, A. R., Choleris, E., & Galea, L. A. (2017). Sex, hormones, and genotype interact to influence psychiatric disease, treatment, and behavioral research. *Journal of Neuroscience Research*, *95*(1–2), 50–64.

Goddard, M. R., Godfray, H. C. J., & Burt, A. (2005). Sex increases the efficacy of natural selection in experimental yeast populations. *Nature*, *434*(7033), 636–640.

Goldfoot, D. A., Westerborg-van Loon, H., Groeneveld, W., & Slob, A. K. (1980). Behavioral and physiological evidence of sexual climax in the female stump-tailed macaque (Macaque arctoides). *Science, N.Y.,* *208,* 1477–1479.

Gosling, S. D., Vazire, S., Srivastava, S., & John, O. P. (2004). Should we trust web-based studies? A comparative analysis of six preconceptions about internet questionnaires. *American Psychologist, 59*(2).

Gott, B. (2002). Fire-making in Tasmania: Absence of evidence is not evidence of absence. *Current Anthropology, 43*(4), 650–656.

Gottschall, J. (2012). *The storytelling animal: How stories make us human.* Houghton Mifflin Harcourt.

Gould, S. J. (1987). Freudian slip. *Natural History, 96*(2), 14.

Gould, S. J. (1991). *Bully for brontosaurus. Further reflections in natural history.* W.W. Norton.

Gould, S. J. (1995). A task for paleobiology at the threshold of majority. *Paleobiology, 21*(1), 1–14.

Gould, S. J. (1997). *The mismeasure of man* (Rev. and expanded ed.). W.W. Norton.

Gould, S. J. (2000). The spice of life. *Leader to Leader, 15,* 14–19.

Gould, S. J. (2008). A biological homage to Mickey Mouse. *Ecotone, 4*(1), 333–340.

Gozes, I. (2017). Sexual divergence in activity-dependent neuroprotective protein impacting autism, schizophrenia, and Alzheimer's disease. *Journal of Neuroscience Research, 95*(1–2), 652–660.

Grafenberg, E. (1950). The role of the urethra in female orgasm. *International Journal of Sexology, 3,* 145–148.

Grammer, K., Renninger, L., & Fischer, B. (2004). Disco clothing, female sexual motivation, and relationship status: Is she dressed to impress? *Journal of Sex Research, 41,* 66–74.

Grant, M., & Mulas, A. (1982). *Eros in Pompeii. The secret rooms of the national museum of naples.* Bonanza Books.

Gravina, G. L., Brandetti, F., Martini, P., Carosa, E., Di Stasi, S. M., Morano, S., . . . Jannini, E. A. (2008). Measurement of the thickness of the urethrovaginal space in women with or without vaginal orgasm. *The Journal of Sexual Medicine, 5*(3), 610–618.

Greer, G. (1971). *The female eunuch.* MacGibbon and Kee.

Greer, G. (2014). Interview: *The Rebels of Oz: Germaine, Clive, Barry and Bob* [TV Programme]. *BBC Four.* Aired 1st July 2014, Serendipity Productions.

Greer, G. (2019). TV violence and women. Article for *Radio Times* [Magazine]. www.radiotimes.com/news/tv/2018-05-05/tv-violence-women-the-bridge-germaine-greer/ last accessed 09/10/2019.

Gregory, R. L. (2015). *Eye and brain: The psychology of seeing* (Vol. 38). Princeton University Press.

Greiling, H., & Buss, D. M. (2000). Women's sexual strategies: The hidden dimension of short-term mating. *Personality and Individual Differences, 28,* 929–963.

Grimble, A. F. (1952). *A pattern of islands.* J. Murray.

Gross, S. A., & Didio, L. J. (1987). Comparative morphology of the prostate in adult male and female Praomys (Mastomys) Natalensis studied with electron microscopy. *Journal of Submicroscopic Cytology, 19*(1), 77–84.

Gruenbaum, E. (1996). The cultural debate over female circumcision: The Sudanese are arguing this one out for themselves. *Medical Anthropological Quarterly, 10,* 455–475.

Guéguen, N. (2009). Menstrual cycle phases and female receptivity to a courtship solicitation: An evaluation in a nightclub. *Evolution and Human Behavior, 30,* 351–355.

Gupta, A., Mayer, E. A., Fling, C., Labus, J. S., Naliboff, B. D., Hong, J. Y., & Kilpatrick, L. A. (2017). Sex-based differences in brain alterations across chronic pain conditions. *Journal of Neuroscience Research, 95*(1–2), 604–616.

Gur, R. C., & Gur, R. E. (2017). Complementarity of sex differences in brain and behavior: From laterality to multimodal neuroimaging. *Journal of Neuroscience Research, 95*(1–2), 189–199.

Gurven, M. D., & Lieberman, D. E. (2020). WEIRD bodies: Mismatch, medicine and missing diversity. *Evolution and Human Behavior, 41*(5), 330–340.

Gwynne, D. T. (1981). Sexual difference theory: Mormon crickets show role reversal in mate choice. *Science, 213,* 14.

Hadith 5229 (2007). Abu Dawud, al-Bayhaq Umm 'Atiyyah. *Mustadrak Hakim, 3,* 525. https://islamqa.org/hanafi/hadithanswers/120067/is-there-a-hadith-regarding-female-circumcision/ last accessed 27/10/2023.

Haeffel, G. J., Gibb, B. E., Metalsky, G. I., Alloy, L. B., Abramson, L. Y., Hankin, B. L., . . . Swendsen, J. D. (2008). Measuring cognitive vulnerability to depression: Development and validation of the cognitive style questionnaire. *Clinical Psychology Review, 28*(5), 824–836.

Haeffel, G. J., & Howard, G. S. (2010). Self-report: Psychology's four-letter word. *American Journal of Psychology, 123*(2), 181–188.

Hahn-Holbrook, J., Holt-Lunstad, J., Holbrook, C., Coyne, S. M., & Lawson, E. T. (2011). Maternal defense: Breast feeding increases aggression by reducing stress. *Psychological Science*, 22(10), 1288–1295.

Haidt, J. (2001). The emotional dog and its rational tail: A social intuitionist approach to moral judgment. *Psychological Review*, 108(4), 814.

Hald, G. M., & Høgh-Olesen, H. (2010). Receptivity to sexual invitations from strangers of the opposite gender. *Evolution and Human Behavior*, 31(6), 453–458.

Haldane, J. B. S. (1927). A mathematical theory of natural and artificial selection, part V: Selection and mutation. In *Mathematical proceedings of the Cambridge philosophical society* (Vol. 23, No. 7, pp. 838–844). Cambridge University Press.

Halpern, C. T., Joyner, K., Udry, J. R., & Suchindran, C. (2000). Smart teens don't have sex (or kiss much either). *Journal of Adolescent Health*, 26(3), 213–225.

Hamerton, J. L., Canning, N., Ray, M., & Smith, S. (1975). A cytogenetic survey of 14,069 newborn infants: I. Incidence of chromosome abnormalities. *Clinical Genetics*, 8(4), 223–243.

Harris, E. A., Hornsey, M. J., Larsen, H. F., & Barlow, F. K. (2019). Beliefs about gender predict faking orgasm in heterosexual women. *Archives of Sexual Behavior*, 48, 2419–2433.

Harris, R. F. (1996). Gender bender. *Current Biology*, 6(7), 765.

Harrison, L. M., Noble, D. W., & Jennions, M. D. (2022). A meta-analysis of sex differences in animal personality: No evidence for the greater male variability hypothesis. *Biological Reviews*, 97(2), 679–707.

Harrow, J., Frankish, A., Gonzalez, J. M., Tapanari, E., Diekhans, M., Kokocinski, F., . . . Hubbard, T. J. (2012). GENCODE: The reference human genome annotation for the ENCODE project. *Genome Research*, 22(9), 1760–1774.

Hartman, C. G., & Ball, J. (1931). On the almost instantaneous transport of spermatozoa through the cervix and uterus of the rat. *Proceedings of the Society of Experimental Biological Medicine*, 28, 312–314.

Haselton, M. G., & Buss, D. M. (2000). Error management theory: A new perspective on biases in cross-sex mind reading. *Journal of Personality and Social Psychology*, 78(1), 81.

Haselton, M. G., & Miller, G. F. (2006). Women's fertility across the cycle increases the short-term attractiveness of creative intelligence. *Human Nature*, 17, 50–73.

Hausmann, M. (2017). Why sex hormones matter for neuroscience: A very short review on sex, sex hormones, and functional brain asymmetries. *Journal of Neuroscience Research*, 95(1–2), 40–49.

Hayes, R. O. (1975). Female genital mutilation, fertility control, women's roles, and the patrilineage in modern Sudan: A functional analysis 1. *American Ethnologist*, 2(4), 617–633.

Hazen, H. (1983). *Endless rapture rape, romance, and the female imagination*. Charles Scribbener.

Heidegger, M. (1996). *Being and time: A translation of Sein und Zeit*. SUNY Press.

Heinrichs, M., & Domes, G. (2008). Neuropeptides and social behaviour: Effects of oxytocin and vasopressin in humans. *Progress in Brain Research*, 170, 337–350.

Henrich, J., Heine, S. J., & Norenzayan, A. (2010). The weirdest people in the world. *Behavioral and Brain Sciences*, 33(2–3), 61–83.

Herdt, G. H. (2017). Fetish and fantasy in Sambia initiation. In *Rituals of manhood* (pp. 44–98). Routledge.

Hess, N. H., & Hagen, E. H. (2006). Sex differences in indirect aggression: Psychological evidence from young adults. *Evolution and Human Behavior*, 27(3), 231–245.

Hevesi, K., Horvath, Z., Sal, D., Miklos, E., & Rowland, D. L. (2021). Faking orgasm: Relationship to orgasmic problems and relationship type in heterosexual women. *Sexual Medicine*, 9(5), 100419.

Hicks, E. K. (1996). *Infibulation: Female mutilation in Islamic northeastern Africa*. Transaction.

Hill, K. (1993). Life history theory and evolutionary anthropology. *Evolutionary Anthropology: Issues, News, and Reviews*, 2(3), 78–88.

Hill, S. (2019). *This is your brain on birth control: The surprising science of women, hormones, and the law of unintended consequences*. Penguin.

Hinde, R. A., & Barden, L. A. (1985). The evolution of the teddy bear. *Animal Behaviour*, 33(4), 1371–1373.

Hite, S. (1979). The Hite report. *Journal of School Health*, 49(5), 251–254.

Hite, S. (1981). *The Hite report on male sexuality*. Ballantine Books.

Hodes, G. E., Walker, D. M., Labonté, B., Nestler, E. J., & Russo, S. J. (2017). Understanding the epigenetic basis of sex differences in depression. *Journal of Neuroscience Research*, 95(1–2), 692–702.

Holloway, R. L. (2017). In the trenches with the corpus callosum: Some redux of redux. *Journal of Neuroscience Research*, 95(1–2), 21–23.

Holmes, M. M., Putz, O., Crews, D., & Wade, J. (2005). Normally occurring intersexuality and testosterone induced plasticity in the copulatory system of adult leopard geckos. *Hormones and Behavior*, 47, 439–445.

Hood, J. C. (2007). Orthodoxy vs. power: The defining traits of grounded theory. In *The Sage handbook of grounded theory* (pp. 151–164). SAGE.

Hooven, C. (2021). *T: The story of testosterone, the hormone that dominates and divides us.* Henry Holt and Company.

Hosken, D. J. (2008). Clitoral variation says nothing about female orgasm. *Evolution & Development, 10,* 393–395.

Hoving, H. J., Bush, S. L., & Robison, B. H. (2012). A shot in the dark: Same-sex sexual behaviour in a deep-sea squid. *Biology Letters, 8*(2), 287–290.

Howell, C. J. (1999). Epidural versus non-epidural analgesia for pain relief in labour. *The Cochrane Database of Systematic Reviews, 3.*

Hrdy, S. B. (2009a). *Mothers and others: The evolutionary origins of mutual understanding.* Belknap Press of Harvard University Press.

Hrdy, S. B. (2009b). *The woman that never evolved.* Harvard University Press.

Hughes, S. M., Farley, S. D., & Rhodes, B. C. (2010). Vocal and physiological changes in response to the physical attractiveness of conversational partners. *Journal of Nonverbal Behavior, 34,* 155–167.

Humphrey, L. T., Dean, M. C., & Stringer, C. B. (1999). Morphological variation in great ape and modern human mandibles. *The Journal of Anatomy, 195*(4), 491–513.

Huynh, H. K., Willemsen, A. T., & Holstege, G. (2013). Female orgasm but not male ejaculation activates the pituitary. A PET-neuro-imaging study. *Neuroimage, 76,* 178–182.

Huynh, H. K., Willemsen, A. T., Lovick, T. A., & Holstege, G. (2013). Pontine control of ejaculation and female orgasm. *The Journal of Sexual Medicine, 10*(12), 3038–3048.

Imperato-McGinley, J., Peterson, R. E., Gautier, T., & Sturla, E. (1979). Androgens and the evolution of male-gender identity among male pseudohermaphrodites with 5α-reductase deficiency. *New England Journal of Medicine, 300*(22), 1233–1237.

Insel, T. R. (2010). The challenge of translation in social neuroscience: A review of oxytocin, vasopressin, and affiliative behavior. *Neuron, 65*(6), 768–779.

Irish Times. (2016). *[Op ed unattributed] Mystery of the female orgasm revealed.* www.irishtimes.com/life-and-style/health-family/mystery-of-the-female-orgasm-revealed-1.2742453 last accessed 25/08/2023.

Irwig, M. S. (2017). Testosterone therapy for transgender men. *The Lancet Diabetes & Endocrinology, 5*(4), 301–311.

Jacob, F. (1977). Evolution and tinkering. *Science, 196*(4295), 1161–1166.

Jacobs, P. A., Melville, M., Ratcliffe, S., Keay, A. J., & Syme, J. (1974). A cytogenetic survey of 11,680 new-born infants. *Annals of Human Genetics, 37*(4), 359–376.

Jahanshad, N., & Thompson, P. M. (2017). Multimodal neuroimaging of male and female brain structure in health and disease across the life span. *Journal of Neuroscience Research, 95*(1–2), 371–379.

James, E. L. (2012). *Fifty shades of grey.* Arrow.

James, W. (1890). The importance of individuals. *The Open Court, 4*(154), 24–37.

James, W. H. (1993). The incidence of superfecundation and of double paternity in the general population. *AMG Acta geneticae medicae et gemellologiae: Twin Research, 42*(3–4), 257–262.

Jani, N. N., & Wise, T. N. (1988). Antidepressants and inhibited female orgasm: A literature review. *Journal of Sex & Marital Therapy, 14*(4), 279–284.

Janicke, T., Häderer, I. K., Lajeunesse, M. J., & Anthes, N. (2016). Darwinian sex roles confirmed across the animal kingdom. *Science Advances, 2*(2), e1500983.

Jannini, E. A., Rubio-Casillas, A., Whipple, B., Buisson, O., Komisaruk, B. R., & Brody, S. (2012). Female orgasm (s): One, two, several. *The Journal of Sexual Medicine, 9*(4), 956–965.

Janus, S. S., & Janus, C. L. (1993). *The Janus report on sexual behavior.* Wiley.

Jardine, R., & Martin, N. G. (1983). Spatial ability and throwing accuracy. *Behavior Genetics, 13,* 331–340.

Jayne, C. (1984). Freud, Grafenberg, and the neglected vagina: Thoughts concerning an historical omission in sexology. *Journal of Sex Research,* 212–215.

Jennions, M. D., & Kokko, H. (2010). Sexual selection. *Evolutionary Behavioral Ecology,* 343–364.

John, O. P., & Srivastava, S. (1999). The Big Five trait taxonomy: History, measurement, and theoretical perspectives. *Handbook of Personality: Theory and Research, 2,* 102–138.

Jones, B. C., Feinberg, D. R., Watkins, C. D., Fincher, C. L., Little, A. C., & DeBruine, L. M. (2013). Pathogen disgust predicts women's preferences for masculinity in men's voices, faces, and bodies. *Behavioral Ecology, 24*(2), 373–379.

Jones, E. (1953). *Sigmund Freud: Life and work.* Hogarth Press.

Jones, T. (1983). *Monty Python's the meaning of life* [Motion picture]. Celandine Films.

Joshi, P. (2014). The best vibrators and sex toys to buy now (Magazine article, updated August 2023). *Good Housekeeping.* www.goodhousekeeping.co.uk/health/health-advice/the-great-good-housekeeping-vibrator-test%20 accessed 18/4/2014.

Jozifkova, E., & Flegr, J. (2006). Dominance, submissivity (and homosexuality) in general population. Testing of evolutionary hypothesis of sadomasochism by internet-trap-method. *Neuroendocrinology Letters, 27*(6), 711–718.

Jozifkova, E., & Konvicka, M. (2009). Sexual arousal by higher-and lower-ranking partner: Manifestation of a mating strategy? *The Journal of Sexual Medicine, 6*(12), 3327–3334.

Judson, O. (2002). *Dr. Tatiana's sex advice to all creation: The definitive guide to the evolutionary biology of sex.* Macmillan.

Judson, O. P., & Normark, B. B. (1996). Ancient asexual scandals. *Trends in Ecology & Evolution, 11*(2), 41–46.

Kaiser, T., Del Giudice, M., & Booth, T. (2020). Global sex differences in personality: Replication with an open online dataset. *Journal of Personality, 88*(3), 415–429.

Kant in Kant, I. (1987). *Critique of judgment. 1790* (Trans. W. S. Pluhar). Hackett.

Kaplan, H. (1974). *The new sex therapy: Active treatment of sexual dysfunctions.* Brunner-Routledge.

Kassindja, F., & Bashir, L. M. (1999). *Do they hear you when you cry?* Dell Publishing.

Kavoussi, R. (1996, November). Clinical profile/safety and efficacy data. In *Wellbutrin advisory panel meeting.*

Kell, C. A., von Kriegstein, K., Rosler, A., Kleinschmidt, A., & Laufs, H. (2005). The sensory cortical representation of the human penis: Revisiting somatotopy in the male homunculus. *Journal of Neuroscience, 25*, 5984–5987.

Kenrick, D. T., Groth, G. E., Trost, M. R., & Sadalla, E. K. (1993). Integrating evolutionary and social exchange perspectives on relationships: Effects of gender, self-appraisal, and involvement level on mate selection criteria. *Journal of Personality and Social Psychology, 64*, 951–969.

Kenrick, D. T., Sadalla, E. K., Groth, G., & Trost, M. R. (1990). Evolution, traits, and the stages of human courtship: Qualifying the parental investment model. *Journal of Personality, 58*, 97–116.

Kerschbaum, H. H., Hofbauer, I., Gföllner, A., Ebner, B., Bresgen, N., & Bäuml, K. H. T. (2017). Sex, age, and sex hormones affect recall of words in a directed forgetting paradigm. *Journal of Neuroscience Research, 95*(1–2), 251–259.

Khattab, H. (1996). *Women's perceptions of sexuality in rural Giza* (Monographs in Reproductive Health, No. 1). The Population Council.

Kierkegaard, S. (2013). *Kierkegaard's writings, VII, volume 7: Philosophical fragments, or a fragment of philosophy/Johannes Climacus, or De omnibus dubitandum est* (Two books in one volume) (Vol. 22). Princeton University Press.

Kilgallon, S. J., & Simmons, L. W. (2005). Image content influences men's semen quality. *Biology Letters, 1*(3), 253–255.

Kimberly, C., & Hans, J. D. (2017). From fantasy to reality: A grounded theory of experiences in the swinging lifestyle. *Archives of Sexual Behavior, 46*(3), 789–799.

Kimura, M. (1994). *Population genetics, molecular evolution, and the neutral theory: Selected papers.* University of Chicago Press.

King, R. (2013). Fists of furry: At what point did human fists part company with the rest of the hominid lineage? *Journal of Experimental Biology, 216*(12), 2361–2361.

King, R. (2016). I can't get no (boolean) satisfaction: A reply to Barrett et al. (2015). *Frontiers in Psychology, 7*, 1880.

King, R. (2022). "Nose job": Possible side effects of Sars-Cov-2. *Archives of Sexual Behavior, 51*(2), 705–705.

King, R., & Belsky, J. (2012). A typological approach to testing the evolutionary functions of human female orgasm. *Archives of Sexual Behavior, 41*(5), 1145–1160.

King, R., Belsky, J., Mah, K., & Binik, Y. (2011). Are there different types of female orgasm? *Archives of Sexual Behavior, 40*, 865–875.

King, R., Dempsey, M., & Valentine, K. A. (2016). Measuring sperm backflow following female orgasm: A new method. *Socioaffective Neuroscience & Psychology, 6*(1), 31927.

King, R., & O'Riordan, C. (2019). Near the knuckle: How evolutionary logic helps explain Irish Traveller bare-knuckle contests. *Human Nature, 30*(3), 272–298.

Kinsey, A. C., Pomeroy, W. B., Martin, C. E., & Gebhard, P. H. (1953). *Sexual behavior in the human female.* Saunders.

Kinsey, A. C., Pomeroy, W. B., Martin, C. E., & Sloan, S. (1948). *Sexual behavior in the human male* (Vol. 1). Saunders.

Knaus, H. H. (1950). *Die Physiologie der Zeugung des Menschen*. Maudrich. (Republished as *the physiology of human reproduction*. Wilhelm Maudlach).

Knobloch, H. S., & Grinevich, V. (2014). Evolution of oxytocin pathways in the brain of vertebrates. *Frontiers in Behavioral Neuroscience, 8*, 31.

Knox, R. V. (2001). *Artificial insemination of swine: Improving reproductive efficiency of the breeding herd*. www.gov.mb.ca/agriculture/livestock/pork/pdf/bab13s04.pdf last accessed 25/08/2023.

Kobayashi, H., & Kohshima, S. (1997). Unique morphology of the human eye. *Nature, 387*(6635), 767.

Kobelt, G. L. (1851). *De l'appareil du sens génital des deux sexes dans l'espèce humaine et dans quelques mammifères, au point de vue anatomique et physiologique, traduit de l'allemand par Kaula H, Berger-Levrault et fils, Strasbourg et Paris (1851)* (1re éd). allemande (Original work published 1844).

Koedt, A. (1968). *The myth of the vaginal orgasm* (First Published in *Notes from the First Year*). New York Radical Women.

Kokko, H., Booksmythe, I., & Jennions, M. D. (2013). Causality and sex roles: Prejudice against patterns? A reply to Ah-King. *Trends in Ecology & Evolution, 28*(1), 2–4.

Kolko, D. J. (Ed.). (2002). *Handbook on firesetting in children and youth*. Elsevier.

Komisaruk, B., Wise, N., Frangos, E., Liu, W. C., Allen, K., & Brody, S. (2011). Women's clitoris, vagina, and cervix mapped on the sensory cortex: fMRI evidence. *Journal of Sexual Medicine, 8*, 2822–2830.

Komisaruk, B. R., & Sansone, G. (2003). Neural pathways mediating vaginal function: The Vagus nerves and spinal cord oxytocin. *Scandinavian Journal of Psychology, 44*, 241–250.

Komisaruk, B. R., Bianca, R., Sansone, G., Go, L. E., Cueva-Rolo, R., Beyer, C., & Whipple, B. (1996). Brain-mediated responses to vaginocervical stimulation in spinal cord-transected rats: Role of the Vagus nerves. *Brain Research, 708*(1–2), 128–134.

Komisaruk, B. R., Gerdes, C. A., & Whipple, B. (1997). Complete spinal cord injury does not block perceptual responses to genital self-stimulation in women. *Archives of Neurology, 54*(12), 1513–1520.

Komisaruk, B. R., & Whipple, B. (2005). Functional MRI of the brain during orgasm in women. *Annual Review of Sex Research, 16*(1), 62–86.

Komisaruk, B. R., Whipple, B., Crawford, A., Grimes, S., Liu, W. C., Kalnin, A., & Mosier, K. (2004). Brain activation during vaginocervical self-stimulation and orgasm in women with complete spinal cord injury: fMRI evidence of mediation by the Vagus nerves. *Brain Research, 1024*(1–2), 77–88.

Komisaruk, B. R., Whipple, B., Gerdes, C. A., Harkness, B., & Keyes, J. W., Jr. (1997). Brainstem responses to cervical self-stimulation: Preliminary PET-scan analysis. *International Behavioral Neuroscience Society Annual Conference Abstract Book, 6*, 38.

Kompanje, E. J. O. (2006). Painful sexual intercourse caused by a disproportionately long penis: An historical note on a remarkable treatment devised by Guilhelmius Fabricius Hildanus (1560–1634). *Archives of Sexual Behavior, 35*, 603–605.

Kontula, O., & Miettinen, A. (2016). Determinants of female sexual orgasms. *Socioaffective Neuroscience & Psychology, 6*(1), 31624.

Koopman, P., Gubbay, J., Vivian, N., Goodfellow, P., & Lovell-Badge, R. (1991). Male development of chromosomally female mice transgenic for Sry. *Nature, 351*(6322), 117.

Koss, W. A., & Frick, K. M. (2017). Sex differences in hippocampal function. *Journal of Neuroscience Research, 95*(1–2), 539–562.

Kothe, E. (1996). Tetrapolar fungal mating types: Sexes by the thousands. *FEMS Microbiology Review, 18*(1), 65–87.

Krehbiel, R. H., & Carstens, H. P. (1939). Roentgen studies of the mechanism involved in sperm transportation in the female rabbit. *American Journal of Physiology, 125*, 571–577.

Krüger, T. H. C., Haake, P., Hartmann, U., Schedlowski, M., & Exton, M. S. (2002). Orgasm induced prolactin secretion: Feedback control of sexual drive? *Neuroscience & Biobehavioral Reviews, 26*, 31–44.

Kruuk, H., & Kruuk, H. (1972). *The spotted hyena: A study of predation and social behavior*. University of Chicago Press.

Kubrick, S. (1999). *Eyes wide shut* [Film]. Warner Bros.

Kuhle, B. X., & Radtke, S. (2013). Born both ways: The alloparenting hypothesis for sexual fluidity in women. *Evolutionary Psychology, 11*(2).

Kurth, F., Jancke, L., & Luders, E. (2017). Sexual dimorphism of Broca's region: More gray matter in female brains in Brodmann areas 44 and 45. *Journal of Neuroscience Research, 95*(1–2), 626–632.

Ladas, A. K., Whipple, B., & Perry, J. D. (1982). *The G spot and other recent discoveries about human sexuality*. McDougal.

Lai, M. C., Lerch, J. P., Floris, D. L., Ruigrok, A. N., Pohl, A., Lombardo, M. V., & Baron-Cohen, S. (2017). Imaging sex/gender and autism in the brain: Etiological implications. *Journal of Neuroscience Research*, *95*(1–2), 380–397.

Laman-Maharg, A., & Trainor, B. C. (2017). Stress, sex, and motivated behaviors. *Journal of Neuroscience Research*, *95*(1–2), 83–92.

Landripet, I., Štulhofer, A., & Diamond, M. (2006). Pornography, sexual violence and crime statistics: A cross-cultural perspective. In *IASR thirty-second annual meeting. Book of Abstracts*.

Larson, C. M., Pillsworth, E. G., & Haselton, M. G. (2012). Ovulatory shifts in women's attractions to primary partners and other men: Further evidence of the importance of primary partner sexual attractiveness. *PLoS One*, *7*, e44456. https://doi.org/10.1371/journal.pone.0044456.

Lassek, W. D., & Gaulin, S. J. (2009). Costs and benefits of fat-free muscle mass in men: Relationship to mating success, dietary requirements, and native immunity. *Evolution and Human Behavior*, *30*(5), 322–328.

Lassek, W. D., & Gaulin, S. J. (2022). Substantial but misunderstood human sexual dimorphism results mainly from sexual selection on males and natural selection on females. *Frontiers in Psychology*, *13*, 859931.

Laumann, E. O., Nicolosi, A., Glasser, D. B., Paik, A., Gingell, C., Moreira, E., & Wang, T. (2005). Sexual problems among women and men aged 40–80 y: Prevalence and correlates identified in the Global Study of Sexual Attitudes and Behaviors. *International Journal of Impotence Research*, *17*(1), 39.

Lavie, M., & Willig, C. (2005). "I don't feel like melting butter": An interpretative phenomenological analysis of the experience of Inorgasmia. *Psychology & Health*, *20*(1), 115–128.

Lax, R. F. (2000). Socially sanctioned violence against women: Female genital mutilation is its most brutal form. *Clinical Social Work Journal*, *28*(4), 403–412.

Leary, R. F., Allendorf, F. W., & Knudsen, K. L. (1984). Superior developmental stability of heterozygotes at enzyme loci in salmonid fishes. *The American Naturalist*, *124*(4), 540–551.

Lee, A. J., & Zietsch, B. P. (2011). Experimental evidence that women's mate preferences are directly influenced by cues of pathogen prevalence and resource scarcity. *Biology Letters*, rsbl20110454.

Lehmiller, J. J. (2018). *Tell me what you want: The science of sexual desire and how it can help you improve your sex life*. Hachette.

Lehtonen, J., & Kokko, H. (2011). Two roads to two sexes: Unifying gamete competition and gamete limitation in a single model of anisogamy evolution. *Behavioral Ecology and Sociobiology*, *65*(3), 445–459.

Leipold, E., Liebmann, L., Korenke, G. C., Heinrich, T., Gießelmann, S., Baets, J., . . . Kurth, I. (2013). A de novo gain-of-function mutation in SCN11A causes loss of pain perception. *Nature Genetics*, *45*(11), 1399–1404.

Lessing, D. (1962). *The golden notebook*. Simon & Schuster.

Levin, R. (1981). The female orgasm-a current appraisal. *Journal of Psychosomatic Research*, *25*, 119–133.

Levin, R. (1998). Sex and the human female reproductive tract-what really happens during and after coitus. *International Journal of Impotence Research*, *10*(Suppl. 1), S14–S21.

Levin, R. (2001). Sexual desire and the deconstruction and reconstruction of the human female sexual response model of Masters and Johnson. In W. Everard, E. Laan, & S. Both (Eds.), *Sexual appetite, desire and motivation: Energetics of the sexual system* (pp. 63–93). Royal Netherlands Academy of Arts and Sciences.

Levin, R. (2002). The physiology of sexual arousal in the human female: A recreational and procreational synthesis. *Archives of Sexual Behavior*, *31*, 405–411.

Levin, R. (2004). An orgasm is . . . who defines what an orgasm is? *Sexual and Relationship Therapy*, *19*, 101–107.

Levin, R. (2008). Critically revisiting aspects of the human sexual response cycle of Masters and Johnson: Correcting errors and suggesting modifications. *Sexual and Relationship Therapy*, *23*, 393–399.

Levin, R. J. (2009). Revisiting post-ejaculation refractory time—what we know and what we do not know in males and in females. *The Journal of Sexual Medicine*, *6*(9), 2376–2389.

Levin, R. J. (2011). Can the controversy about the putative role of the human female orgasm in sperm transport be settled with our current physiological knowledge of coitus? *The Journal of Sexual Medicine*, *8*(6), 1566–1578.

Levin, R. J., & Wagner, G. (1985). Orgasm in women in the laboratory—quantitative studies on duration, intensity, latency, and vaginal blood flow. *Archives of Sexual Behavior*, *14*(5), 439–449.

Lewis, J. E., DeGusta, D., Meyer, M. R., Monge, J. M., Mann, A. E., & Holloway, R. L. (2011). The mismeasure of science: Stephen Jay Gould versus Samuel George Morton on skulls and bias. *PLoS Biology*, *9*(6), e1001071.

Lewis, M. (2010). *Liar's poker*. W.W. Norton.

Li, N. (2007). Mate preference necessities in long- and short-term mating: People prioritize in themselves what their mates prioritize in them. *Acta Psychologica Sinica, 39*, 528–535.

Li, N. P., & Kenrick, D. T. (2006). Sex similarities and differences in preferences for short-term mates: What, whether, and why. *Journal of Personality and Social Psychology, 90*, 468–489.

Libby, R. W., Gray, L., & White, M. (1978). A test and reformulation of reference group and role correlates of premarital sexual permissiveness theory. *Journal of Marriage and the Family, 40*, 79–92.

Lidborg, L. H., Cross, C. P., & Boothroyd, L. G. (2020). *Masculinity matters (but mostly if you're muscular): A meta-analysis of the relationships between sexually dimorphic traits in men and mating/reproductive success*. BioRxiv.

Lieberman, H. (2017). *Buzz: The stimulating history of the sex toy*. Pegasus Books.

Lightfoot-Klein, H. (1984). *Prisoners of ritual: An odyssey into female genital circumcision in Africa*. Haworth Press, Inc. (Reprinted 1989)

Lightfoot-Klein, H. (1989). The sexual experience and marital adjustment of genitally circumcised and infibulated females in the Sudan. *The Journal of Sex Research, 26*(3), 375–392.

Lippa, R. A. (2007). The preferred traits of mates in a cross-national study of heterosexual and homosexual men and women: An examination of biological and cultural influences. *Archives of Sexual Behavior, 36*, 193–208.

Little, A. C., Jones, B. C., Penton-Voak, I. S., Burt, D. M., & Perrett, D. I. (2002). Partnership status and the temporal context of relationships influence human female preferences for sexual dimorphism in male face shape. *Proceedings of the Royal Society of London B, 269*, 1095–1103.

Lively, C. M., & Dybdahl, M. F. (2000). Parasite adaptation to locally common host genotypes. *Nature, 405*(6787), 679–681.

Lloyd, E. A. (2005). *The case of the female orgasm: Bias in the science of evolution*. Harvard University Press.

Lonsdorf, E. V. (2017). Sex differences in nonhuman primate behavioral development. *Journal of Neuroscience Research, 95*(1–2), 213–221.

Loomba-Albrecht, L. A., & Styne, D. M. (2009). Effect of puberty on body composition. *Current Opinion in Endocrinology, Diabetes and Obesity, 16*(1), 10–15.

Lorenz, K. (1937). Imprinting. *Auk, 54*(1), 245–273.

Lorenz, K. Z. (1966). A. The psychobiological approach: Methods and results-evolution of ritualization in the biological and cultural spheres. *Philosophical Transactions of the Royal Society of London. Series B, Biological Sciences, 251*(772), 273–284.

Love, B. (1992). *The encyclopedia of unusual sex practices*. Barricade Books.

Lovecraft, H. P. (1999). *The call of Cthulhu and other weird stories*. Penguin.

Low, B. S. (1979). Sexual selection and human ornamentation. In N. Chagnon & W. Irons (Eds.), *Evolutionary theory and human social behavior*. Duxbury.

Low, B. S., Alexander, R. D., & Noonan, K. M. (1987). Human hips, breasts, and buttocks: Is fat deceptive? *Ethology and Sociobiology, 8*, 249–257.

Lu, Q., Lai, J., Du, Y., Huang, T., Prukpitikul, P., Xu, Y., & Hu, S. (2019). Sexual dimorphism of oxytocin and vasopressin in social cognition and behavior. *Psychology Research and Behavior Management*, 337–349.

Luoto, S., Krams, I., & Rantala, M. J. (2019). A life history approach to the female sexual orientation spectrum: Evolution, development, causal mechanisms, and health. *Archives of Sexual Behavior, 48*(5), 1273–1308.

Lynch, V. J. (2008). Clitoral and penile size variability are not significantly different: Lack of evidence for the byproduct theory of the female orgasm. *Evolution & Development, 10*, 396–397.

MacHale, D., & Cohen, Y. (2018). *New light on George Boole*. Cork University Press.

Magnanti, B. (2012). *The sex myth: Why everything we're told is wrong*. Hachette UK.

Mah, K., & Binik, Y. M. (2001). The nature of the human orgasm: A critical review of major trends. *Clinical Psychology Review, 21*, 823–856.

Mah, K., & Binik, Y. M. (2002). Do all orgasms feel alike? Evaluating a two-dimensional model of the orgasm experience across gender and sexual context. *Journal of Sex Research, 39*(2), 104–113.

Maier, T. (2013). *Masters of sex: The life and times of William Masters and Virginia Johnson, the couple who taught America how to love*. Basic Books.

Maines, R. (1999). *The technology of orgasm. Hysteria", the vibrator, and women's sexual satisfaction*. Johns Hopkins Press.

Maister, L., Fotopoulou, A., Turnbull, O., & Tsakiris, M. (2020). The erogenous mirror: Intersubjective and multisensory maps of sexual arousal in men and women. *Archives of Sexual Behavior, 49*(8), 2919–2933.

Malamuth, N. M. (Ed.). (2014). *Pornography and sexual aggression*. Elsevier.

Manson, J. H. (1997). Primate consortships: A critical review. *Current Anthropology, 38*(3), 353–374.

Margulis, L. (1981). *Symbiosis in cell evolution: Life and its environment on the early earth*. Boston University Press.

Marlowe, F. W. (2005). Hunter-gatherers and human evolution. *Evolutionary Anthropology: Issues, News, and Reviews, 14*(2), 54–67.

Marshall, A. J. (1954). Bower-birds. *Biological Reviews, 29*(1), 1–45.

Martin, G. R. (2003). *A storm of swords, Part 2*. Blood and Gold.

Martin, L., & Pullman, G. K. (1991). *The great Eskimo vocabulary hoax*. University of Chicago Press.

Martin, L. J., & Sollars, S. I. (2017). Contributory role of sex differences in the variations of gustatory function. *Journal of Neuroscience Research, 95*(1–2), 594–603.

Martin-Wintle, M. S., Shepherdson, D., Zhang, G., Zhang, H., Li, D., Zhou, X., . . . Swaisgood, R. R. (2015). Free mate choice enhances conservation breeding in the endangered giant panda. *Nature Communications, 6*, 10125.

Mason, G. J. (2010). Species differences in responses to captivity: Stress, welfare and the comparative method. *Trends in Ecology and Evolution, 25*(12), 713–721.

Masters, W. H. (1959). The sexual response cycle of the human female: Vaginal lubrication. *Annals of the New York Academy of Sciences, 83*(1), 301–317.

Masters, W. H., & Johnson, V. E. (1965). The sexual response cycles of the human male and female: Comparative anatomy and physiology. In F. A. Beach (Ed.), *Sex and behavior* (pp. 512–534). Wiley.

Masters, W. H., & Johnson, V. E. (1966). *Human sexual response*. Churchill.

Matteo, S., & Rissman, E. F. (1984). Increased sexual activity during the midcycle portion of the human menstrual cycle. *Hormones and Behavior, 18*(3), 249–255.

Mautz, B. S., Wong, B. B., Peters, R. A., & Jennions, M. D. (2013). Penis size interacts with body shape and height to influence male attractiveness. *Proceedings of the National Academy of Sciences, 110*(17), 6925–6930.

Maxwell, S. E., Lau, M. Y., & Howard, G. S. (2015). Is psychology suffering from a replication crisis? What does "failure to replicate" really mean? *American Psychologist, 70*(6), 487.

Maynard-Smith, J. (1978). *Models in ecology*. CUP Archive.

Mayr, E. (1961). Cause and effect in biology: Kinds of causes, predictability, and teleology are viewed by a practicing biologist. *Science, 134*(3489), 1501–1506.

Mayr, E. (1974). Behavior programs and evolutionary strategies: Natural selection sometimes favors a genetically "closed" behavior program, sometimes an" open" one. *American Scientist, 62*(6), 650–659.

McCaughey, M., & French, C. (2001). Women's sex-toy parties: Technology, orgasm, and commodification. *Sexuality & Culture, 5*, 77–96.

McCook, A. (2006). Is peer review broken? Submissions are up, reviewers are overtaxed, and authors are lodging complaint after complaint about the process at top-tier journals. What's wrong with peer review? *The Scientist, 20*(2), 26–35.

McCoy, N. L., & Pitino, L. (2002). Pheromonal influences on sociosexual behavior in young women. *Physiology & Behavior, 75*(3), 367–375.

McEwen, B. S., & Milner, T. A. (2017). Understanding the broad influence of sex hormones and sex differences in the brain. *Journal of Neuroscience Research, 95*(1–2), 24–39.

McNeil, E. (1986). *9 ½ weeks*. Signet.

McNeil, M. (2019). *[Online resource] Honest Courtesan*. https://maggiemcneill.wordpress.com/2019/01/31/time-management/#comments last accessed 01/02/2019.

McRae-Clark, A. L., Cason, A. M., Kohtz, A. S., Moran Santa-Maria, M., Aston-Jones, G., & Brady, K. T. (2017). Impact of gender on corticotropin-releasing factor and noradrenergic sensitivity in cocaine use disorder. *Journal of Neuroscience Research, 95*(1–2), 320–327.

McWhorter, J. H. (2014). *The language hoax: Why the world looks the same in any language*. Oxford University Press.

Mead, M. (1963). *Sex and temperament in three primitive societies*. Morrow.

Merrick, T. (2019). From 'intersex' to 'DSD': A case of epistemic injustice. *Synthese, 196*(11), 4429–4447.

Merz, C. J., & Wolf, O. T. (2017). Sex differences in stress effects on emotional learning. *Journal of Neuroscience Research, 95*(1–2), 93–105.

Messenger, J. C. (1971). Sex and repression in an Irish folk community. In *Human sexual behavior: Variations in the ethnographic spectrum* (pp. 3–37). Prentice Hall.

Meston, C. M., Levin, R. J., Sipski, M. L., Hull, E. M., & Heiman, J. R. (2004). Women's orgasm. *Annual Review of Sex Research*, *15*(1), 173–257.

Mezalira, A., Dallanora, D., Bernardi, M. L., Wentz, I., & Bortolozzo, F. P. (2005). Influence of sperm cell dose and post-insemination backflow on reproductive performance of intrauterine inseminated sows. *Reproduction in Domestic Animals*, *40*(1), 1–5.

Millar, R. (1952). Forces observed during coitus in thoroughbreds. *Australian Ethology Journal*, *28*, 127–128.

Miller, G. (2009). *Spent*. Viking Penguin.

Miller, G. (2011). *The mating mind: How sexual choice shaped the evolution of human nature*. Anchor.

Miller, G. (2019). Polyamory is growing and we need to get serious about it. *Quillette* (Online article in popular science journal). https://quillette.com/2019/10/29/polyamory-is-growing-and-we-need-to-get-serious-about-it/ last accessed 07/11/2019.

Miller, G. F. (1998). How mate choice shaped human nature: A review of sexual selection and human evolution. In *Handbook of evolutionary psychology: Ideas, issues, and applications* (pp. 87–129). Psychology Press.

Miller, G., Tybur, J. M., & Jordan, B. D. (2007). Ovulatory cycle effects on tip earnings by lap dancers: Economic evidence for human estrus? *Evolution and Human Behavior*, *28*(6), 375–381.

Miller, G. F. (2007). Sexual selection for moral virtues. *The Quarterly Review of Biology*, *82*(2), 97–125.

Miner, H. (1956). Body ritual among the Nacirema. *American Anthropologist*, *58*(3), 503–507.

Møller, A. P., & Eriksson, M. (1994). Patterns of fluctuating asymmetry in flowers: Implications for sexual selection in plants. *Journal of Evolutionary Biology*, *7*(1), 97–113.

Møller, A. P., & Pomiankowski, A. A. (1993). Fluctuating asymmetry and sexual selection. *Genetica*, *89*(1–3), 267–279.

Monbiot, G. (2019). Newspaper article about English public school system. *Guardian Newspaper*. www.theguardian.com/commentisfree/2019/nov/07/boarding-schools-boris-johnson-bullies last accessed 06/11/2019.

Money, J., Hampson, J. G., & Hampson, J. L. (1955). Hermaphroditism: Recommendations concerning assignment of sex, change of sex and psychologic management. *Bulletin of the Johns Hopkins Hospital*, *97*(4), 284–300.

Moore, H. D. M., Martin, M., & Birkhead, T. R. (1999). No evidence for killer sperm or other selective interactions between human spermatozoa in ejaculates of different males *in vitro*. *Proceedings of the Royal Society of London B*, *266*, 2343–2350.

Moore, M. M. (1985). Nonverbal courtship patterns in women: Context and consequences. *Ethology and Sociobiology*, *6*(4), 237–247.

Moore, M. M., & Butler, D. L. (1989). Predictive aspects of nonverbal courtship behavior in women. *Semiotica*, *76*(3–4), 205–216.

Morgan, M. H., & Carrier, D. R. (2013). Protective buttressing of the human fist and the evolution of hominin hands. *Journal of Experimental Biology*, *216*(2), 236–244.

Morris, D. (1967). *The naked ape*. Jonathan Cape.

Mostafa, T., El Khouly, G., & Hassan, A. (2012). Pheromones in sex and reproduction: Do they have a role in humans? *Journal of Advanced Research*, *3*(1), 1–9.

Mukherjee, S. (2017). *The gene: An intimate history*. Simon & Schuster.

Murdock, G. P. (1967). *Ethnographic atlas*. Pittsburgh University Press.

Murphy, M. R., Seckl, J. R., Burton, S., Checkley, S. A., & Lightman S. L. (1987). Changes in oxytocin and vasopressin secretion during sexual activity in men. *Journal of Clinical Endocrinology and Metabolism*, *65*, 738–741.

Nagel, T. (1989). *The view from nowhere*. Oxford University Press.

Nash, P. (1961). Training an elite: The prefect-fagging system in the English public school. *History of Education Quarterly*, *1*(1), 14–21.

Nathanson, P., & Young, K. K. (2001). *Spreading misandry: The teaching of contempt for men in popular culture*. McGill-Queen's Press-MQUP.

National Safety Council. *Website with details on adoption and risks*. National Safety Council. www.nsc.org/learn/safety-knowledge/Pages/injury-facts.aspx last accessed 29/09/2017.

Nelson, L. H., & Lenz, K. M. (2017). The immune system as a novel regulator of sex differences in brain and behavioral development. *Journal of Neuroscience Research*, *95*(1–2), 447–461.

Nesse, R. M. (2019). *Good reasons for bad feelings: Insights from the frontier of evolutionary psychiatry*. Dutton.

Nesse, R. M., & Williams, G. C. (2012). *Why we get sick: The new science of Darwinian medicine.* Vintage.

Nesse, R. M., Williams, G. C., & Mysterud, I. (1995). Why we get sick. *Trends in Ecology and Evolution,* *10*(7), 300–301.

Nettle, D. (2002). Height and reproductive success in a cohort of British men. *Human Nature, 13*(4), 473–491.

Nettle, D. (2005). An evolutionary approach to the extraversion continuum. *Evolution and Human Behavior,* *26*(4), 363–373.

Netto, C. A., Sanches, E., Odorcyk, F. K., Duran-Carabali, L. E., & Weis, S. N. (2017). Sex-dependent consequences of neonatal brain hypoxia-ischemia in the rat. *Journal of Neuroscience Research, 95*(1–2), 409–421.

Nielsen, M. W., Andersen, J. P., Schiebinger, L., & Schneider, J. W. (2017). One and a half million medical papers reveal a link between author gender and attention to gender and sex analysis. *Nature Human Behaviour, 1*(11), 791.

Nims, J. P. (1975). Imagery, shaping, and orgasm. *Journal of Sex & Marital Therapy, 1*(3), 198–203.

Nin, A. (1977). *Delta of Venus.* Bantam Books. (Original work published 1940)

Nisbett, R. E., & Wilson, T. D. (1977). Telling more than we can know: Verbal reports on mental processes. *Psychological Review, 84*(3), 231.

Nishimori, K., Young, L. J., Guo, Q., Wang, Z., Insel, T. R., & Matzuk, M. M. (1996). Oxytocin is required for nursing but is not essential for parturition or reproductive behavior. *Proceedings of National Academy of Science U S A, 93,* 11699–11704.

Nolin, M. J., & Petersen, K. K. (1992). Gender differences in parent-child communication about sexuality: An exploratory study. *Journal of Adolescent Research, 7*(1), 59–79.

Obermeyer, C. M. (1999). Female genital surgeries: The known, the unknown, and the unknowable. *Medical Anthropology Quarterly, 13,* 79–106.

Ochs, E. P., Mah, K., & Binik, Y. M. (2002). Obtaining data about human sexual functioning from the Internet. In A. Cooper (Ed.), *Sex and the Internet: A guidebook for clinicians* (pp. 245–262). Brunner-Routledge.

Odent, M. (1999). *The scientification of love.* Free Association Books.

O'Donovan, D., Hindle, J. E., McKeown, S., & O'Donovan, S. (1993). Effect of visitors on the behaviour of female Cheetahs Acinonyx jubutus and cubs. *International Zoo Yearbook, 32*(1), 238–244.

Ogas, O., & Gaddam, S. (2011). *A billion wicked thoughts: What the world's largest experiment reveals about human desire.* Dutton, Penguin Books.

Ogilvie, M. B., & Choquette, C. J. (1981). Nettie Maria Stevens (1861–1912): Her life and contributions to cytogenetics. *Proceedings of the American Philosophical Society, 125*(4), 292–311.

Ogura, A., Ikeo, K., & Gojobori, T. (2004). Comparative analysis of gene expression for convergent evolution of camera eye between octopus and human. *Genome Research, 14*(8), 1555–1561.

Ohno, M., Maeda, T., & Matsunobu, A. (1991). A cytogenetic study of spontaneous abortions with direct analysis of chorionic villi. *Obstetrics & Gynecology, 77*(3), 394–398.

Oliveira-Pinto, A. V., Santos, R. M., Coutinho, R. A., Oliveira, L. M., Santos, G. B., Alho, A. T., . . . Pasqualucci, C. A. (2014). Sexual dimorphism in the human olfactory bulb: Females have more neurons and glial cells than males. *PLoS One, 9*(11), e111733.

OMGYES.com-[Website collecting and sharing orgasm experiences]. *The science of women's pleasure.* https://start.omgyes.com/join last accessed 14/08/2023.

Oosterbeek, H., Sloof, R., & Van De Kuilen, G. (2004). Cultural differences in ultimatum game experiments: Evidence from a meta-analysis. *Experimental Economics, 7,* 171–188.

Opperman, E., Braun, V., Clarke, V., & Rogers, C. (2014). "It feels so good it almost hurts": Young adults' experiences of orgasm and sexual pleasure. *The Journal of Sex Research, 51*(5), 503–515.

Orsini, C. A., & Setlow, B. (2017). Sex differences in animal models of decision making. *Journal of Neuroscience Research, 95*(1–2), 260–269.

Orwell, G. (2002). Politics and the English language. *British Army Review-London-Ministry of Defence,* 52–56.

Ostrzenski, A., Krajewski, P., Ganjei-Azar, P., Wasiutynski, A. J., Scheinberg, M. N., Tarka, S., & Fudalej, M. (2014). Verification of the anatomy and newly discovered histology of the G-spot complex. *BJOG: An International Journal of Obstetrics & Gynaecology, 121*(11), 1333–1340.

Owens, L., Shute, R., & Slee, P. (2000a). "Guess what I just heard!": Indirect aggression among teenage girls in Australia. *Aggressive Behavior, 26,* 67–83.

Owens, L., Shute, R., & Slee, P. (2000b). "I'm in and you're out . . . :" Explanations for teenage girls' indirect aggression. *Psychology, Evolution, and Gender, 2,* 19–46.

Parish, A. R. (1996). Female relationships in bonobos (Pan paniscus). *Human Nature, 7*(1), 61–96.

Pasteur, L. (1861). *Sur les corpuscles organisés qui existent dans l'atmosphère: Examen de la doctrine des générations spontanées.* Leçon Professée a la Société Chimique de Paris.

Paus, T., Wong, A. P. Y., Syme, C., & Pausova, Z. (2017). Sex differences in the adolescent brain and body: Findings from the saguenay youth study. *Journal of Neuroscience Research, 95*(1–2), 362–370.

Pavlicev, M., & Wagner, G. (2016). The evolutionary origin of female orgasm. *Journal of Experimental Zoology Part B. Molecular and Developmental Evolution. 326*(6), 326–337.

Pavlova, M. A. (2017). Sex and gender affect the social brain: Beyond simplicity. *Journal of Neuroscience Research, 95*(1–2), 235–250.

Pavlovic, J. M., Akcali, D., Bolay, H., Bernstein, C., & Maleki, N. (2017). Sex-related influences in migraine. *Journal of Neuroscience Research, 95*(1–2), 587–593.

Pawlowski, B., & Jasienska, G. (2005). Women's preferences for sexual dimorphism in height depend on menstrual cycle phase and expected duration of relationship. *Biological Psychology, 70*, 38–43.

Pawlowski, W. (2009). BDSM: The ultimate expression of healthy sexuality. In W. J. Taverner & R. W. McKee (Eds.), *Taking sides: Clashing views in human sexuality* (11th ed., pp. 70–75). McGraw-Hill.

Pazhoohi, F., Garza, R., Doyle, J. F., Macedo, A. F., & Arantes, J. (2019). Sex differences for preferences of shoulder to hip ratio in men and women: An eye tracking study. *Evolutionary Psychological Science, 5*, 405–415.

Penfield, W., & Boldrey, E. (1937). Somatic motor and sensory representation in the cerebral cortex of man as studied by electrical stimulation. *Brain, 60*(4), 389–443.

Penton-Voak, I. S., Little, A. C., Jones, B. C., Burt, D. M., Tiddeman, B. P., Perrett, D. I. (2003). Female condition influences preferences for sexual dimorphism in faces of male humans (Homo sapiens). *Journal of Comparative Psychology, 117*, 264–271.

Perel, E. (2006). *Mating in captivity* (p. 272). HarperCollins.

Perry, J. D., & Whipple, B. (1981). Pelvic muscle strength of female ejaculators: Evidence in support of a new theory of orgasm. *Journal of Sex Research, 17*(1), 22–39.

Perry, J. F. (1988). Do men have a G-spot? *Australian Forum, 2*, 37–41.

Pickles, V. R. (1967). Uterine suction during orgasm. Letter to the *British Medical Journal, 1*(5537), 427.

Pierotti, R., Annett, C. A., & Hand, J. L. (1997). Male and female perceptions of pair-bond dynamics: Monogamy in western gulls, larus occidentalis. In *Feminism and evolutionary biology* (pp. 261–275). Springer.

Pike, C. J. (2017). Sex and the development of Alzheimer's disease. *Journal of Neuroscience Research, 95* (1–2), 671–680.

Pillsworth, E. G., & Haselton, M. G. (2006). Male sexual attractiveness predicts differential ovulatory shifts in female extra-pair attraction and male mate retention. *Evolution & Human Behavior, 27*, 247–258.

Pinker, S. (1995). *The language instinct: The new science of language and mind.* Penguin.

Pinker, S. (2003). *The blank slate: The modern denial of human nature.* Penguin.

Pinker, S. (2007). *The stuff of thought: Language as a window into human nature.* Penguin.

Pinker, S. (2011). *The better angels of our nature: The decline of violence in history and its causes.* Penguin.

Pinker, S. (2019). *Enlightenment now: The case for reason, science, humanism, and progress.* Penguin Books.

Plato. (1928). *The theaetetus of Plato* (Trans. M. J. Levett). Wylie & Company.

Playà, E., Vinicius, L., & Vasey, P. L. (2017). Need for alloparental care and attitudes toward homosexuals in 58 countries: Implications for the kin selection hypothesis. *Evolutionary Psychological Science, 3*(4), 345–352.

Plomin, R., Caspi, A., Pervin, L. A., & John, O. P. (1990). Behavioral genetics and personality. In *Handbook of personality: Theory and research* (Vol. 2, pp. 251–276). Guildford Press.

Plomin, R., DeFries, J. C., & McClearn, G. E. (2008). *Behavioral genetics.* Macmillan.

Pogun, S., Yararbas, G., Nesil, T., & Kanit, L. (2017). Sex differences in nicotine preference. *Journal of Neuroscience Research, 95*(1–2), 148–162.

Popper, K. (1957). *The poverty of historicism.* Routledge.

Popper, K. (2014). *Conjectures and refutations: The growth of scientific knowledge.* Routledge.

Porpora, D. V. (2015). *Reconstructing sociology: The critical realist approach.* Cambridge University Press.

Poulsen, H. (1970). Nesting behaviour of the Black-Casqued Hornbill Ceratogymna atrata (Temm.) and the Great Hornbill Buceros bicornis L. *Ornis Scandinavica (Scandinavian Journal of Ornithology), 1*(1), 11–15.

Pound, N. (2002). Male interest in visual cues of sperm competition risk. *Evolution and Human Behavior, 23*(6), 443–466.

Prause, N. (2011). The human female orgasm: Critical evaluations of proposed psychological sequelae. *Sexual and Relationship Therapy, 26*(4), 315–328.

Prause, N., Kuang, L., Lee, P. M., & Miller, G. F. (2016). Clitorally stimulated orgasms are associated with better control of sexual desire, and not associated with depression or anxiety, compared with vaginally stimulated orgasms. 'Clitoral' versus 'vaginal' orgasms: False dichotomies and differential effects. *Journal of Sexual Medicine, 13*(11), 1676–1685.

Printzlau, F., Wolstencroft, J., & Skuse, D. H. (2017). Cognitive, behavioral, and neural consequences of sex chromosome aneuploidy. *Journal of Neuroscience Research, 95*(1–2), 311–319.

Proverbio, A. M. (2017). Sex differences in social cognition: The case of face processing. *Journal of Neuroscience Research, 95*(1–2), 222–234.

Provost, M. P., Kormos, C., Kosakoski, G., & Quinsey, V. L. (2006). Sociosexuality in women and preference for facial masculinization and somatotype in men. *Archives of Sexual Behavior, 35*, 305–312.

Provost, M. P., Troje, N. F., & Quinsey, V. L. (2008). Short-term mating strategies and attraction to masculinity in point-light walkers. *Evolution and Human Behavior, 29*, 65–69.

Prüfer, K., Munch, K., Hellmann, I., Akagi, K., Miller, J. R., Walenz, B., . . . Pääbo, S. (2012). The bonobo genome compared with the chimpanzee and human genomes. *Nature, 486*(7404), 527–531.

Prum, R. O. (2017). *The evolution of beauty: How Darwin's forgotten theory of mate choice shapes the animal world-and us.* Anchor.

Puts, D. A. (2006). Cyclic variation in women's preferences for masculine traits: Potential hormonal causes. *Human Nature, 17*, 114–127.

Puts, D. A., Hill, A. K., Bailey, D. H., Walker, R. S., Rendall, D., Wheatley, J. R., . . . Ramos-Fernandez, G. (2016). Sexual selection on male vocal fundamental frequency in humans and other anthropoids. *Proceedings of the Royal Society B: Biological Sciences, 283*(1829), 20152830.

Queller, D. C. (1997). Why do females care more than males? *Proceedings of the Royal Society of London B: Biological Sciences, 264*(1388), 1555–1557.

Quine, W. V. O. (1953). *From a logical point of view.* Harvard University Press.

Quist, M. C., Watkins, C. D., Smith, F. G., Little, A. C., DeBruine, L. M., & Jones, B. C. (2012). Sociosexuality predicts women's preferences for symmetry in men's faces. *Archives of Sexual Behavior, 41*, 1415–1421.

Radtke, S. (2013). Sexual fluidity in women: How feminist research influenced evolutionary studies of same-sex behavior. *Journal of Social, Evolutionary, and Cultural Psychology, 7*(4), 336.

Ramachandran, V. S., & Blakeslee, S. (1999). *Phantoms in the brain: Human nature and the architecture of the mind.* Fourth Estate Ltd.

Rancour-Laferriere, D. (1983). Four adaptive aspects of the female orgasm. *Journal of Social and Biological Structures, 6*, 319–333.

Reage, P. (1954). *The story of O.* Pauvet Press.

Reber, J., & Tranel, D. (2017). Sex differences in the functional lateralization of emotion and decision making in the human brain. *Journal of Neuroscience Research, 95*(1–2), 270–278.

Regan, P. C. (1998a). Minimum mate selection standards as a function of perceived mate value, relationship context, and gender. *Journal of Psychology and Human Sexuality, 10*, 53–73.

Regan, P. C. (1998b). What if you can't get what you want? Willingness to compromise ideal mate selection standards as a function of sex, mate value, and relationship context. *Personality and Social Psychology Bulletin, 24*, 1294–1303.

Regan, P. C., & Berscheid, E. (1997). Gender differences in characteristics desired in a potential sexual and marriage partner. *Journal of Psychology and Human Sexuality, 9*, 25–37.

Regan, P. C., Levin, L., Sprecher, S., Christopher, F. S., & Cate, R. (2000). Partner preferences: What characteristics do men and women desire in their short-term and long-term romantic partners? *Journal of Psychology and Human Sexuality, 12*, 1–21.

Regan, P. C., Medina, R., & Joshi, A. (2001). Partner preferences among homosexual men and women: What is desirable in a sex partner is not necessarily desirable in a romantic partner. *Social Behavior and Personality, 29*, 625–633.

Reschke-Hernández, A. E., Okerstrom, K. L., Bowles Edwards, A., & Tranel, D. (2017). Sex and stress: Men and women show different cortisol responses to psychological stress induced by the Trier social stress test and the Iowa singing social stress test. *Journal of Neuroscience Research, 95*(1–2), 106–114.

Reynolds, J. J., & McCrea, S. M. (2016). Life history theory and exploitative strategies. *Evolutionary Psychology, 14*(3).

Ricci, D., Cesarini, L., Groppo, M., De Carli, A., Gallini, F., Serrao, F., . . . Mosca, F. (2008). Early assessment of visual function in full term newborns. *Early Human Development, 84*(2), 107–113.

Richters, J., de Visser, R. O., Rissel, C. E., Grulich, A. E., & Smith, A. M. (2008). Demographic and psychoso-cial features of participants in bondage and discipline, "sadomasochism" or dominance and submission (BDSM): Data from a national survey. *The Journal of Sexual Medicine*, *5*(7), 1660–1668.

Richters, J., de Visser, R., Rissel, C., & Smith, A. (2006). Sexual practices at last heterosexual encounter and occurrence of orgasm in a national survey. *Journal of Sex Search*, *43*(3), 217–226.

Riddell, F. (2014). *No, no, no! Victorians didn't invent the vibrator*. www.theguardian.com/commentis-free/2014/nov/10/victorians-invent-vibrator-orgasms-women-doctors-fantasy last accessed 16/02/2018.

Rippon, G. (2019). *Gender and our brains: How new neuroscience explodes the myths of the male and female minds*. Pantheon.

Rizzolo, J. B., & Bradshaw, G. A. (2016). Prevalence and patterns of complex PTSD in Asian elephants (ele-phas maximus). In N. Bandara & T. Wickramaarachchi (Eds.), *Asian elephants in culture and nature*. Centre for Asian Studies University of Kelaniya Sri Lanka.

Robertiello, R. C. (1970). The "clitoral versus vaginal orgasm" controversy and some of its ramifications. *Journal of Sex Research*, *6*(4), 307–311.

Rosen, S., Ham, B., & Mogil, J. S. (2017). Sex differences in neuroimmunity and pain. *Journal of Neuroscience Research*, *95*(1–2), 500–508.

Rosenberg, K., & Trevathan, W. (2002). Birth, obstetrics and human evolution. *BJOG: An International Journal of Obstetrics & Gynaecology*, *109*(11), 1199–1206.

Rosenblatt, P. C. (1967). Marital residence and the functions of romantic love. *Ethnology*, *6*(4), 471–480.

Rosenfeld, C. S. (2017). Sex-dependent differences in voluntary physical activity. *Journal of Neuroscience Research*, *95*(1–2), 279–290.

Rubin, L. H., Yao, L., Keedy, S. K., Reilly, J. L., Bishop, J. R., Carter, C. S., . . . Sweeney, J. A. (2017). Sex differences in associations of arginine vasopressin and oxytocin with resting-state functional brain con-nectivity. *Journal of Neuroscience Research*, *95*(1–2), 576–586.

Russell, B. (1905). On denoting. *Mind*, *14*(56), 479–493.

Russell, J. A., Leng, G., & Douglas, A. J. (2003). The magnocellular oxytocin system, the fount of maternity: Adaptations in pregnancy. *Frontiers in Neuroendocrinology*, *24*, 27–61.

Ryan, C., & Jethá, C. (2010). *Sex at dawn: The prehistoric origins of modern sexuality*. Harper Collins.

Saad, G. (2007). *The evolutionary bases of consumption*. Psychology Press.

Sacco, D. F., Jones, B. C., DeBruine, L. M., & Hugenberg, K. (2012). The roles of sociosexual orienta-tion and relationship status in women's face preferences. *Personality and Individual Differences*, *53*, 1044–1047.

Safron, A. (2016). What is orgasm? A model of sexual trance and climax via rhythmic entrainment. *Socioaffective Neuroscience & Psychology*, *6*(1), 31763.

Salama, S., Boitrelle, F., Gauquelin, A., Malagrida, L., Thiounn, N., & Desvaux, P. (2015). Nature and origin of "squirting" in female sexuality. *The Journal of Sexual Medicine*, *12*(3), 661–666.

Salisbury, C. M., & Fisher, W. A. (2014). "Did you come?" A qualitative exploration of gender differences in beliefs, experiences, and concerns regarding female orgasm occurrence during heterosexual sexual interactions. *The Journal of Sex Research*, *51*(6), 616–631.

Salmon, C. (2012). The pop culture of sex: An evolutionary window on the worlds of pornography and romance. *Review of General Psychology*, *16*(2), 152.

Salmon, C., & Diamond, A. (2012). Evolutionary perspectives on the content analysis of heterosexual and homosexual pornography. *Journal of Social, Evolutionary, and Cultural Psychology*, *6*(2), 193.

Salmon, C., & Symons, D. (2001). *Warrior lovers: Erotic fiction, evolution and female sexuality*. Yale University Press.

Samba Reddy, D. (2017). Sex differences in the anticonvulsant activity of neurosteroids. *Journal of Neuroscience Research*, *95*(1–2), 661–670.

Santos, F. C. A., & Taboga, S. R. (2006). Female prostate: A review about the biological repercussions of this gland in humans and rodents. *Animal Reproduction*, *3*(1), 3–18.

Sartre, J. P. (1948). Existentialism and humanism (1947). *Philosophy: Key Texts*, *115*, 106–116.

Schäfer, L., Mehler, L., Hähner, A., Walliczek, U., Hummel, T., & Croy, I. (2018). Sexual desire after olfac-tory loss: Quantitative and qualitative reports of patients with smell disorders. *Physiology & Behavior*, *201*, 64–69.

Scheib, J. E. (1994). Sperm donor selection and the psychology of female mate choice. *Ethology & Sociobiology*, *15*, 113–129.

Scheib, J. E. (2001). Context-specific mate choice criteria: Women's trade-offs in the contexts of long-term and extra-pair mateships. *Personal Relationships*, *8*, 371–389.

Schlager7 [Youtube Video]. (1967). *Last sword duel*. www.youtube.com/watch?v=e68nuAcSuWQ last accessed 19/08/2023.

Schleifenbaum, L., Driebe, J. C., Gerlach, T. M., Penke, L., & Arslan, R. C. (2021). Women feel more attractive before ovulation: Evidence from a large-scale online diary study. *Evolutionary Human Sciences*, *3*, e47.

Schmidt, J. O., Slessor, K. N., & Winston, M. L. (1993). Roles of Nasonov and queen pheromones in attraction of honeybee swarms. *Naturwissenschaften*, *80*(12), 573–575.

Schmitt, D. P., & Buss, D. M. (1996). Strategic self-promotion and competitor derogation: Sex and context effects on the perceived effectiveness of mate attraction tactics. *Journal of Personality and Social Psychology*, *70*(6), 1185.

Schmitt, D. P., Couden, A., & Baker, M. (2001). Sex, temporal context, and romantic desire: An experimental evaluation of sexual strategies theory. *Personality and Social Psychology Bulletin*, *27*, 833–847.

Schmitt, D. P., Jonason, P. K., Byerley, G. J., Flores, S. D., Illbeck, B. E., O'Leary, K. N., & Qudrat, A. (2012). A reexamination of sex differences in sexuality: New studies reveal old truths? *Current Directions in Psychological Science*, *21*, 135–139.

Schmitt, D. P., Realo, A., Voracek, M., & Allik, J. (2008). Why can't a man be more like a woman? Sex differences in Big Five personality traits across 55 cultures. *Journal of Personality and Social Psychology*, *94*(1).

Schober, J. M., Meyer-Bahlburg, H. F. L., & Ransley, P. G. (2004). Self-assessment of genital anatomy, sexual sensitivity and function in women: Implications for genitoplasty. *British Journal of Urology*, *94*, 589–594.

Schott, G. D. (1993). Penfield's homunculus: A note on cerebral cartography. *Journal of Neurology, Neurosurgery, and Psychiatry*, *56*(4), 329.

Schultz, W. W., van Andel, P., Sabelis, I., & Mooyaart, E. (1999). Magnetic resonance imaging of male and female genitals during coitus and female sexual arousal. *British Medical Journal*, *319*, 1596–1600.

Schulz, J., Bahrami-Rad, D., Beauchamp, J., & Henrich, J. (2018). *The origins of WEIRD psychology*. Available at SSRN 3201031.

Schwambergová, D., Sorokowska, A., Slámová, Ž., Fialová, J. T., Sabiniewicz, A., Nowak-Kornicka, J., . . . Havlíček, J. (2021). No evidence for association between human body odor quality and immune system functioning. *Psychoneuroendocrinology*, *132*, 105363.

Science Daily. (2018). Blog post *Countries with greater gender equality have a lower percentage of female STEM graduates*. www.sciencedaily.com/releases/2018/02/180214150132.htm last accessed 22/08/2023.

Screech, T. (2009). *Sex and the floating world: Erotic imagery in Japan, 1720–1810*. Reaktion Books.

Seal, D. W., Agostinelli, G., & Hannett, C. A. (1994). Extradyadic romantic involvement: Moderating effects of sociosexuality and gender. *Sex Roles*, *31*, 1–22.

Searle, J. R. (1995). *The construction of social reality*. Simon & Schuster.

Sell, A., Lukazsweski, A. W., & Townsley, M. (2017). Cues of upper body strength account for most of the variance in men's bodily attractiveness. *Proceedings of the Royal Society B: Biological Sciences*, *284*(1869), 20171819.

Seo, D., Ahluwalia, A., Potenza, M. N., & Sinha, R. (2017). Gender differences in neural correlates of stress-induced anxiety. *Journal of Neuroscience Research*, *95*(1–2), 115–125.

Shackelford, T. K., Schmitt, D. P., & Buss, D. M. (2005). Universal dimensions of human mate preferences. *Personality and Individual Differences*, *39*(2), 447–458.

Shackelford, T. K., Weekes, V. A., LeBlanc, G. J., Bleske, A. L., Euler, H. A., & Hoier, S. (2000). Female coital orgasm and male attractiveness. *Human Nature*, *11*, 299–306.

Shandall, A. A. (1967). Circumcision and infibulation of females. *Sudan Medical Journal*, *5*, 178–212.

Shearn, D., Bergman, E., Hill, K., Abel, A., & Hinds, L. (1990). Facial coloration and temperature responses in blushing. *Psychophysiology*, *27*(6), 687–693.

Sheehan, E. (1981). Victorian clitoridectomy: Isaac Baker Brown and his harmless operative procedure. *Medical Anthropology Newsletter*, *12*(4), 9–15.

Sheets-Johnstone, M. (1990). Hominid bipedality and sexual selection theory. *Evolutionary Theory*, *9*(1), 57–70.

Shehata, R. (1974). Urethral glands in the wall of the female urethra of rats, mice and closely related rodents. *Cells Tissues Organs*, *90*(3), 381–387.

Shehata, R. (1975). Female prostate in Arvicanthis niloticus and Meriones libycus. *Cells Tissues Organs*, *92*(4), 513–523.

Sherlock, J. M., Sidari, M. J., Harris, E. A., Barlow, F. K., & Zietsch, B. P. (2016). Testing the mate-choice hypothesis of the female orgasm: Disentangling traits and behaviours. *Socioaffective Neuroscience & Psychology*, *6*(1), 31562.

Shors, T. J., Millon, E. M., Chang, H. Y. M., Olson, R. L., & Alderman, B. L. (2017). Do sex differences in rumination explain sex differences in depression? *Journal of Neuroscience Research*, *95*(1–2), 711–718.

Short, R. V. (1980). The origins of human sexuality. In C. R. Austin & R. B. Short (Eds.), *Reproduction in mammals* (Vol. 8, pp. 1–33). Cambridge University Press.

Shresta, A., Dempsey, M., Tuohy-Hamil, S., & King, R. (2022). What does she see in him? Hybristophiles and spree killers. *Journal of Police and Criminal Psychology*, 1–13.

Simler, K., & Hanson, R. (2017). *The elephant in the brain: Hidden motives in everyday life*. Oxford University Press.

Simon, R. J., & Baxter, S. (1989). Gender and violent crime. In N. A. Weiner & M. E. Wolfgang (Eds.), *Violent crime, violent criminals* (pp. 171–197). SAGE.

Simpson, J. A., & Gangestad, S. W. (1992). Sociosexuality and romantic partner choice. *Journal of Personality*, *60*, 31–51.

Simpson, J. A., Gangestad, S. W., Christensen, P., & Niels, K. (1999). Fluctuating asymmetry, sociosexuality, and intrasexual competitive tactics. *Journal of Personality and Social Psychology*, *76*, 159–172.

Singer, I. (1972). Anti-climax: Reviews of *The Nature and Evolution of Female Sexuality*, by Mary Jane Sherfey, and *The Female Orgasm: Psychology, Physiology, Fantasy*, by Seymour Fisher. New York Review of Books (November 30th 1972 Issue). www.nybooks.com/articles/1972/11/30/anti-climax/ last accessed 15/04/2019.

Singer, J., & Singer, I. (1972). Types of female orgasm. *Journal of Sex Research*, *8*, 255–267.

Smith, J. M. (1978). *The evolution of sex*. CUP Archive.

Smith, R. L., Rose, A. J., & Schwartz-Mette, R. A. (2010). Relational and overt aggression in childhood and adolescence: Clarifying mean-level gender differences and associations with peer acceptance. *Social Development*, *19*, 243–269.

Smuts, B., & Nicolson, N. (1989). Reproduction in wild female olive baboons. *American Journal of Primatology*, *19*(4), 229–246.

Soh, D., & Lehmann, C. (2017, April 17). *The rhetorical trap at the heart of the 'neurosexism' debate*. Published (and accessed) Online in *Quillette*. http://quillette.com/2017/04/17/rhetorical-trap-heart-neurosexism-debate/.

Sohrabji, F., Park, M. J., & Mahnke, A. H. (2017). Sex differences in stroke therapies. *Journal of Neuroscience Research*, *95*(1–2), 681–691.

Somerville, M. (1858). *On the connexion of the physical sciences*. J. Murray.

Sorabji, R. (1980). *Necessity, cause and blame*. Cornell University Press.

Sorokowski, P., Sorokowska, A., Karwowski, M., Groyecka, A., Aavik, T., Akello, G., . . . Sternberg, R. J. (2021). Universality of the triangular theory of love: Adaptation and psychometric properties of the triangular love scale in 25 countries. *The Journal of Sex Research*, *58*(1), 106–115.

Speiser, P. W., & White, P. C. (2003). Congenital adrenal hyperplasia. *New England Journal of Medicine*, *349*(8), 776–788.

Spinoza, B. (1992). *Ethics: With the treatise on the emendation of the intellect and selected letters*. Hackett Publishing. (Original work published 1677)

Stearns, S. C. (1976). Life-history tactics: A review of the ideas. *The Quarterly Review of Biology*, *51*(1), 3–47.

Stebbins, G. L. (1982). *Darwin to DNA, molecules to humanity*. W. H. Freeman.

Sternberg, R. J. (1986). A triangular theory of love. *Psychological Review*, *93*(2), 119.

Steverink, D. W. B., Soede, N. M., Bouwman, E. G., & Kemp, B. (1998). Semen backflow after insemination and its effect on fertilisation in sows. *Animal Reproductive Science*, *54*, 109–119.

Stewart, S., Stinnett, H., & Rosenfeld, L. B. (2000). Sex differences in desired characteristics of short-term and long-term relationship partners. *Journal of Social and Personal Relationships*, *17*, 843–853.

Stewart-Williams, S., & Thomas, A. G. (2013). The ape that thought it was a peacock: Does evolutionary psychology exaggerate human sex differences? *Psychological Inquiry*, *24*(3), 137–168.

Stove, D. C. (1991). *The Plato cult and other philosophical follies*. Blackwell Publishers.

Strassmann, B. I. (1997). Polygyny as a risk factor for child mortality among the Dogon. *Current Anthropology*, *38*, 688–695.

Stulhofer, A. (2006). How (un)important is penis size for women with heterosexual experience? Letter to the editor. *Archives of Sexual Behavior*, *35*, 5–6.

Sundahl, D. (2003). *Female ejaculation and the G-spot*. Hunter House.

Sundström Poromaa, I., Comasco, E., Georgakis, M. K., & Skalkidou, A. (2017). Sex differences in depression during pregnancy and the postpartum period. *Journal of Neuroscience Research*, 95(1–2), 719–730.

Sussman, A. (2023). [Online article about Korean feminism] The Cut. *The Cut*. www.thecut.com/2023/03/4b-movement-feminism-south-korea.html last accessed 28/06/2023.

Syed, U. A. M., Davis, D. E., Ko, J. W., Lee, B. K., Huttman, D., Seidl, A., . . . Abboud, J. A. (2017). Quantitative anatomical differences in the shoulder. *Orthopedics*, 40(3), 155–160.

Symons, D. (1979). *The evolution of human sexuality*. Oxford University Press.

Talalaj, J. (1994). *The strangest human sex, ceremonies and customs*. Hill of Content Publishing Company Pty Limited.

Tanner, J. M. (1986). 1 normal growth and techniques of growth assessment. *Clinics in Endocrinology and Metabolism*, 15(3), 411–451.

Tavris, C. (1992). *The mismeasure of woman*. Simon & Schuster.

Taylor, J. (2008). *TV interview for young Turks about his racism with Cenk Uygur*. www.youtube.com/watch?v=oxhsSLos8hk last accessed 03/04/2023.

Terrace, H. S. (2019). *Why chimpanzees can't learn language and only humans can*. Columbia University Press.

Thiessen, D., & Gregg, B. (1980). Human assortative mating and genetic equilibrium: An evolutionary perspective. *Ethology and Sociobiology*, 1(2), 111–140.

Thompson, D. W. (1917). *On growth and form*. Cambridge University Press.

Thornhill, R., & Gangestad, S. W. (1999). The scent of symmetry: A human sex pheromone that signals fitness? *Evolution and Human Behavior*, 20(3), 175–201.

Thornhill, R., & Gangestad, S. W. (2003). Do women have evolved adaptation for extra-pair copulation? In *Evolutionary aesthetics* (pp. 341–368). Springer.

Thornhill, R., & Sauer, P. (1992). Genetic sire effects on the fighting ability of sons and daughters and mating success of sons in a scorpionfly. *Animal Behaviour*, 43(2), 255–264.

Tiefer, L. (2006). Female sexual dysfunction: A case study of disease mongering and activist resistance. *PLoS Medicine*, 3(4), e178.

Tinbergen, N. (1963). On aims and methods of ethology. *Zeitschrift für Tierpsychologie*, 20(4), 410–433.

Toner, J. P., & Adler, N. T. (1986). Influence of mating and vaginocervical stimulation on rat uterine activity. *Journal of Reproduction and Fertility*, 78, 239–249.

Townsend, J. M. (1995). Sex without emotional involvement: An evolutionary interpretation of sex differences. *Archives of Sexual Behavior*, 24(2), 173–206.

Townsend, J. M., & Levy, G. D. (1990). Effects of potential partners' physical attractiveness and socioeconomic status on sexuality and partner selection. *Archives of Sexual Behavior*, 19(2), 149–164.

Trapl, J. (1943). Neue Anschauungen über den Ei- und Samentransport in den inneren Geschlechtsteilen der Frau. *Zentralblatt für Gynäkologie*, 67, 547.

Trivers, R. (1972). *Sexual selection and the descent of man 1871–1971*. Aldine Press.

Trivers, R. (1985). *Social evolution*. Benjamin/Cummings.

Trivers, R. (2011). *The folly of fools: The logic of deceit and self-deception in human life*. Basic Books.

Trivers, R. L. (1974). Parent-offspring conflict. *American Zoologist*, 14(1), 249–264. The quote is from p. 249.

Tronson, N. C., & Collette, K. M. (2017). (Putative) sex differences in neuroimmune modulation of memory. *Journal of Neuroscience Research*, 95(1–2), 472–486.

Twain, M. (1897). *More tramps abroad*. Chatto & Windus.

Tyson, N. G. (2017). *Online statement*. https://twitter.com/neiltyson/statuses/40892790736949249.

Unsworth, C. (2012). *Interview with Guardian [newspaper]*. www.theguardian.com/books/2012/jun/29/cathi-unsworth-on-women-and-noir last accessed 25/03/2023.

Vance, E. B., & Wagner, N. N. (1976). Written descriptions of orgasm: A study of sex differences. *Archives of Sexual Behavior*, 5(1), 87–98.

Van Demark, N. L., & Moeller, A. N. (1951). Speed of spermatozoan transport in the reproductive tract of the estrous cow. *American Journal of Physiology*, 165, 674–679.

van den Berg, P., Fawcett, T. W., Buunk, A. P., & Weissing, F. J. (2013). The evolution of parent–offspring conflict over mate choice. *Evolution and Human Behavior*, 34, 405–411.

Vanston, J. E., & Strother, L. (2017). Sex differences in the human visual system. *Journal of Neuroscience Research*, 95(1–2), 617–625.

Varone, A. (2002). *Erotica Pompeiana. Love inscriptions on the walls of Pompei* (Trans. R. P. Berg). L'Erma di Bretschneider.

van Seters, A. P., & Slob, A. K. (1988). Mutually gratifying heterosexual relationship with micropenis of husband. *Journal of Sex & Marital Therapy, 14*(2), 98–107.

Vehrencamp, S. L., Bradbury, J. W., & Gibson, R. M. (1989). The energetic cost of display in male sage grouse. *Animal Behaviour, 38*(5), 885–896.

Venkataraman, V. (2023). *Blog post addressing Anderson et al., 2023 paper on female contribution to hunting.* www.vivekvenkataraman.com/blog last accessed 11/07/2023.

Vesalius, A. (1564). *Observationum anatomicarum Gabrielis Fallopii examen* (p. 143). Francesco de'Franceschi da Siena.

Victor, T. A., Drevets, W. C., Misaki, M., Bodurka, J., & Savitz, J. (2017). Sex differences in neural responses to subliminal sad and happy faces in healthy individuals: Implications for depression. *Journal of Neuroscience Research, 95*(1–2), 703–710.

von Uexküll, J. (1931). *Der Organismus und die Umwelt.* Verlag nicht ermittelbar.

Voyer, D. (1995). Effect of practice on laterality in a mental rotation task. *Brain and Cognition, 29*(3), 326–335.

Waalen, J., Felitti, V., & Beutler, E. (2002). Haemoglobin and ferritin concentrations in men and women: Cross sectional study. *BMJ, 325*(7356), 137.

Wallen, K. (2006). Commentary on puts' (2006) review of the case of the female orgasm: Bias in the science of evolution. *Archives of Sexual Behavior, 35*(6), 633–636.

Wallen, K., & Lloyd, E. A. (2008a). Clitoral variability compared with penile variability supports nonadaptation of female orgasm. *Evolution & Development, 10*(1), 1–2.

Wallen, K., & Lloyd, E. A. (2008b). Inappropriate comparisons and the weakness of cryptic choice: A reply to Vincent J. Lynch and D. J. Hosken. *Evolution & Development, 10*(4), 398–399.

Wallen, K., & Lloyd, E. A. (2011). Female sexual arousal: Genital anatomy and orgasm in intercourse. *Hormones and Behavior, 59*(5), 780–792.

Warner, J. (1998). Peak of sexual response questionnaire. In *Handbook of sexuality-related measures* (p. 256). SAGE.

Waterman, T. T., & Kroeber, A. L. (1934). *Yurok marriages* (Vol. 35, No. 1). University of California Press.

Watson, N. V., & Kimura, D. (1989). Right-hand superiority for throwing but not for intercepting. *Neuropsychologia, 27*(11–12), 1399–1414.

Webster, M. M., & Rutz, C. (2020). How STRANGE are your study animals? *Nature, 582*(7812), 337–340.

Wedekind, C., & Füri, S. (1997). Body odour preferences in men and women: Do they aim for specific MHC combinations or simply heterozygosity? *Proceedings of the Royal Society of London B: Biological Sciences, 264*(1387), 1471–1479.

Wei, Y. C., Wang, S. R., & Xu, X. H. (2017). Sex differences in brain-derived neurotrophic factor signaling: Functions and implications. *Journal of Neuroscience Research, 95*(1–2), 336–344.

West, S. A., Lively, C. M., & Read, A. F. (1999). A pluralist approach to sex and recombination. *Journal of Evolutionary Biology, 12*(6), 1003–1012.

Wheelock, M. D., Hect, J. L., Hernandez-Andrade, E., Hassan, S. S., Romero, R., Eggebrecht, A. T., & Thomason, M. E. (2019). Sex differences in functional connectivity during fetal brain development. *Developmental Cognitive Neuroscience,* 100632.

Whewell, W. (1834). Mary Somerville: On the connexion of the physical sciences. *Quarterly Review, 51,* 54–68.

Whipple, B. (2008). *Interview with new scientist discussing the Gravina et al. (2008) study of the G-spot.* www.newscientist.com/article/mg19726444.100-ultrasound-nails-location-of-the-elusive-g-spot.html last accessed 20/2/2008.

Whipple, B., Ogden, G., & Komisaruk, B. R. (1992). Physiological correlates of imagery-induced orgasm in women. *Archives of Sexual Behavior, 21*(2), 121–133.

Whiting, B., & Edwards, C. P. (1973). A cross-cultural analysis of sex differences in the behavior of children aged three through 11. *The Journal of Social Psychology, 91*(2), 171–188.

Whiting, J. W., & Whiting, B. B. (1975). Aloofness and intimacy of husbands and wives. *Ethos, 3*(2), 183–207.

Wiederman, M. W., & Dubois, S. L. (1998). Evolution and sex differences in preferences for short-term mates: Results from a policy capturing study. *Evolution and Human Behavior, 19,* 153–170.

Wilder, J. A., Mobasher, Z., & Hammer, M. F. (2004). Genetic evidence for unequal effective population sizes of human females and males. *Molecular Biology and Evolution, 21*(11), 2047–2057.

Wildt, D. E. (1991). Fertilization in cats. In *A comparative overview of mammalian fertilization* (pp. 299–328). Springer US.

Wildt, L., Kissler, S., Licht, P., & Becker, W. (1998). Sperm transport in the human female genital tract and its modulation by oxytocin as assessed by hysterosalpingoscintigraphy, hysterotonography, electrohysterography and Doppler sonography. *Human Reproduction Update, 4,* 655–666.

Williamson, C. (2023). [Podcast interview with Emily Morse, sex therapist]. Why is everyone having such bad sex. *Modern Wisdom 645.* www.youtube.com/results?search_query=Why+Is+Everyone+Having+Such+Bad+Sex last accessed 11/08/2023.

Willis, E. (1981). *Beginning to see the light: Pieces of a decade.* Random House.

Wilson, E. O. (2000). *Consilience.* Alfred A. Knoff.

Wilson, M., & Daly, M. (1985). Competitiveness, risk taking, and violence: The young male syndrome. *Ethology and Sociobiology, 6*(1), 59–73.

Winch, R. F. (1958). *Mate-selection; a study of complementary needs.* Harper.

Winston, R. (2010). *From Britain's leading fertility expert, an intriguing question . . . Is a woman more likely to conceive if she enjoys sex?* www.dailymail.co.uk/femail/article-1279841/From-Britains-leadingfertility-expert-intriguing-question_Is-woman-likely-conceiveenjoys-sex.html last accessed 26/08/2023.

Wise, N. J., Frangos, E., & Komisaruk, B. R. (2017). Brain activity unique to orgasm in women: An fMRI analysis. *The Journal of Sexual Medicine, 14*(11), 1380–1391.

Wisman, A., & Shrira, I. (2020). Sexual chemosignals: Evidence that men process olfactory signals of women's sexual arousal. *Archives of Sexual Behavior, 49,* 1505–1516.

Yang, C. Y., & Lin, C. P. (2017). Magnetoencephalography study of different relationships among low-and high-frequency-band neural activities during the induction of peaceful and fearful audiovisual modalities among males and females. *Journal of Neuroscience Research, 95*(1–2), 176–188.

Yoon, J. H., Abdelmohsen, K., Kim, J., Yang, X., Martindale, J. L., Tominaga-Yamanaka, K., . . . Gorospe, M. (2013). Scaffold function of long non-coding RNA HOTAIR in protein ubiquitination. *Nature Communications, 4*(1), 2939.

Yoshizawa, K., Ferreira, R. L., Kamimura, Y., & Lienhard, C. (2014). Female penis, male vagina, and their correlated evolution in a cave insect. *Current Biology, 24*(9), 1006–1010.

Young, L. C., & VanderWerf, E. A. (2014). Adaptive value of same-sex pairing in Laysan albatross. *Proceedings of the Royal Society of London B: Biological Sciences, 281*(1775), 20132473.

Young, L. C., Zaun, B. J., & VanderWerf, E. A. (2008). Successful same-sex pairing in Laysan albatross. *Biology Letters, 4,* 323–325.

Zahavi, A. (1975). Mate selection—A selection for a handicap. *Journal of Theoretical Biology, 53*(1), 205–214.

Zak, P. J., Kurzban, R., & Matzner, W. T. (2005). Oxytocin is associated with human trustworthiness. *Hormones and Behaviour, 48,* 522–527.

Zallinger, R. (1970). *Illustration of the march of progress.* Time Life.

Zaviačič, M. (1994). Sexual asphyxiophilia (Koczwarism) in women and the biological phenomenon of female ejaculation. *Medical Hypotheses, 42*(5), 318–322.

Zaviačič, M., & Ablin, R. J. (1999). *The human female prostate: From vestigial Skene's paraurethral glands and ducts to woman's functional prostate.* Slovak Academic Press.

Zaviačič, M., Jakubovská, V., Belošovič, M., & Breza, J. (2000). Ultrastructure of the normal adult human female prostate gland (Skene's gland). *Anatomy and Embryology, 201*(1), 51–61.

Zaviačič, M., Zajíčková, M., Blažeková, J., Donárová, L., Stvrtina, S., Mikulecký, M., . . . Breza, J. (2000). Weight, size, macroanatomy, and histology of the normal prostate in the adult human female: A minireview. *Journal of Histotechnology, 23*(1), 61–69.

Zervomanolakis, I., Ott, H. W., Hadziomerovic, D., Mattle, V., Seeber, B. E., Virgolini, I., . . . Wildt, L. (2007). Physiology of upward transport in the human female genital tract. *Annals of the New York Academy of Science, 1101,* 1–20.

Zhana. *Casual sex project [online resource].* www.drzhana.com/about/ last accessed 25/03/2023.

Zhao, N., Zhang, X., Noah, J. A., Tiede, M., & Hirsch, J. (2023). Separable processes for live "In-person" and live "Zoom-like" faces. *Imaging Neuroscience.* https://doi.org/10.1162/imag_a_00027.

Zietsch, B. P., & Santtila, P. (2013). No direct relationship between human female orgasm rate and number of offspring. *Animal Behaviour, 86*(2), 253–255.

Zinner, D. P., Nunn, C. L., van Schaik, C. P., & Kappeler, P. M. (2004). Sexual selection and exaggerated sexual swellings of female primates. In *Sexual selection in primates: New and comparative perspectives* (pp. 71–89).

Zubrow, E. (1989). The demographic modelling of Neanderthal extinction. In *The human revolution* (pp. 212–231). Edinburgh University Press.

Zucker, K. (2002). From the editor's desk: Receiving the torch in the era of sexological renaissance. *Archives of Sexual Behavior, 31,* 1–6.

Index

Taylor & Francis eBooks

www.taylorfrancis.com

A single destination for eBooks from Taylor & Francis
with increased functionality and an improved user
experience to meet the needs of our customers.

90,000+ eBooks of award-winning academic content in
Humanities, Social Science, Science, Technology, Engineering,
and Medical written by a global network of editors and authors.

TAYLOR & FRANCIS EBOOKS OFFERS:

A streamlined
experience for
our library
customers

A single point
of discovery
for all of our
eBook content

Improved
search and
discovery of
content at both
book and
chapter level

REQUEST A FREE TRIAL
support@taylorfrancis.com

 Routledge
Taylor & Francis Group

 CRC Press
Taylor & Francis Group